W9-CCI-177

Workbook
to Accompany
Anatomy & Physiology | Revealed® Version 2.0

Robert B. Broyles, Jr.
Butler County Community College

 **McGraw-Hill**
Higher Education

Boston   Burr Ridge, IL   Dubuque, IA   New York   San Francisco   St. Louis
Bangkok   Bogotá   Caracas   Kuala Lumpur   Lisbon   London   Madrid   Mexico City
Milan   Montreal   New Delhi   Santiago   Seoul   Singapore   Sydney   Taipei   Toronto

The **McGraw·Hill** Companies

McGraw-Hill
Higher Education

Workbook to accompany *Anatomy & Physiology Revealed*® Version 2.0

Published by McGraw-Hill, a business unit of The McGraw-Hill Companies, Inc., 1221 Avenue of the Americas, New York, NY 10020, Copyright © 2009 by The McGraw-Hill Companies, Inc. All rights reserved. No part of this publication may be reproduced or distributed in any form or by any means, or stored in a database or retrieval system, without the prior written consent of The McGraw-Hill Companies, Inc., including, but not limited to, in any network or other electronic storage or transmission, or broadcast for distance learning.

Some ancillaries, including electronic and print components, may not be available to customers outside the United States.

 This book is printed on recycled, acid-free paper containing 10% postconsumer waste.

4 5 6 7 8 9 0 BAN/BAN 0 9 8

ISBN 978–0–07–337814–5
MHID 0–07–337814–3

Publisher: *Michelle Watnick*
Senior Sponsoring Editor: *James F. Connely*
Senior Developmental Editor: *Kathleen R. Loewenberg*
Outside Developmental Services: *Robin Reed/Barb Tucker*
Marketing Manager: *Lynn M. Breithaupt*
Project Manager: *April R. Southwood*
Lead Production Supervisor: *Sandy Ludovissy*
Designer: *John Albert Joran*
Senior Photo Research Coordinator: *John C. Leland*
Images: *All images owned by McGraw-Hill Companies, Inc.*
Compositor: *Carlisle Publishing Services*
Typeface: *10/12 Stone Serif*
Printer: *R. R. Donnelley Menasha, WI*

www.mhhe.com

PREFACE

When I first became aware of *Anatomy & Physiology | Revealed®*, it immediately became obvious to me that this was the ultimate hands-on learning tool for any student studying the human body. Through the use of layer-by-layer dissection photographs of actual cadavers, students can gain insight into the intricate design of the human body. Radiological images and animations offer their unique perspectives as well. Not every student has access to a cadaver, and even those that do, cannot take it home for review. Now, however, students can take a cadaver home with them, through this accurate and detailed study of the human body.

The updated version of *Anatomy & Physiology | Revealed®* now includes the integumentary system, expanded skeletal system coverage, a separate histology section, and a powerful search engine that quickly locates information across all body systems.

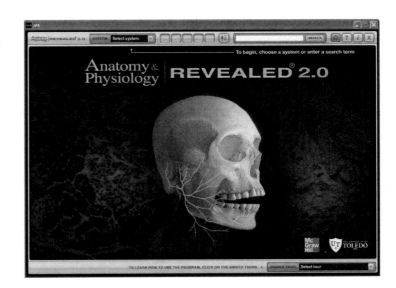

Audience

This workbook is designed to complement *Anatomy & Physiology | Revealed®*, a powerful learning tool. Regardless of their computer skills, students will be up and running in a short time as they learn how to master the concepts within the user-friendly interface. Developed to support one-semester or two-semester anatomy and physiology courses, human anatomy and human physiology courses, and a variety of health-related career programs, this workbook, combined with *Anatomy & Physiology | Revealed®*, will ensure students the very best opportunity for getting the most out of their classes. For many schools, the combination of the two will take the place of their laboratory manual.

Key Features

- The inside front cover offers key screen captures as reminders for basic navigation steps.

- The first chapter walks students through an introduction to the format and important sections of *Anatomy & Physiology | Revealed®*.

- Exercises follow the arrangement of *Anatomy & Physiology | Revealed®* requiring students to complete steps and answer follow-up questions.

- Perforated pages allow exercises to be easily removed and handed in to instructors.

- "Heads-Up!" sections offer tips and reminders about the program.

- "Check-Point" questions serve as brief self-tests to check comprehension.

- Useful tables include bonus questions on related topics.

- To help students master terminology, the inside back cover lists key parts of anatomical and physiological terms, their meanings and example words.

Acknowledgements

I offer my heartfelt thanks to colleagues who've offered
suggestions and/or reviewed the workbook to
accompany the initial version of *Anatomy & Physiology |
Revealed®*. Their contributions are much appreciated.
I'd also like to recognize the team at McGraw-Hill:
Michelle Watnick, Publisher; Jim Connely, Sponsoring
Editor; Lynn Breithaupt, Marketing Manager; Mary
Powers, Managing Editor; April Southwood, Project
Manager; Mary Jane Lampe, Project Coordinator;
John Joran, Designer; Sandy Ludovissy, Production
Supervisor; and Nikki Puccio, Editorial Assistant.

I would like to give a special thank you to my
Developmental Editor, Kathy Loewenberg, and to the
Editors at Carlisle Communication, Robin Reed,
Barb Tucker, and Wendy Langerud. This revision would
not have been possible without their guidance and
counsel.

Dr. Ann Stalheim Smith, Finally, I thank my
Professor Emeritus of Human Body at Kansas State
University taught me how to teach while I was an
undergrad in her classroom so many years ago; I thank
her for being such a remarkable role model.

Finally, I thank my beloved bride, Patricia, again for
her loving support and sacrifices.

Reviewers

Shaheem Abrahams
Thomas Nelson Community College

Isaac Barjis
New York City College of Technology

Jerry Barton
Tarrant County College—South Campus

David Bastedo
San Bernardino Valley College

Valerie A. Bennett
Clarion University of Pennsylvania

J. Gordon Betts
Tyler Junior College

Lois Brewer Borek
Georgia State University

Scott Dunham
Illinois Central College

Kathryn Durham
Lorain County Community College

Adam Eiler
San Jacinto College South

Deanna Ferguson
Gloucester County College

Cynthia R. Hartzog
Catawba Valley Community College

Amy Harwell
Oregon State University

Chris Herman
Eastern Michigan University

Mark F. Hoover
Penn State Altoona

Julie Huggins
Arkansas State University

Susanne Kalup
Westmoreland County Community College

Michael Kopenits
Amarillo College

Ellen Lathrop-Davis
Community College of Baltimore County Essex

Elisabeth C. Martin
College of Lake County

Alice McAfee
University of Toledo

Patrick McArthur
Okaloosa Walton College

Elizabeth M. Meyer
College of Lake County

Alfredo Munoz
University of Texas at Brownsville

Margaret (Betsy) Ott
Tyler Junior College

Mark Paternostro
Pennsylvania College of Technology

Robert L. Pope
Miami Dade College

Gregory K. Reeder
Broward Community College

Dawn E. Roberts
University of Massachusetts Amherst

Walied Samarrai
New York City College of Technology

Brad Sarchet
Manatee Community College

Kelly Sexton
North Lake College

Jeff Simpson
Metropolitan State College of Denver

Mitzie Sowell
Pensacola Junior College

Eric L. Sun
Macon State College

Anupama Trzaska
St. Louis Community College

Cinnamon VanPutte
Southwestern Illinois College

Hugo Rodriguez Uribe
University of Texas at Brownsville

Anthony Weinhaus
University of Minnesota

Judith Moon Wolff
Delaware Technical and Community College

Linda Wooter
Bishop State Community College

A Special Note to Students

Anatomy & Physiology | Revealed® is not a learning tool that is exhausted with a single use, nor is it any less valuable the more times you use it. The high quality cadaver photos in *Anatomy & Physiology | Revealed®* are designed to provide you a unique opportunity to match visuals of the human body with what you are learning from your instructor. With each use, you will glean more information to add to your knowledge base, and as you do, you will also gain the confidence that you need at exam time. With this in mind, use *Anatomy & Physiology | Revealed®* on a daily basis to hone your skills, refresh your memory, and stay sharp.

Be sure to take advantage of the Self-Test aspect of *Anatomy & Physiology | Revealed®*. It allows you to determine how well you understand the information presented and will reveal areas that you may need to revisit.

This workbook is written for YOU, the student. If you have any comments on how it may be improved, please let me know—I am always open to suggestions! My desire is that this workbook, in combination with *Anatomy & Physiology | Revealed®*, will help you to excel in your study of the human body.

Wishing you the best . . .

Bob Broyles
bbroyles@butlercc.edu

Bob embarked on his teaching career while in the third grade, bringing critters from his rural home to the kindergarten classroom. While in graduate school, he began teaching undergraduate anatomy and physiology labs. This set the stage for his current position at Butler County Community College, where he teaches anatomy and physiology lecture and lab and cadaver dissection.

Bob has a passion for teaching; desiring to guide others to understand the intricacies of the natural world as experienced in their daily lives. His other passions include organic gardening (he is a Master Gardener), landscaping for wildlife (primarily birds), prairie restoration, photography, and bird studies.

Bob's previous projects with McGraw-Hill Publishing include the Instructor's PowerPoint Presentations for Saladin's *Human Anatomy* text, multiple textbook reviews, and Online Tutor for Kenneth S. Saladin's *Anatomy and Physiology* textbook.

Bob resides with his wife Patricia on a small farm, where they grow prairie wildflowers, as well as trees and shrubs that benefit birds, bees, and butterflies.

CONTENTS

Chapter 7: The Lymphatic System   331

Chapter 8: The Respiratory System   353

Chapter 12: The Endocrine System   465

Overview: The Endocrine System   465

CHAPTER **1**

Introduction: Becoming Familiar with *Anatomy and Physiology | Revealed*®

Overview: Getting Started

You have before you a workbook to accompany the most powerful learning tool available for learning human anatomy and physiology. *Anatomy and Physiology | Revealed*® will guide you through the human body as no other software or textbook has ever done. Nowhere else will you be able to peel back layers of the body in a virtual dissection of an actual human cadaver. Nowhere else will you have virtual pins on the dissected human form with the ability to not only inform you of the name of a structure, but also to explain in detail that structure's anatomy and physiology. These same identification characteristics are also incorporated into tissue slides (histology) and radiological imaging (X-ray, MRI, CT Scan, etc.). Nowhere else will you have correlated animations to explain in great detail the form and function of your body. And, as the capstone to your learning experience, you are able to quiz yourself through self-testing sessions designed to reinforce what you have learned.

With such a powerful learning tool, it could be a daunting task simply getting started. Fear not! You will never find a more user-friendly interface. And, to make it even less intimidating, this chapter will walk you through each of the many facets of *Anatomy and Physiology | Revealed*®, allowing you to gain hands-on experience and build your confidence before you begin to study your first system. With this chapter under your belt, the balance of this workbook will be a breeze.

Let's get started.

Once you have opened the program, you will see the screen to the right:

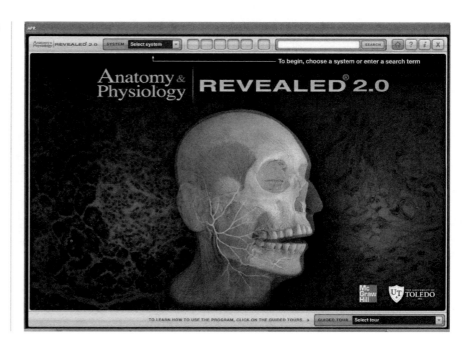

This is the **Home screen.**

HEADS UP! ————

You will find, as you go through this workbook, that all terms referring to items on the screen of Anatomy and Physiology | Revealed® *will be in bold type. For example, the opening screen is referred to as the* **Home screen.** *Later, you will see the names of the different buttons in* Anatomy and Physiology | Revealed® *listed in bold in the same case in this workbook as they appear on your screen. An example of this is the* **DISSECTION** *button. This is intended to make it easier for you to refer to these terms in this workbook while viewing* Anatomy and Physiology | Revealed® *on your computer screen.*

Let's navigate around the **Home screen,** starting at the top left. The software version of *Anatomy and Physiology | Revealed*® is listed in the top left-hand corner. To the right of this is the **SYSTEM** menu. **Select system** is visible in the drop-down box in this menu. Click the downward arrow and you will see the eleven systems covered in *Anatomy and Physiology | Revealed*®.

Reclick the downward arrow to close this menu.

To the right of the **SYSTEM** menu are five blank boxes. These are buttons that don't become activated until a system is chosen from the **SYSTEM** menu. We will view these shortly.

To the right of the blank boxes is the **ANATOMY TERMS** button.

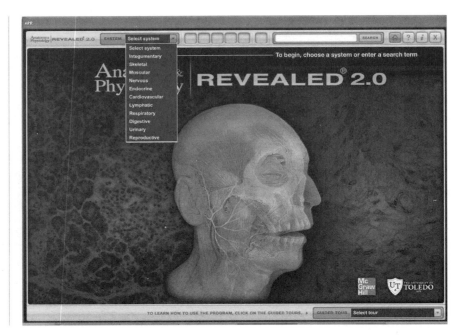

By clicking this button you are presented with the screen below:

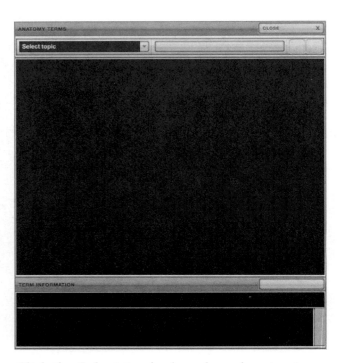

Click the **Select topic** drop-down box to view your topic options:

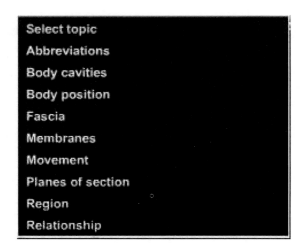

Click on **Body position,** and the **Select term** menu becomes available. Click the drop-down box and select **Anatomical position,** and you will see the screen below:

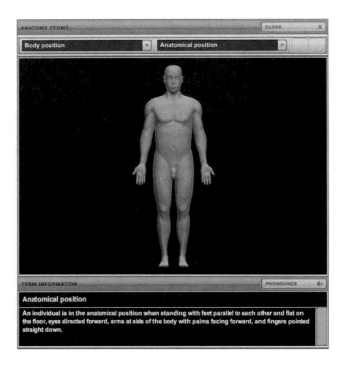

You are presented with an illustration depicting the term **Anatomical position,** and a definition is provided below the illustration. Notice the **PRONOUNCE** button located between the illustration and the definition. Click on this button to hear the correct pronunciation of the selected term. Explore the options available to you by selecting several different topics and terms. After you are finished, click the **CLOSE** button to the top right of the screen to return to the **Home screen.**

To the right of the **ANATOMY TERMS** button is the **SEARCH** menu. Enter a term and click the **SEARCH** button. Your results will be listed in four areas—**Structures, Animations, General Topics/Views,** and **Anatomy Terms.**

Four buttons or icons are located in the top-right corner of *Anatomy and Physiology | Revealed*®. These are the **SUB NAVIGATION** buttons. Mouse-over each one to reveal its identity:

The **HOME** button will return you to the **Home screen.**

The **HELP** button provides access to information specific to the section you are using.

The **INFO** button provides information about the creation of *Anatomy and Physiology | Revealed*®.

The **EXIT** button allows you to end your *Anatomy and Physiology | Revealed*® session.

At the bottom right of the **Home screen** is the **GUIDED TOUR** button. Click on the drop-down menu to see your options for this narrated descriptive tour of the many facets of *Anatomy and Physiology | Revealed*®:

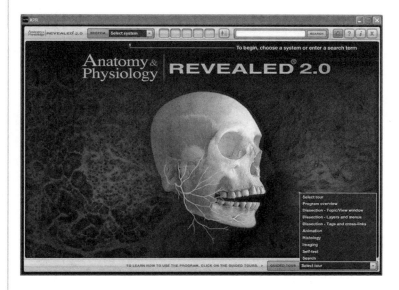

Click on **Program overview** to see a summary of *Anatomy and Physiology | Revealed*®. When finished, click the **CLOSE** button at the top-right to return to the **Home screen.**

We will use the **Muscular system** first to walk through the many facets of this powerful learning tool called ***Anatomy and Physiology | Revealed*®.** After you have mastered this information, you will be able to apply these skills to all of the body systems.

Return to the **Home screen,** if you are not there already. In the **SYSTEM** menu, click the drop-down box indicated by **Select system:**

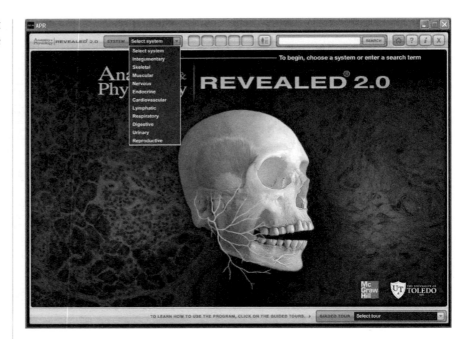

Click on **Muscular,** and you will see the *Muscular system* opening screen:

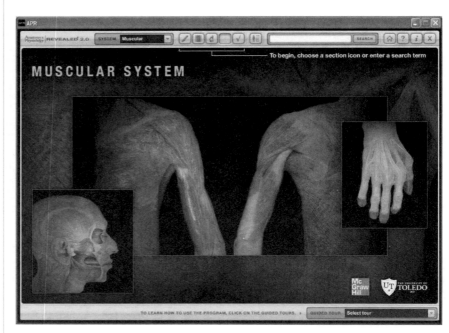

We can now explore the five buttons at the top center of the screen:

These five buttons, the **MAIN SECTIONS,** are the heart and soul of *Anatomy and Physiology | Revealed*®. It is through the use of these buttons that we can access the five sections of *Anatomy and Physiology | Revealed*® and gain a deeper understanding of the human body.

The five **MAIN SECTIONS** and their respective buttons are:

DISSECTION

ANIMATIONS

HISTOLOGY

IMAGING

SELF-TEST

We will look at these one at a time, in the next five sections of this chapter.

CHECK POINT:

Becoming Familiar with *Anatomy and Physiology | Revealed*®

1. What is the name for the first screen you see when opening *Anatomy and Physiology | Revealed*®?
2. Why is the text of this workbook often in bold letters?
3. What are the functions of the four **SUB NAVIGATION** buttons?
4. What button would you click on for a helpful guided tour of *Anatomy and Physiology | Revealed*®?
5. What are the five **MAIN SECTIONS** in *Anatomy and Physiology | Revealed*®?

The DISSECTION Section

The first **MAIN SECTIONS** button or icon is shaped like a scalpel. Mouse-over the button and it will identify itself as **DISSECTION** as seen in the image below:

Click the **DISSECTION** button, and you will see the image below:

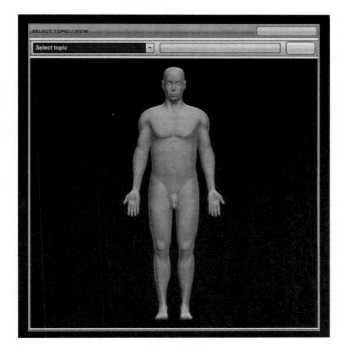

The large area on the right of the screen is the **IMAGE AREA.** It contains the **SELECT TOPIC/VIEW** window

with two menus, the **Select topic** menu and the **Select view** menu, plus the model area. Click the **Select topic menu** to view the available options:

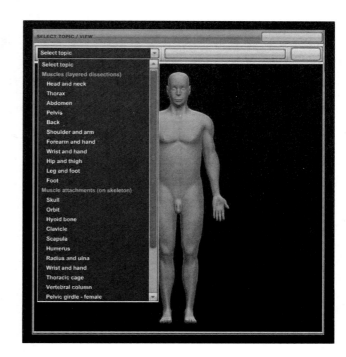

There is a slider bar to the right of the options listed, indicating more options are available than will fit on the screen. Place the mouse curser over the slider, left click but do not release, move your mouse downward, and then release the mouse button. This will move the slider downward and reveal the remaining options in this menu.

Move the slider back up to its original position. On this menu there are major groupings of structures listed in blue text and specific structures in these regions listed under them in white text. By clicking on one of these listed structures you will be able to view a dissection of that structure.

Click on **Head and neck** (white text), listed under **Muscles (layered dissections)** in blue text:

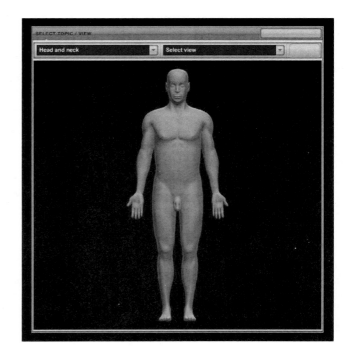

Head and neck is visible in the **Select topic** menu and the **Select view** menu becomes available. Click on the **Select view** menu:

Select **Anterior** from the menu.

The anterior portion of the head and neck is highlighted in blue, and to the right of the **Select view** menu the **GO** button is flashing green. Click the **GO** button:

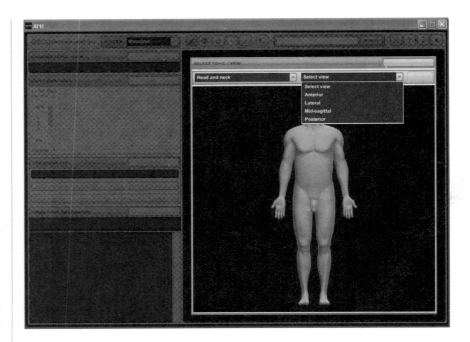

Visible in the **IMAGE AREA** to the right of the screen is the **Anterior** view of the **Head and neck.** You may ask, "Why don't I see the muscles, if this is the **Muscular System?**" Great question. The answer is in the left one-third of the screen. This area contains control boxes to manipulate the image and a **STRUCTURE INFORMATION** window where information about the specific structures is presented. These controls allow you to peel layers away from the cadaver to view the deeper structures, such as the muscles, blood vessels, organs, and bones. Let's discuss these control boxes from top to bottom:

The **CURRENT TOPIC: VIEW** window tells you which topic and view you have chosen for the current image. This example shows the **CURRENT TOPIC** as **Head and neck,** and the **CURRENT VIEW** as **Anterior.** In the top right corner of this window is a **CHANGE TOPIC/VIEW** button, which will take you back to the **Select topic** and **Select view** menus allowing you to change the image in the **IMAGE AREA.**

The **LAYER CONTROLS** window contains two methods to view the dissected image of the cadaver in the **IMAGE AREA.** First, the row of **LAYER TAGS ON/OFF** buttons will provide sequentially deeper views of the cadaver's anatomy with small colored "pins" or tags to indicate specific structures that you should learn. Feel free to click on these buttons to see the layers they represent and the pins presented for you. Some of the pins are blue while others are green. The blue pins indicate structures that are *not* a part of the system being studied—in this case, they are *not* a part of the muscular system. The green pins indicate structures that *are* a part of the system being

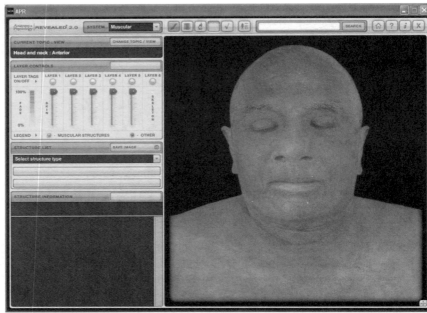

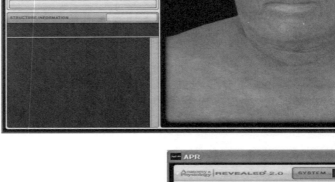

CURRENT TOPIC:VIEW
window

LAYER CONTROLS
window

STRUCTURE LIST
window

STRUCTURE INFORMATION
window

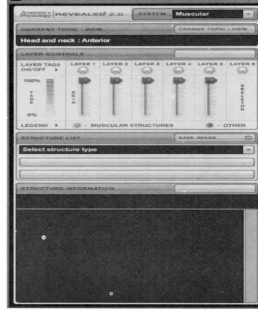

studied, that is, they *are* a part of the muscular system. In case you don't remember which pins are which, the **LEGEND** in the **LAYER CONTROLS** window will clarify this for you.

The second method for viewing the sequential dissected layers of the cadaver is the row of red sliders located between the **LAYER TAGS ON/OFF** buttons and the **LEGEND** in the **LAYER CONTROLS** window. These sliders can be moved downward to reveal progressively deeper layers of dissection. The **FADE** scale to the left of the sliders indicates the amount of fading between layers provided as you move the sliders downward. The first slider is labeled **SKIN.** This is the most superficial layer, and as the slider is moved downward, the skin is removed to reveal the muscles deep to it. To do this, place the mouse curser on the first slider. Click and hold the left mouse button, and move the mouse backward, toward you. As you do, the slider moves downward and the view in the **IMAGE AREA** changes to reveal the deeper anatomy. Repeat this process with the remaining sliders, to explore how the sliders function.

Return all of the sliders in the up position, and we will revisit the row of **LAYER TAGS ON/OFF** buttons at the top of the **LAYER CONTROLS** window. Click **LAYER 1,** and you will see blue pins on the skin of the cadaver.

Mouse-over the individual blue pins. They will turn red, and the identification of that structure will be visible:

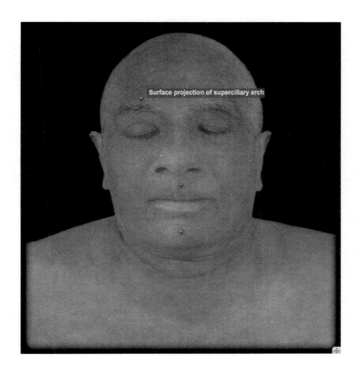

Left-click on the pin. The structures become highlighted, and information about that structure is displayed in the **STRUCTURE INFORMATION** window at the bottom left of the screen:

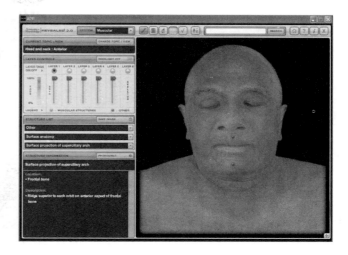

This information in the **STRUCTURE INFORMATION** window is valuable for your studies of the human body. In addition, it will be used to answer the review questions presented to you in this workbook. The **PRONOUNCE** button is located in the **STRUCTURE INFORMATION** window as well. Click on this button to hear the correct pronunciation of the structure.

When a structure is highlighted in the **IMAGE AREA,** the **HIGHLIGHT OFF** button becomes available in the top-right corner of the **LAYER CONTROLS** window. Once it is activated, the **HIGHLIGHT OFF** button can be toggled off and on to reveal the highlighted structure. Also, when a structure is highlighted, it will remain visible even when all of the sliders are moved downward. This allows you to see the correlation of the skeleton with any of the other structures.

Whenever one of the **LAYER TAGS ON/OFF** buttons is on, all of the sliders to the left of that button will be in the down position:

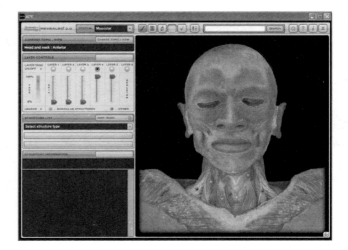

The **STRUCTURE LIST** window is located below the **LAYER CONTROLS** window. To the right of the window title is the **SAVE IMAGE** button. Anytime you want to save an image in the **IMAGE AREA,** click this button.

The **STRUCTURE LIST** window contains a **Select structure type** menu, which allows you to view the same structures as you would using the **LAYER TAGS** or **SLIDERS.** Click the **Select structure type** menu, and you will have the option of selecting **Muscles** or **Other:**

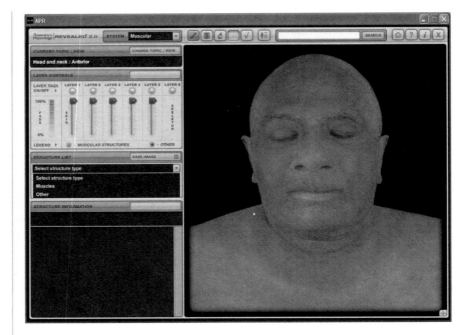

Select **Muscles,** and the **Select structure group** menu becomes available. Click on it:

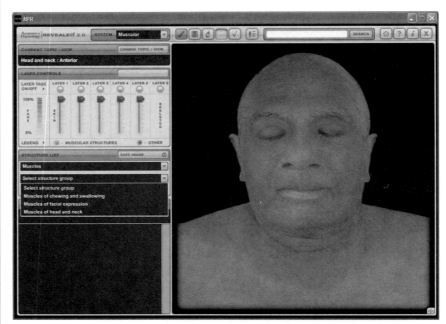

Select **Muscles of facial expression,** and the **Select structure** menu becomes available. Click on it:

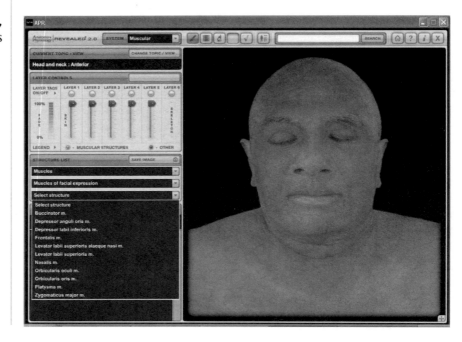

Select **Orbicularis oculi m.,** and this muscle will be highlighted on the cadaver in the **IMAGE AREA:**

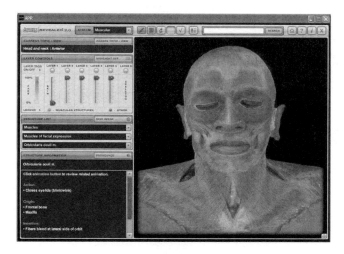

The **STRUCTURE INFORMATION** window contains details concerning this muscle, and, near the top of this window is a message highlighted in green indicating that an animation is available for this structure. Notice also at the top-center of the screen that the second button from the left, the **ANIMATIONS** button, is highlighted in green. Click this button to view the animation. We will discuss the **ANIMATIONS** button in more detail in the next section.

C H E C K   P O I N T :

The DISSECTION Section

1. What icon or button do you click on to enter the dissection section of *Anatomy and Physiology | Revealed®?*
2. What is the name for the area of the screen where the images are?
3. When clicking the **LAYER** buttons, pins of different colors are located on the dissected cadaver. What are those colors, and what do they signify?
4. What happens when you mouse-over one of the pins? What happens when you left-click on them?
5. What button do you select to change the image you are viewing in *Anatomy and Physiology | Revealed®?*

The ANIMATIONS Section

Once you have selected a system from the **Home screen,** the **ANIMATIONS** button becomes active. By clicking this button, a list of animations specific to the current system becomes available for you to choose from. Let's walk through the process.

If you are in the **Home screen,** choose **Muscular** from the **SYSTEM** menu. Once you are in a particular system, you may access the **ANIMATIONS** button at any time by clicking on it. You are now in the **Muscular** system, so click on the **ANIMATIONS** button. When you do,

you will be presented with a mostly black screen with an **ANIMATION LIST** menu in the top left:

Click on **Select animation** or its associated drop-down menu to show the list of available animations:

There is a slider on the right side of the list. Slide it downward to view the complete list of animations. There are two main headings in blue type for the muscular system, **Anatomy and Physiology** and **Muscle actions.** Click on **Skeletal muscle** under the **Anatomy and Physiology** heading to load the animation:

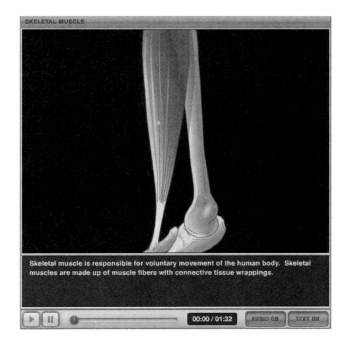

The **IMAGE AREA** is the area where the animation is viewed. The **ANIMATION INFORMATION** window, located below the **IMAGE AREA,** contains the text of the animation as it plays. Located in the lower-left area of the animation screen are the **ANIMATION CONTROLS,** and in the lower-right area are the **ANIMATION OPTIONS.** The **ANIMATION CONTROLS** consist of the **PLAY** button (arrow), the **PAUSE** button (two bars), and the red **SLIDER.** The red **SLIDER** moves from left to right as the animation plays and can be grasped by the mouse and moved right and left to advance the animation or to replay a part of it. The **ANIMATION OPTIONS** allow you to control whether the **AUDIO** or **TEXT** is **ON** or **OFF.** The default is **ON** for both.

Click the **PLAY** button, and view the animation. After viewing this animation, click the drop-down box in the **ANIMATION LIST.** Under the **Muscle actions** heading, select **Biceps brachii m.**

Click the **PLAY** button and view the animation. Familiarize yourself with the **ANIMATIONS** section of *Anatomy and Physiology | Revealed*® by viewing several animations under each heading.

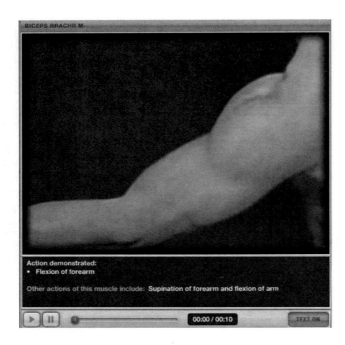

The HISTOLOGY Section

Just like the previous sections of *Anatomy and Physiology | Revealed*®, the **HISTOLOGY** section can be accessed from any screen once you have entered a specific body system. If you are just starting *Anatomy and Physiology | Revealed*®, the **Home screen** will be the first screen you see. From the **SYSTEM** menu, select **Muscular.** Click on the third button from the left in the **MAIN SECTIONS** area, the **HISTOLOGY** button. If you are already in the **Muscular System,** click on the **HISTOLOGY** button in the **MAIN SECTIONS** area.

Click the **Select topic** menu:

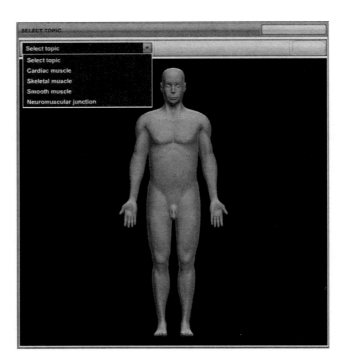

Select **Cardiac muscle.**

CHECK POINT:

The ANIMATIONS Section

1. What steps must be completed before the **ANIMATIONS** button becomes active?
2. Name the headings for the animations available for the muscular system.
3. If you were in the **DISSECTION** section, and decided that you wanted to view an animation pertaining to the system you are viewing, what steps would you take to view it?

Click the flashing green **GO** button to see the image to the right:

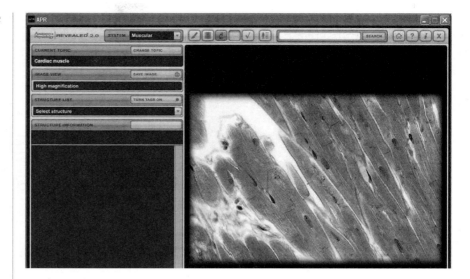

You are now viewing the **Cardiac muscle** tissue in the **IMAGE AREA.** Click the **TURN TAGS ON** button:

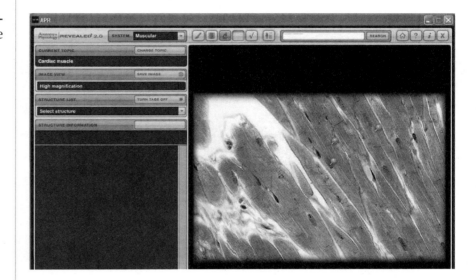

Just like you did in the **DISSECTION** section, you can mouse-over the pins to activate the identification labels:

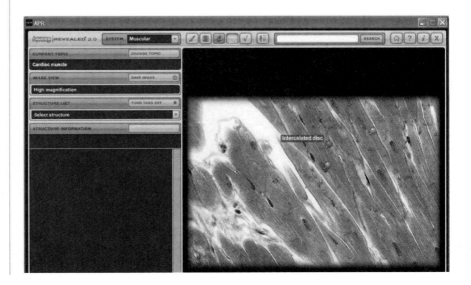

You can also left-click the pins to highlight that structure and activate the **STRUCTURE IDENTIFICATION** window:

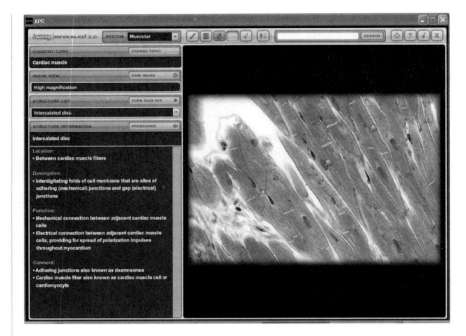

If you want to view structures without activating the pins on the image, you can click on the **STRUCTURE LIST** menu and click on, for example, **Intercalated disc:**

This view is identical to the one where you clicked the pin for Intercalated disc, showing the structures highlighted in the **IMAGE AREA** and the **STRUCTURE INFORMATION** window activated.

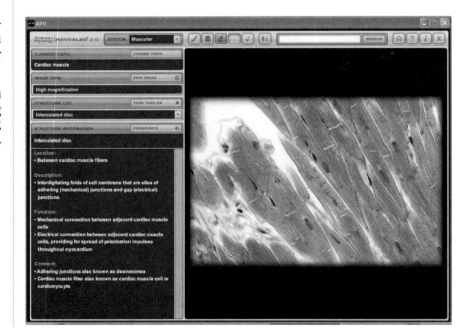

C H E C K   P O I N T :

The HISTOLOGY Section

1. Describe the sequence of steps that you would take to access the **HISTOLOGY** section of the **Nervous System.** Assume that you are starting from the **Home screen.**
2. What steps would you then take to view the **Skeletal muscle** tissue?
3. What two different methods could you use to access the **STRUCTURE IDENTIFICATION** window for the **Skeletal muscle** tissue?
4. What steps would you take to view the **Neuromuscular junction** tissue? Include steps to highlight structures and activate the **STRUCTURE IDENTIFICATION** window.

The IMAGING Section

Just like the previous sections of *Anatomy and Physiology | Revealed*®, the **IMAGING** section can be accessed from any screen once you have entered a specific body system. We have been using the muscular system to demonstrate the many facets of *Anatomy and Physiology | Revealed*®, but the muscular system has no **IMAGING** section, so we will visit the skeletal system to explore this section.

If you are in the muscular system of *Anatomy and Physiology | Revealed*®, click the **SYSTEM** menu and select **Skeletal.** If you are just starting *Anatomy and Physiology | Revealed*®, the **Home screen** will be the first screen you see. From the **SYSTEM** menu, select **Skeletal.** Click the fourth button from the left in the **MAIN SECTIONS** area, the **IMAGING** button.

Click the **Select topic** menu:

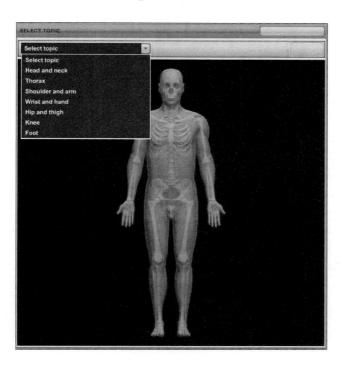

Select **Head and neck.** The head and neck of the model will be highlighted in blue, and the **GO** button will flash green. Click the **GO** button.

In the **CURRENT TOPIC** window **Head and neck** is listed. To the right of the window title is a **CHANGE TOPIC** button. By clicking this button you return to the **SELECT TOPIC** menu. In the **IMAGE TYPE/VIEW** window, **X-ray** is listed as the image type, indicating that there are no other image types available for the **Head and neck.** With other systems, some of the other options include CT-Scan, MRI, angiogram, and bronchogram.

Click the **Select view** menu, and select **Lateral** and you will see the image below:

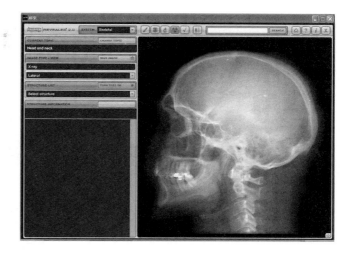

In the **STRUCTURE LIST** window you have two options. First, you can click the **TURN TAGS ON** button to place pins on the X-ray in the **IMAGE AREA:**

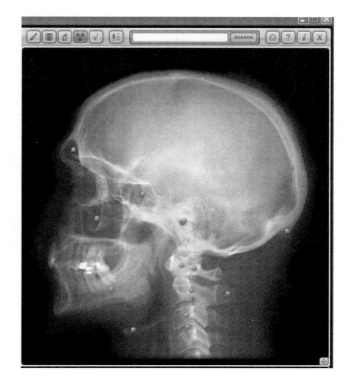

You can now mouse-over or left-click the pins like you do in the **DISSECTION** section to identify the structures . . .

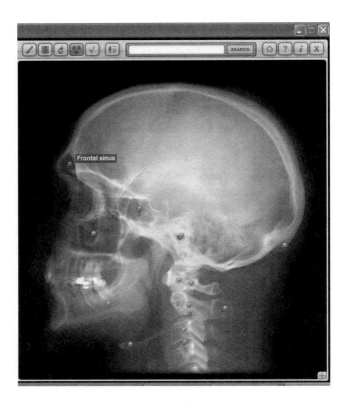

. . . or left-click the pin to highlight the structure and activate the **STRUCTURE INFORMATION** window:

The second option for the **STRUCTURE LIST** window is to click the **Select structure** menu, and click on a structure from the list:

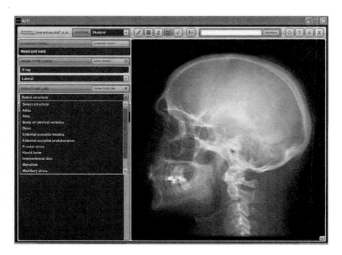

The menu will close with that structure remaining in the menu window, the structure will be highlighted on the X-ray in the **IMAGE AREA,** and the **STRUCTURE INFORMATION** window will be activated. The **PRONOUNCE** button is also activated, allowing you to hear the correct pronunciation of the structure name.

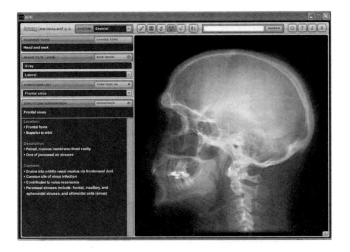

You can also save the images in this section by clicking the **SAVE IMAGE** button.

You can change the view in the **IMAGING** section by clicking the drop-down box in the **IMAGE TYPE/ VIEW** window:

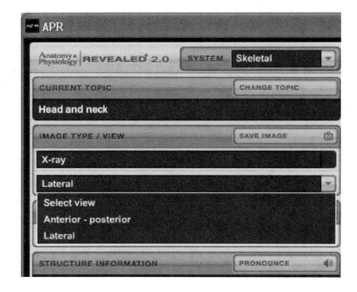

Select **Anterior - posterior:**

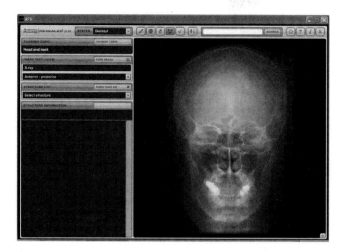

Click the **TURN TAGS ON** button, like you did with the previous image. Also, as an alternate approach, click the **STRUCTURE LIST** menu to click and view individual structures.

CHECK POINT:

The IMAGING Section

1. List the images available in the **Skeletal System** for the **Head and neck.**
2. List the images available in the **Skeletal System** for the **Hip and thigh.**
3. List the images available in the **Skeletal System** for the **Knee.**

The SELF-TEST Section

The **SELF-TEST** section of *Anatomy and Physiology | Revealed*® provides an opportunity for you to assess your understanding of the structures and concepts that have been presented. These self-tests are valuable for you to use in preparation for upcoming examinations or quizzes. They allow you to find your strengths *and* your weaknesses, and thus you will be able to fine-tune your study time as needed. Be sure to use this section regularly before you move on to new material, so that you can increase your understanding of the subject.

Just like the previous sections of *Anatomy and Physiology | Revealed*®, the **SELF-TEST** section can be accessed from any screen once you have entered a specific body system. If you are just starting *Anatomy and Physiology | Revealed*®, the **Home screen** will be the first screen you see. From the **SYSTEM** menu, select **Skeletal.** Click on the fifth button from the left in the **MAIN SECTIONS** area, the **SELF-TEST** button. If you are already in the **Skeletal System,** click on the **SELF-TEST** button in the **MAIN SECTIONS** area:

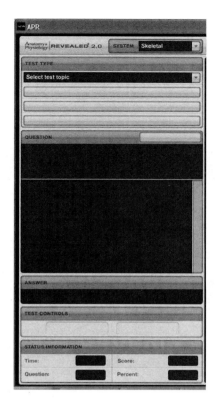

Click the **Select test topic** menu:

You can take a self test over the current system, the **Skeleton,** or over the correlated **Histology** or **Animations** provided with this system.

Click on **Skeleton:**

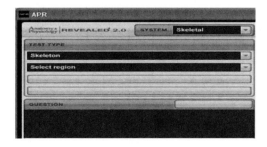

Open the **Select region** menu:

Select **Skull and associated bones:**

Click the **Select test type** menu:

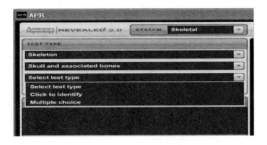

There are two options for the type of test you take. Click the first option, **Click to identify:**

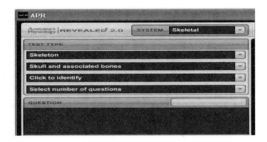

Now click the next menu, **Select number of questions:**

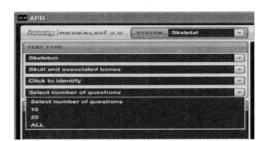

You have a choice for the number of questions that you will answer for this **SELF-TEST.** If you have a limited amount of time, choosing **10** or **25** questions will give you a brief test. But, for the most benefit from this section, choose the **ALL** option, which will ask you questions from every image in the section. Either way, the self tests will look identical:

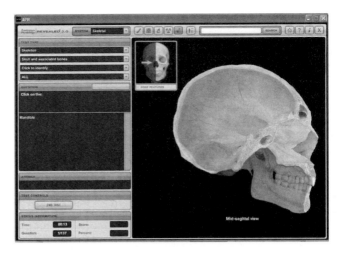

In the **QUESTION** window, the instructions say **"Click on the:"** followed by **"Mental tubercle."** Using your mouse, click on the structure asked for in this question, the mental tubercle. If you are correct in identifying the structure . . .

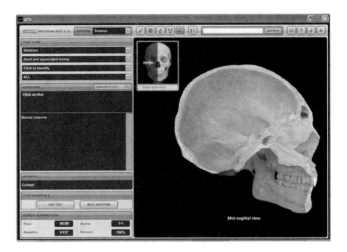

. . . the **ANSWER** window displays a green **"Correct,"** and the structure you clicked on is highlighted in the **IMAGE AREA.** If you answered the question incorrectly, **"Incorrect"** will appear in red in the **ANSWER** window, and **"correct structure shown"** in green. The correct structure will be highlighted on the image in the **IMAGE AREA.**

In the **TEST CONTROLS** window, you now have the option to **END TEST** or to answer the **NEXT QUESTION.**

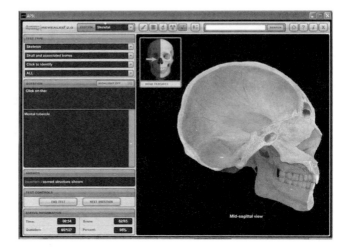

Your score is kept in the **STATUS INFORMATION** window, showing you **Time** taken for that question, the **Question** number, your **Score,** and your **Percent.** The **time** counts down from 60 seconds, and if you do not answer the question in the 60 seconds allowed, the question is counted as **"Incorrect."** At any time after a question is answered, you have the option to **END TEST** or to answer the **NEXT QUESTION.**

Click the **END TEST** button.

In the **IMAGE AREA** is your score card. At the top is your **Score** and your **Percent.** To the right of these is the **SAVE RESULTS** button. By clicking here, you can save your score to a file for future reference. In the **STRUCTURES** window is a list of all of the structures you identified incorrectly and correctly. If you did not complete the test, there is a note in green to **"Click to view more"** questions.

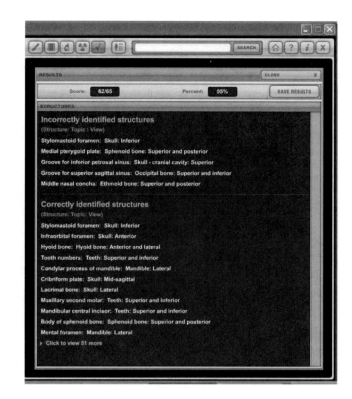

In the top-left corner of the screen, in the **TEST TYPE** window, is the option to **"Take new test and clear results."** Click on this menu item:

Click **Skeleton.** In the **Select region** menu, select **Skull and associated bones.** In the **Select test type** menu, select **Multiple choice.** In the **Select number of questions** menu, select **ALL:**

For the **Multiple choice** questions in the **QUESTION** window, click the correct answer with your mouse. These tests are completed and scored in the same manner as the **Click to identify** tests, with the exception of incorrect answers, where the correct answer is given to you in the **ANSWER** window. Your results can be viewed and saved in the same manner as the **Click to identify** tests.

Be sure to explore the **SELF-TEST** available for the other **Topics, Histology** and **Animation.**

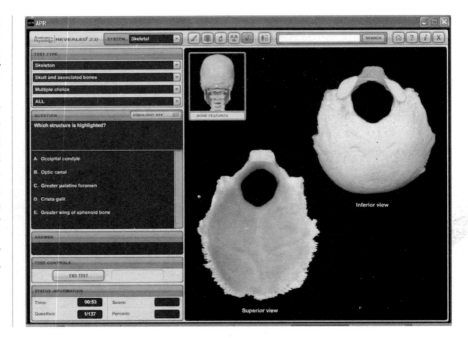

CHECK POINT:

The SELF-TEST Section

1. Describe the sequence of steps that you would take to access the **SELF-TEST** section of the **Nervous System.** Assume that you are starting from the **Home screen.**
2. List the **Test topics** available for the **Muscular System.**
3. How much time is allowed for each question in a **Self-Test**?
4. What are the advantages of selecting **ALL** from the **Select number of questions** menu?
5. Describe the process of saving your results from a **Self-Test.**

By viewing the Quick-View Tabs under the headings for each **DISSECTION, IMAGING,** and **HISTOLOGY** exercise in this workbook, you will be able to easily see which menus to select for that particular exercise. As shown in the example below, from the **Select Topic** menu, you first would click an **"Muscles"** and then **"Head and Neck."** From the **Select View** menu, you would select **"Anterior."** Finally, you would press the flashing green **GO!** button. Other than just a few exceptions, all exercises will begin with this Quick-View Tab Bar to direct you to the correct place in the *Anatomy and Physiology | Revealed*® program.

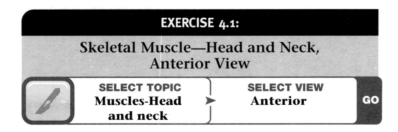

EXERCISE 4.1:

Skeletal Muscle—Head and Neck, Anterior View

SELECT TOPIC	SELECT VIEW	
Muscles-Head and neck	Anterior	GO

I N   R E V I E W

What Have I Learned?

1. Name the five **MAIN SECTIONS** in *Anatomy and Physiology | Revealed®*.

2. List the steps you would go through from the *Anatomy and Physiology | Revealed®* opening screen to viewing a deep dissection of the posterior view of the arm and hand.

3. List the steps you would go through from the *Anatomy and Physiology | Revealed®* opening screen to viewing the animation of **Appositional bone growth.**

4. Name the function of each of the following buttons:

   a)    b)    c)

   d)    e)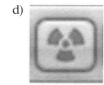

5. Name the function of each of these six buttons:

   (a)   (b)   (c) (d) (e) (f)

a)

b)

c)

d)

e)

f)

6. Using *Anatomy and Physiology | Revealed®*, define **Protraction.**

7. Using *Anatomy and Physiology | Revealed®*, define **Oblique plane.**

8. Using *Anatomy and Physiology | Revealed®*, define **Anatomical position.**

9. Using *Anatomy and Physiology | Revealed®*, define **Proximal** and **Distal.**

10. Using *Anatomy and Physiology | Revealed®*, define **Abdomen** and **Pelvis.**

CHAPTER 2

The Integumentary System

Overview: The Integumentary System

*Your skin (or integument) is your body's largest organ, weighing in at around 15 percent of your body weight. The **Integumentary System** consists not only of your skin, but also your hair, nails, sweat, and sebaceous glands—the so-called accessory organs. In our exploration of the **Integumentary System,** we will focus our attention on the nails and the layers of the skin in the dissection views and the detailed structure of the skin and its accessory organs in the histology views.*

HEADS UP!

*There are only three sections available for the Integumentary System: **DISSECTION, HISTOLOGY,** and **SELF-TEST.** There are no **ANIMATION** or **IMAGE** sections for this system.*

- *From the **Home screen,** click the drop-down box on the **Select system** menu.*

- *From the systems listed, click on **Integumentary** and you will see the image to the right:*

- *This is the opening screen for the **Integumentary System.***

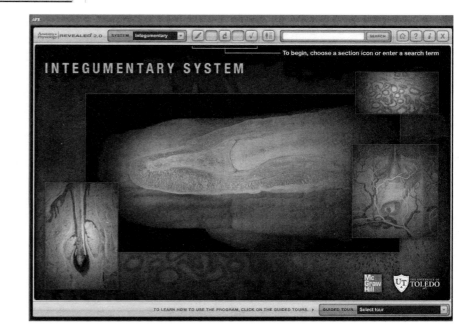

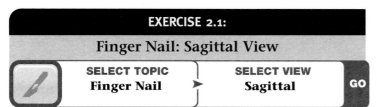

EXERCISE 2.1:
Finger Nail: Sagittal View

| | SELECT TOPIC **Finger Nail** | ▶ | SELECT VIEW **Sagittal** | GO |

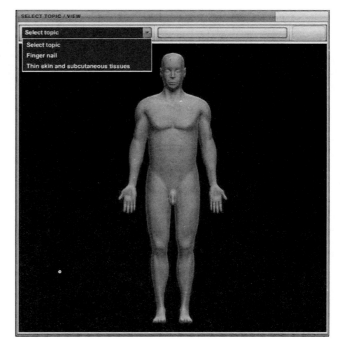

HEADS UP!

*The **View** menu displays **Sagittal**. This indicates that this is the only view available for the finger nail. Otherwise, as you will recall from Chapter 1, the **Select view** menu would have a drop-down box with different views to choose from.*

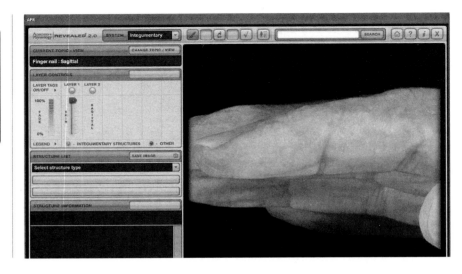

- *Click on **LAYER 1** in the **LAYER CONTROLS** window, and you will see the following image:*

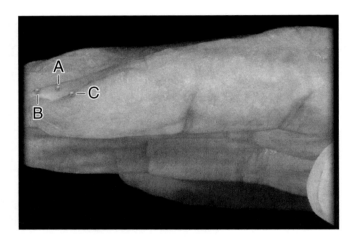

- *Mouse-over the green pins on the screen to find the information necessary to fill in the following blanks.*

A. _____

B. _____

C. _____

CHECK POINT:

Finger Nail

- *Left-click the pins on the screen to find the information necessary to answer the following questions:*
  1. Name the structures that are scalelike modifications of epidermis.
  2. What are the two functions of these structures?
  3. What is the name for the distal edge of these structures?

- *Click on **LAYER 2** in the **LAYER CONTROLS** window, and you will see the following image:*

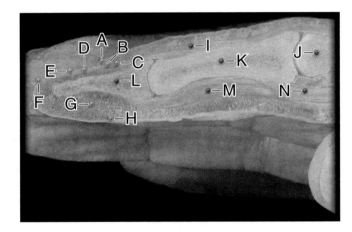

- *Mouse-over the green pins on the screen to find the information necessary to fill in the following blanks:*

A. _____

B. _____

C. _____

D. _____

E. _____

F. _____

G. _____

H. _____

HEADS UP!

Some of the structures in the dissections are not part of the integumentary system, but are important structures in the vicinity. These nonintegumentary structures are tagged with blue pins to distinguish them from the green pins of the integumentary system.

Non-integumentary System Structures (blue pins)

I. _____

J. _____

K. _____

L. _____

M. _____

N. _____

CHECK POINT:

Finger Nail, cont'd

4. Name the structure that consists of the stratum corneum of the proximal nail fold.
5. What is another name for this structure?
6. What structure does this structure overlie?

IN REVIEW

What Have I Learned?

The following questions cover the material that you just learned—the **Integumentary System.** Use the information in the **STRUCTURE INFORMATION** window for these structures to answer the following questions on a separate piece of paper:

1. Name the structures that are scale-like modifications of the epidermis.

2. What are the two functions of these structures?

3. What is the name for the distal edge of these structures?

4. Name the part of the nail plate that consists of layers of compacted, highly keratinized epithelial cells.

5. What layer of the epidermis does this structure correspond to?

6. What is the name for the epidermal fold along the lateral edge of the nail plate?

7. Name the structure that consists of the stratum corneum of the proximal nail fold.

8. What is another name for this structure?

9. What structure does this structure overlie?

10. What is the name for the proximal part of the nail plate?

11. What is its function?

12. Name the growth zone of the nail that contains mitotic cells.

13. Normal nail growth is _____ mm/day.

14. Which grow faster, finger nails or toe nails?

15. Where do the muscles of facial expression insert?

16. Name the four functions of this structure.

17. Name the skin layer deep to the epidermis.

18. Name the two functions of the epidermis.

19. Name the five epidermal layers of thick skin, from deep to superficial.

20. Which of these five layers is absent in thin skin?

21. Name the structure that provides nutrients for the epidermis.

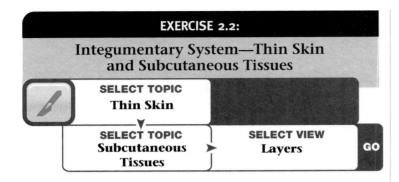

EXERCISE 2.2:

Integumentary System—Thin Skin and Subcutaneous Tissues

SELECT TOPIC
Thin Skin

SELECT TOPIC
Subcutaneous Tissues

SELECT VIEW
Layers

GO

HEADS UP!

If, during these exercises, you are already in the section indicated by the first tab under the Exercise title, click on the **CHANGE TOPIC/VIEW** *button above the* **LAYER CONTROLS** *window.*

- *Click on* **LAYER 1** *in the* **LAYER CONTROLS** *window, and you will see the following image:*

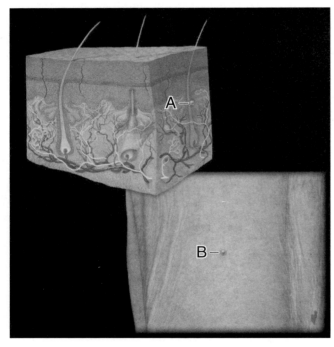

- *Mouse-over the green pins on the screen to find the information necessary to fill in the following blanks:*

A. _____

B. _____

CHECK POINT:

Thin Skin and Subcutaneous Tissues

1. Name the two layers of the epidermis consisting of cells without nuclei.
2. Name the major cell type of the epidermis.
3. Name the accessory organ of the skin that consists of a filament of keratinized cells.

- *Click on* **LAYER 2** *in the* **LAYER CONTROLS** *window, and you will see the following image:*

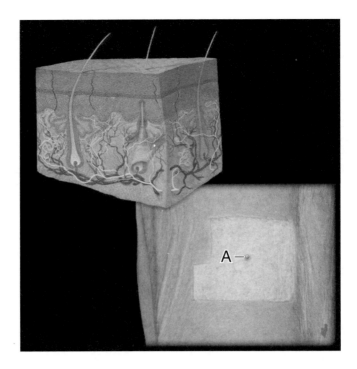

- *Mouse-over the green pin on the screen to find the information necessary to fill in the following blank:*

A. _____

CHECK POINT:

Thin Skin and Subcutaneous Tissues, cont'd

4. Name the skin layer that lies deep to the epidermis.
5. What tissue-type does this layer consist of?
6. This layer consists of two structural layers. What are they?

- *Click on* **LAYER 3** *in the* **LAYER CONTROLS** *window, and you will see the following image:*

- *Mouse-over the green pin on the screen to find the information necessary to fill in the following blank:*

A. _____

CHECK POINT:
_____

Thin Skin and Subcutaneous Tissues, cont'd

7. Name the integumentary layer located deep to the skin.
8. What tissue types are found in this layer?

- *Click on **LAYER 4** in the **LAYER CONTROLS** window, and you will see the following image:*

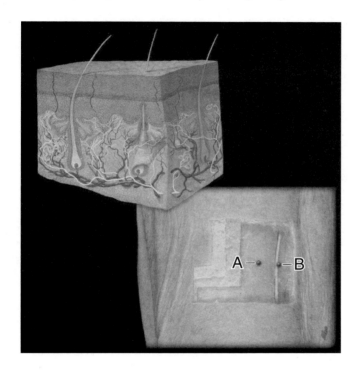

- *Mouse-over the blue pins on the screen to find the information necessary to fill in the following blanks:*

Non-integumentary System Structures (blue pins)

A. _____

B. _____

CHECK POINT:
_____

Thin Skin and Subcutaneous Tissues, cont'd

9. What structures are located in the integumentary layer located deep to the skin?
10. Name the deep fascia of the thigh.

- *Click on **LAYER 5** in the **LAYER CONTROLS** window, and you will see the following image:*

- *Mouse-over the blue pin on the screen to find the information necessary to fill in the following blank:*

Non-integumentary System Structure (blue pin)

A. _____

IN   REVIEW

What Have I Learned?

The following questions cover the material that you just learned—the **Integumentary System.** Use the information in the **STRUCTURE INFORMATION** window for these structures to answer the following questions on a separate piece of paper:

1. Name the two layers of the epidermis consisting of cells without nuclei.

2. Name the major cell type of the epidermis.

3. Other than the cells listed in question 2, what other cells are found in the epidermis?

4. Name the accessory organ of the skin that consists of a filament of keratinized cells.

5. What skin type contains these structures?

6. What is the name of the oblique tube in the skin where these structures are located?

7. Where on the body surface are these structures not found?

8. Name the skin layer that lies deep to the epidermis.

9. What tissue type does this layer consist of?

10. This layer consists of two structural layers. What are they?

11. What fibers located in the dermis give strength to the skin?

12. What general sensory reception occurs through the dermis?

13. How are these stimuli received?

14. Name two other functions of the dermis not mentioned in these questions.

15. Name the integumentary layer located deep to the skin.

16. What tissue types are found in this layer?

17. What structures are located in this layer?

18. Name three functions of this integumentary layer.

19. What are two other names for this layer?

20. Name the deep fascia of the thigh.

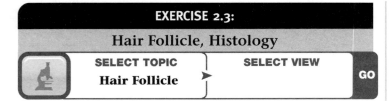

EXERCISE 2.3:

Hair Follicle, Histology

SELECT TOPIC   SELECT VIEW

Hair Follicle   GO

- *Click the* **TURN TAGS ON** *button in the* **STRUCTURE LIST** *panel, and you will see the following image:*

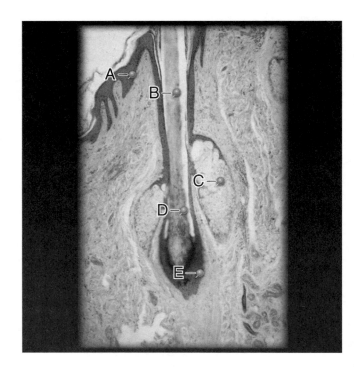

- *Mouse-over the green pins on the screen to find the information necessary to fill in the following blanks:*

A. _____

B. _____

C. _____

D. _____

E. _____

C H E C K   P O I N T:

Hair Follicle, Histology

1. Name the structure responsible for hair growth.
2. Where is this structure located?
3. Name the filamentous, pigmented, and keratinized structure that projects from the epidermal surface.

EXERCISE 2.4:

Thick Skin, Histology (Low Magnification)

  SELECT TOPIC **Thick Skin** ▶ SELECT VIEW   **GO**

H E A D S   U P!

*If you are already in the **HISTOLOGY** section, click the **CHANGE TOPIC** button at the top of the screen.*

- *In the **IMAGE VIEW** panel, click on the **Select magnification** drop-down menu.*

- *Select **Low magnification.***

- *Click the **TURN TAGS ON** button located in the **STRUCTURE LIST** panel, and you will see the following image:*

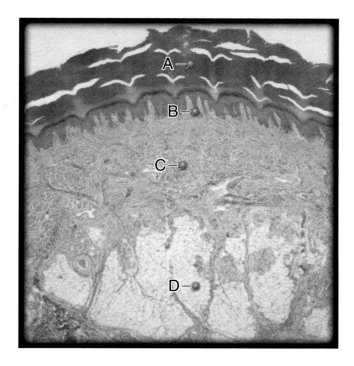

- *Mouse-over the green pins on the screen to find the information necessary to fill in the following blanks:*

A. _____

B. _____

C. _____

D. _____

C H E C K   P O I N T:

Thick Skin, Histology (Low Magnification)

1. Where on the body surface would you find thick skin? Thin skin?
2. Name the structures found at the interface between the dermis and the epidermis.
3. What is the function of these structures?

EXERCISE 2.5:

Thick Skin, Histology (High Magnification)

SELECT TOPIC	SELECT VIEW	
Thick Skin ▸		**GO**

- *From the* **Select magnification** *menu, select* **High magnification.**

- *Click the* **TURN TAGS ON** *button, and you will see the following image:*

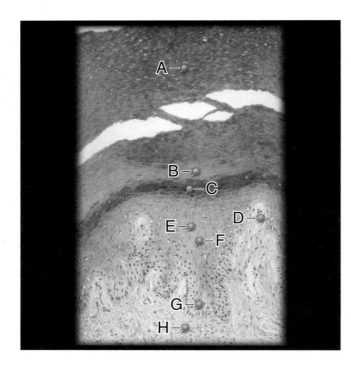

- *Mouse-over the green pins on the screen to find the information necessary to fill in the following blanks:*

A. _____

B. _____

C. _____

D. _____

E. _____

F. _____

G. _____

H. _____

CHECK POINT:

Thick Skin, Histology (High Magnification)

1. Name the layer of the epidermis that creates a barrier to liquids.
2. What relationship does your skin have with household dust?
3. Where do the replacement cells for the sloughed stratum corneum cells originate?

EXERCISE 2.6:

Thin Skin, Histology (Low Magnification)

SELECT TOPIC	SELECT VIEW	
Thin Skin ▸		**GO**

- *The* **Select magnification** *menu will read* **Low magnification.**

- *Click the* **TURN TAGS ON** *button, and you will see the following image:*

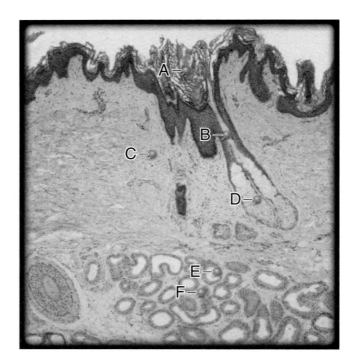

- *Mouse-over the green pins on the screen to find the information necessary to fill in the following blanks:*

A. _____

B. _____

C. _____

D. _____

E. _____

F. _____

CHECK POINT:

Thin Skin, Histology (Low Magnification)

1. Name the layer of the skin consisting of stratified squamous epithelium.
2. The cells from which epidermal layers lack nuclei?
3. Name the two types of sweat glands. What is the difference between the two?

IN REVIEW

What Have I Learned?

The following questions cover the material that you just learned—**Histology of the Integumentary System.** Use the information in the **STRUCTURE INFORMATION** window for these structures to answer the following questions on a separate piece of paper.

1. Name the structure responsible for hair growth.

2. Where is this structure located?

3. Name the filamentous, pigmented, and keratinized structure that projects from the epidermal surface.

4. What is the function of this structure?

5. Name the angulated tubular invagination of the epidermis containing an inner epidermic and an outer dermic coat.

6. What is its function?

7. List the characteristic parts of this structure.

8. What structures are associated with a hair follicle?

9. What is another name for the arrector muscle of the hair?

10. What is included with the apocrine gland secretions?

11. Name the simple, saccular holocrine gland with ducts opening into the hair follicle or onto the skin.

12. Name two locations where these glands are *not* found.

13. Where on the body surface would you find thick skin? Thin skin?

14. Name the structures found at the interface between the dermis and the epidermis.

15. What is the function of these structures?

16. Describe these structures.

17. Define "avascular."

18. Name the layer of the epidermis that creates a barrier to liquids.

19. What relationship does your skin have with household dust?

20. What is the meaning of the Latin word "cornu"?

21. Where do the replacement cells for the sloughed stratum corneum cells originate?

22. The stratum corneum consists of _____ layers of cornified dead cells.

23. Name the layer of the epidermis present only in thick skin.

24. What is keratohyalin? Where is it located?

25. What is the function of the stratum spinosum?

26. The cells from which two layers of the epidermis are responsible for the turnover of epidermal cells?

27. Name the deepest layer of the epidermis.

28. What is another name for this layer?

29. What is the function of this layer?

30. Name the layer of the skin consisting of stratified squamous epithelium.

31. The cells from which epidermal layers lack nuclei?

32. Name the two types of sweat glands. What is the difference between the two?

33. In severe heat stress, the body may sweat _____ per hour.

Self-Test

Take this opportunity to quiz yourself by taking the **SELF-TEST.** See page 15 for a reminder on how to access the self-test for this section.

HEADS UP!

The **Select test type** *menu will not always have the option to choose between* **Click to identify** *and* **Multiple choice** *tests. Sometimes, as is the case for this chapter, only one test option will be available.*

CHAPTER 3

The Skeletal System

Overview: Skeletal System

We are born with 270 bones in our bodies, and even more bones form during childhood. By the time we reach adulthood though, several separate bones have fused together so that the number of our bones has decreased to around 206[1], which make up the adult skeletal system. An example of this reduction occurs in each half of our pelvis, where three separate bones—the ilium, the ischium, and the pubis—fuse into one single bone called the *os coxa.*

The skeletal system is further divided into the **axial skeleton,** consisting of the bones of the skull, vertebral column, and the thoracic cage; and the **appendicular skeleton,** which consists of the bones of the upper and lower extremities along with their associated girdles (Table 3.1).

[1]Around 206—some people develop varying numbers of miscellaneous bones, either **sesamoid bones,** which form within some tendons as a response to stress (such as the patella) or **sutural bones,** which develop within the sutures of the skull.

CHECK POINT:

Overview: Skeletal System

1. The average human adult has _____ bones in their body, while the average newborn has _____ bones in theirs.
2. Explain the difference.

Naming Bony Processes and Other Landmarks

Bony landmarks are various ridges, spines, depressions, pores, bumps, grooves, and articulating structures on the surface of bones. These structures allow for the passage of blood vessels and nerves; for joints between bones; and for the attachment of ligaments, muscles, and tendons. Therefore, a working knowledge of the names of these structures will be a valuable asset when considering the structure and function of these surface features.

Some of the most commonly encountered bony landmarks are listed in Table 3.2.

Table 3.1	Summary of the Bones of the Adult Skeletal System		
Axial Skeleton—80 Bones		**Appendicular Skeleton—126 Bones**	
Skull and Hyoid	23 bones	Pectoral Girdle	4 bones
Inner Ear Ossicles	6 bones	Upper Extremities	60 bones
Vertebral Column	26 bones	Pelvic Girdle	2 bones
Sternum and Ribs	25 bones	Lower Extremities	60 bones
Total – 206 Bones			

Bonus Question: Using this information as a reference, how would this table appear if it was a list of the bones for a newborn?

Table 3.2	Common Bony Landmarks	
FEATURE	**DESCRIPTION**	**EXAMPLE**
Landmarks of Articulation:		
Condyle:	Smooth, rounded knob	occipital condyle of skull
Facet:	Smooth, flat, slightly concave or convex articular surface	articular facets of vertebrae
Head:	Prominent expanded end of a bone, sometimes rounded	head of the femur
Elevated Landmarks:		
Process:	Any bony prominence	mastoid process of the skull
Spine:	Sharp, slender, or narrow process	spine of the scapula

Continued

Table 3.2	Continued	
FEATURE	**DESCRIPTION**	**EXAMPLE**
Elevated Landmarks: (continued)		
Crest:	Narrow ridge	iliac crest of the pelvis
Line:	Slightly raised, elongated ridge	nuchal lines of the skull
Tuberosity:	Rough surface	tibial tuberosity
Tubercle:	Small, rounded process	greater tubercle of the humerus
Trochanter:	Massive processes unique to the femur	greater trochanter of the femur
Epicondyle:	Projection superior to a condyle	medial epicondyle of the femur
Depressions or Flat Surfaces:		
Alveolus:	Pit or socket	tooth socket
Fossa:	Shallow; broad, or elongated basin	mandibular fossa
Fovea:	Small pit	fovea capitis of the femur
Sulcus:	Groove for a tendon, nerve, or blood vessel	intertubercular sulcus of the humerus
Spaces or openings:		
Foramen:	Hole through a bone, usually round	foramen magnum of the skull
Fissure:	Slit through a bone	orbital fissure behind the eye
Meatus or canal:	Tubular passage or tunnel through a bone	auditory meatus of the ear
Sinus:	Space or cavity within a bone	frontal sinus of the skull

Bonus Question: Alveolus is a common term in human anatomy. How many other examples of alveoli can you find in *Anatomy & Physiology | Revealed®️ (APR)?*

CHECK POINT:

Naming Bone Processes and Other Landmarks

1. Explain the difference between a crest and a line.
2. Explain the difference between a condyle and an epicondyle.
3. Explain the difference between a foramen and a fissure.

EXERCISE 3.1:

Coloring Exercise

Color in the structures with colored pens or pencils.

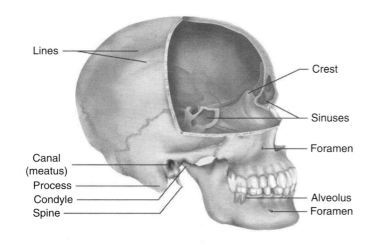

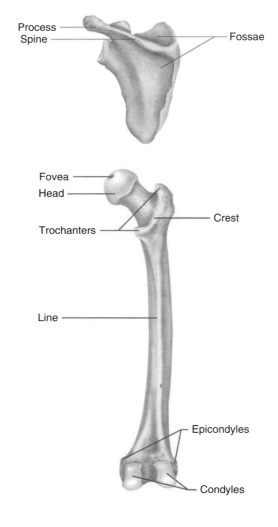

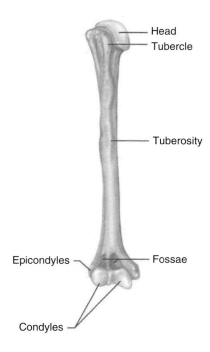

Head
Tubercle

Tuberosity

Epicondyles — — Fossae

Condyles

A Few Notes About Naming Processes

A process is any bony prominence—that is, a piece of bone that sticks out from the rest of the bone. When it comes to naming these processes, there are a few rules that need to be followed to minimize confusion. For example, let's consider when a process articulates with another bone, such as the zygomatic process of the temporal bone. This process is a structure on the **temporal bone** that articulates with the **zygomatic bone.** The formula for naming one of these processes is:

the x process of the y bone

where x = the name of bone articulated with
and y = the name of the bone it is part of

So, with the ZYGOMATIC process of the TEMPORAL bone, x = ZYGOMATIC (the bone articulated with) and y = TEMPORAL (the bone it is part of). Now, how does the *zygomatic* process of the *temporal* bone compare to the *temporal* process of the *zygomatic* bone? If you are not sure, don't worry, we will cover these structures shortly.

Let's look at the styloid process of the temporal bone. *This* styloid process does *not* articulate with any other bone, but it shares its name with the styloid processes of both the ulna and radius bones of the forearm. Therefore, it must be named in reference to the bone that it is part of to prevent confusion—hence the name the styloid process of the temporal bone. What problems would you predict could occur if this distinction is not made in an emergency room scenario?

Some processes, the mastoid process for example, are unique in name and do not articulate with any other bones. These require no further clarification when naming them.

CHECK POINT:

A Few Notes About Naming Processes

1. Consider the *temporal* process of the *zygomatic* bone.
   a) This process articulates with which bone?
   b) This process is part of which bone?
2. What is the correct way to say the two styloid processes of the forearm?
3. Why is it correct to refer to the mastoid process as the mastoid process and not the mastoid process of the temporal bone?

- *From the* **Home screen,** *click the drop-down box on the* **Select system** *menu.*

- *From the systems listed, click on* **Skeletal,** *and you will see the image to the right:*

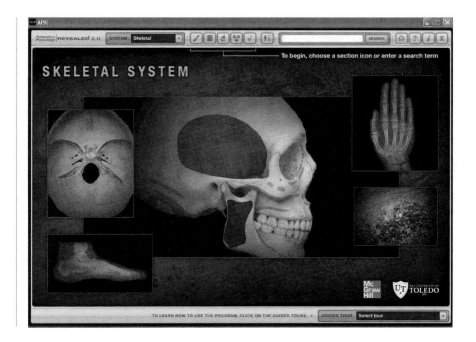

- This is the opening screen for the **Skeletal System.**

Animations: Joint Movements and Appositional Bone Growth

We cannot discuss the bones of the skeleton without first understanding their movements in reference to each other and with the rest of the body. The location of this movement or articulation between bones is the **joint.** In the **ANIMATIONS** section of the **Skeletal System,** several of these joint movements are listed for you. By selecting a specific joint movement, a definition and short animated example will explain each one.

EXERCISE 3.2a:

Joint Movements

SELECT ANIMATION
See Movements at Right
(or below, above, etc.)

PLAY

- *After viewing the animation and reading the definition, define the following joint movements. Click the **Select animation** menu in the **ANIMATION LIST** window to select the next term:*

Flexion:

Extension:

Abduction:

Adduction:

Pronation:

Supination:

Elevation:

Depression:

Circumduction:

Rotation:

Dorsiflexion:

Plantar flexion:

Eversion:

Inversion:

Protraction:

Retraction:

CHECK POINT:

Joint Movements

1. A movement that raises a body part is. . . .
2. A movement of a body part away from the main axis of the body or structure is. . . .
3. A movement at the ankle so that the dorsum of the foot is elevated is. . . .
4. A movement that decreases the angle between two bones at a joint is. . . .
5. A movement that increases the angle between two bones at a joint is. . . .

EXERCISE 3.2b:

Appositional Bone Growth

SELECT ANIMATION
Appositional Bone Growth

PLAY

- *After viewing the animation, answer the following questions:*

1. Define appositional bone growth.

2. Which cells produce bone material?

3. How is a tunnel formed around a blood vessel?

4. How is the tunnel filled in to produce a new osteon?

5. What is the name for the concentric rings that form the osteon?

- *When you are finished with the animation, click on the* **Dissection** *button at the top of the screen to begin the following exercises, or click on the* **Exit** *button at the top right of the screen to exit* **Anatomy & Physiology | Revealed®.**

HEADS UP!

*Anatomy & Physiology | Revealed® has several animations available to aid your study of different systems. Watch for the **Animation** button at the bottom of the screen to be highlighted green, which indicates that an animation is available for the specific structure(s) you are viewing.*

Skeleton (Regions)

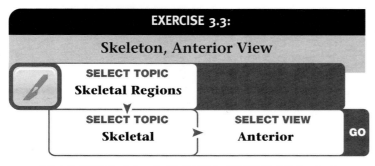

EXERCISE 3.3:

Skeleton, Anterior View

SELECT TOPIC
Skeletal Regions

SELECT TOPIC
Skeletal

SELECT VIEW
Anterior

GO

- *Click **LAYER 1** in the **LAYER CONTROLS** window, and you will see the following image:*

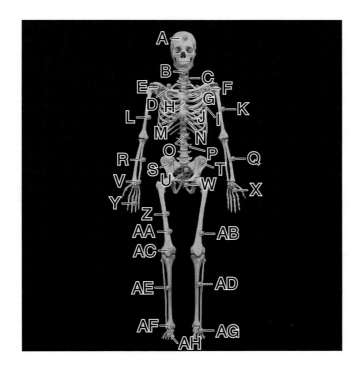

- *Mouse-over the pins on the screen to find the information necessary to identify the following structures:*

A. _____

B. _____

C. _____

D. _____

E. _____

F. _____

G. _____

H. _____

I. _____

J. _____

K. _____

L. _____

M. _____

N. _____

O. _____

P. _____

Q. _____

R. _____

S. _____

T. _____

U. _____

V. _____

W. _____

X. _____

Y. _____

Z. _____

AA. _____

AB. _____

AC. _____

AD. _____

AE. _____

AF. _____

AG. _____

AH. _____

CHECK POINT:

Skeleton, Anterior View

1. Name the bones that make up the cranial group. How many of each bone are in this group?
2. Name the bones that make up the facial group. How many of each bone are in this group?
3. Name the bones that form the pelvic girdle.
4. Name the bones that form the wrist.
5. The vertebral column is composed of _____ vertebrae. How many of each group of vertebrae are there?

EXERCISE 3.4:

Skeleton, Posterior View

SELECT TOPIC	SELECT VIEW	
Skeleton	**Posterior**	GO

• *Click* **LAYER 1** *in the* **LAYER CONTROLS** *window, and you will see the following image:*

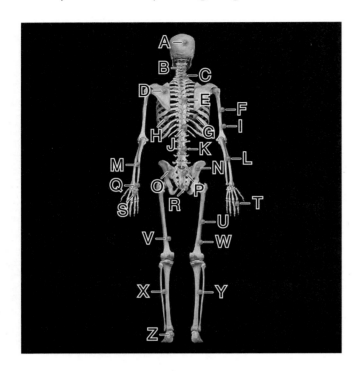

• *Mouse-over the pins on the screen to find the information necessary to identify the following structures:*

A. _____

B. _____

C. _____

D. _____

E. _____

F. _____

G. _____

H. _____

I. _____

J. _____

K. _____

L. _____

M. _____

N. _____

O. _____

P. _____

Q. _____

R. _____

S. _____

T. _____

U. _____

V. _____

W. _____

X. _____

Y. _____

Z. _____

CHECK POINT:

Skeleton, Posterior View

1. Name the group of bones that form the palm of the hand. What structures form the knuckles?
2. Name the group of bones that form the foot. How many are there?
3. Name the group of bones that form the fingers. How many compose each finger?
4. Name the lateral bone of the forearm. Name the medial bone of the forearm.
5. What is another name for the tailbone?

EXERCISE 3.5:

Locating Structures of the Head and Neck

There are 30 bones in the human adult head and neck, which can present a daunting task when it comes to learning each bone. With this in mind, the following tables are designed to allow you to discover which bones are visible in each view of *Anatomy & Physiology | Revealed*®. These tables are not meant to be tedious, but rather to help you become more familiar with each bone.

On the left of the following tables are names of structures found on the head and neck. The names of specific bones are aligned to the left, and the structures found on those bones are listed under them indented to the right. Using *Anatomy & Physiology | Revealed*®, open each dissection view listed across the top right of the table and put an "X" in the columns under the views where you find the structures listed in the left column. Not all bones or structures will be visible in all views.

Table 3.3	Structures of the Head and Neck—Cranial Bones								
Structures	**Dissection Views**								
	Individual Bone	**Anterior**	**Inferior Skull**	**Lateral**	**Mid-sagittal**	**Posterior**	**Superior**	**Skull Cranial Cavity**	**Orbit**
Frontal bone									
Frontal sinus									
Supraorbital notch									
Parietal bone									
Occipital bone									
External occipital protuberance									
Foramen magnum									
Occipital condyle									
Temporal bone									
Carotid canal									
External acoustic meatus									
Internal acoustic meatus									
Mastoid process									
Squamous part of temporal bone									
Styloid process of temporal bone									
Zygomatic process of temporal bone									
Sphenoid bone									
Body									
Foramen ovale									
Foramen spinosum									
Greater wing									
Sella turcica									
Sphenoidal sinus									
Ethmoid bone									
Crista galli									
Ethmoid air cells									
Nasal concha—middle									
Nasal concha—superior									
Perpendicular plate of the ethmoid bone									

Bonus Question: What is the name for the bone shaped like a "butterfly"?

Table 3.4	Structures of the Head and Neck—Facial Bones								
Structures	**Dissection Views**								
	Individual Bone	**Anterior**	**Inferior Skull**	**Lateral**	**Mid-sagittal**	**Posterior**	**Superior**	**Skull Cranial Cavity**	**Orbit**
Maxilla									
Alveolar process of maxilla									
Infraorbital foramen									
Palatine bone									
Zygomatic bone									
Temporal process of zygomatic bone									
Nasal bone									
Vomer bone									
Mandible									
Alveolar process of mandible									
Angle of mandible									
Body of mandible									
Condylar process of mandible									
Coronoid process of mandible									
Mandibular foramen									
Mental foramen									
Ramus of mandible									

Bonus Question: What is the anatomical term that refers to the chin area? How do you suppose it received this name?

Table 3.5	Structures of the Head and Neck—Other Skull Structures								
Structures	**Dissection Views**								
	Individual Bone	**Anterior**	**Inferior Skull**	**Lateral**	**Mid-sagittal**	**Posterior**	**Superior**	**Skull Cranial Cavity**	**Orbit**
Coronal suture									
Cranial fossa—anterior									
Cranial fossa—middle									
Cranial fossa—posterior									
Foramen lacerum									
Hard palate									
Hyoid bone									
Jugular foramen									
Lambda									
Lambdoid suture									
Nasal concha (inferior)									
Pterion									
Sagittal suture									
Septal cartilage									
Temporomandibular joint (TMJ)									
Articular disk of the TMJ									
Zygomatic arch									

Bonus Question: What is the name for the "cheekbone"?

Table 3.6	Structures of the Head and Neck—Cervical Vertebrae								
Structures	**Dissection Views**								
	Individual Bone	**Anterior**	**Inferior Skull**	**Lateral**	**Mid-sagittal**	**Posterior**	**Superior**	**Skull Cranial Cavity**	**Orbit**
Atlas (C1 vertebra)									
Axis (C2 vertebra)									
Cervical vertebrae									
Spinous process (cervical)									
Transverse process (cervical)									
Vertebral body (cervical)									
Intervertebral disk									

Bonus Question: Which bones are characterized by the presence of "transverse foramina"?

EXERCISE 3.6:

Head and Neck, Anterior View

SELECT TOPIC	SELECT VIEW	
Head and Neck ▶	**Anterior**	GO

- *Click* **LAYER 1** *in the* **LAYER CONTROLS** *window, and you will see the following image:*

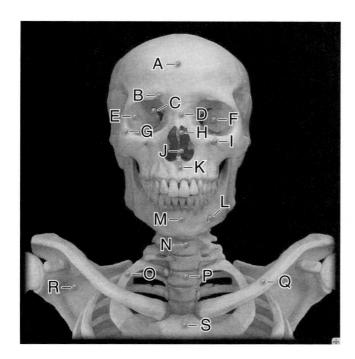

- *Mouse-over the pins on the screen to find the information necessary to identify the following structures:*

A. _____

B. _____

C. _____

D. _____

E. _____

F. _____

G. _____

H. _____

I. _____

J. _____

K. _____

L. _____

M. _____

N. _____

O. _____

P. _____

Q. _____

R. _____

S. _____

CHECK POINT:

Head and Neck, Anterior View

1. Name the bones contributing to the orbit.
2. Name the bone known as the "collar bone."
3. What structures are formed by the frontal bone?
4. Which is the shortest rib?
5. Name the superior part of the sternum.

EXERCISE 3.7:

Head and Neck, Lateral View

SELECT TOPIC	SELECT VIEW	
Head and Neck ▶	**Lateral**	**GO**

- *Click* **LAYER 1** *in the* **LAYER CONTROLS** *window, and you will see the following image:*

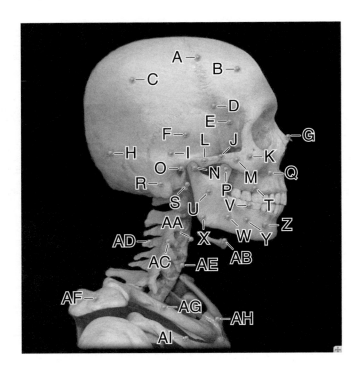

- *Mouse-over the pins on the screen to find the information necessary to identify the following structures:*

A. _____

B. _____

C. _____

D. _____

E. _____

F. _____

G. _____

H. _____

I. _____

J. _____

K. _____

L. _____

M. _____

N. _____

O. _____

P. _____

Q. _____

R. _____

S. _____

T. _____

U. _____

V. _____

W. _____

X. _____

Y. _____

Z. _____

AA. _____

AB. _____

AC. _____

AD. _____

AE. _____

AF. _____

AG. _____

AH. _____

AI. _____

CHECK POINT:

Head and Neck, Lateral View

1. Name the only bone in the body that does not articulate with any other bone.
2. Name the two bone processes that make up the zygomatic arch.
3. Name the "sockets" for the teeth.
4. Describe the pterion.
5. What structures of which bones form the temporomandibular joint?

EXERCISE 3.8:

Thorax, Anterior View

SELECT TOPIC	SELECT VIEW	
Thorax ▸	**Anterior**	**GO**

• *Click **LAYER 1** in the **LAYER CONTROLS** window, and you will see the following image:*

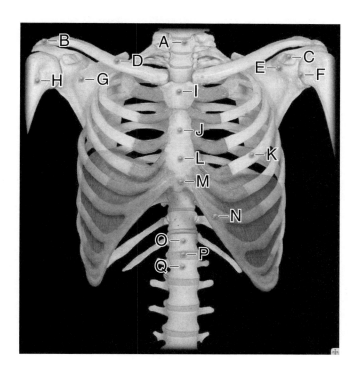

• *Mouse-over the pins on the screen to find the information necessary to identify the following structures:*

A. _____

B. _____

C. _____

D. _____

E. _____

F. _____

G. _____

H. _____

I. _____

J. _____

K. _____

L. _____

M. _____

N. _____

O. _____

P. _____

Q. _____

CHECK POINT:

Thorax, Anterior View

1. Name the three parts of the sternum.
2. Name a landmark for intramuscular injections.
3. Name the structures that attach the ribs to the sternum.
4. Name the two bones that form the glenohumeral joint.
5. Name the structure found between the bodies of all but two vertebrae. This structure is lacking between which two vertebrae?

EXERCISE 3.9:

Abdomen, Anterior View

SELECT TOPIC	SELECT VIEW	
Abdomen ▸	**Anterior**	**GO**

• *Click **LAYER 1** in the **LAYER CONTROLS** window, and you will see the following image:*

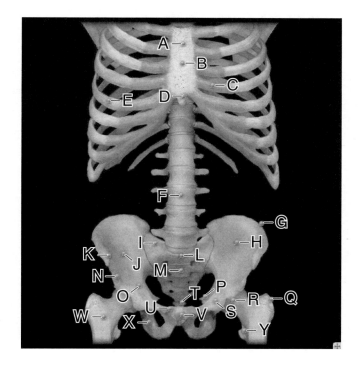

• *Mouse-over the pins on the screen to find the information necessary to identify the following structures:*

A. _____

B. _____

C. _____

D. _____

E. _____

F. _____

G. _____

H. _____

I. _____

J. _____

K. _____

L. _____

M. _____

N. _____

O. _____

P. _____

Q. _____

R. _____

S. _____

T. _____

U. _____

V. _____

W. _____

X. _____

Y. _____

CHECK POINT:

Abdomen, Anterior View

1. What process occurs to the pubic symphysis during late pregnancy?
2. What Latin term means "wing"? Where is a structure with this name located?
3. Name a landmark for intramuscular injections.
4. What term refers to the hip joint socket? Which bones contribute to this structure?
5. Name the structure formed by five fused vertebrae.

EXERCISE 3.10:

Back, Posterior View

SELECT TOPIC	SELECT VIEW	
Back	**Posterior**	GO

- *Click **LAYER 1** in the **LAYER CONTROLS** window, and you will see the following image:*

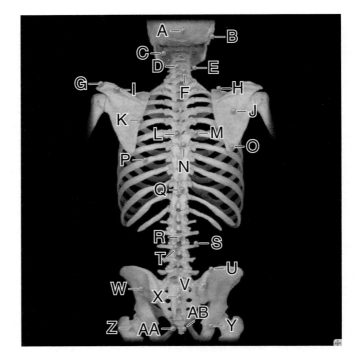

- *Mouse-over the pins on the screen to find the information necessary to identify the following structures:*

A. _____

B. _____

C. _____

D. _____

E. _____

F. _____

G. _____

H. _____

I. _____

J. _____

K. _____

L. _____

M. _____

N. _____

O. _____

P. _____

Q. _____

R. _____

S. _____

T. _____

U. _____

V. _____

W. _____

X. _____

Y. _____

Z. _____

AA. _____

AB. _____

CHECK POINT:

Back, Posterior View

1. Name the large, triangular flat bone of the superior back.
2. Name the three coxal bones.
3. Name the vertebrae of the lower back. How many are there?
4. Which intervertebral discs most commonly herniate?
5. Name the structure marked by a "dimple" on the lower back.

Skull and Associated Bones

Animation: The Skull

| | SELECT ANIMATION **Skull** | PLAY |

• *After viewing the animation, answer these questions:*

1. The bones that surround and protect the brain are referred to as the _____ bones.

2. The bones that form the underlying structure of the face are referred to as the _____ bones.

3. With one exception, the bones of the skull articulate with each other through joints known as _____. The exception is the _____.

4. There are numerous holes in the skull called _____.

5. What are the functions of these holes?

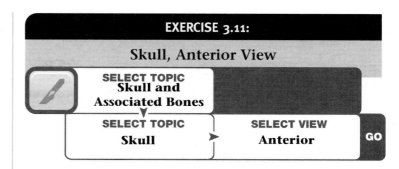

EXERCISE 3.11:

Skull, Anterior View

| SELECT TOPIC **Skull and Associated Bones** |
| SELECT TOPIC **Skull** | ▶ | SELECT VIEW **Anterior** | GO |

HEADS UP!

Many of the views we will see of the Skeletal System will indicate the locations of muscle attachments. These attachments are of two types – **Origins** *and* **Insertions.** *Origins, as indicated by red shading, are the relatively stationary or immobile points of skeletal muscle attachment. The* **Insertions,** *indicated by blue shading, are the points where a muscle attaches to a bone and produces movement. One way to remember the distinction between the two is to consider your life. Your* **Origin,** *or location of your birth, never changes or moves as you go about your life. But, your* **Insertion,** *where you are inserted on the earth at this time may be different than your origin. Thus, your* **Origin never moves,** *but when you move, your* **Insertion moves** *to another location.*

• *Click* **LAYER 1** *in the* **LAYER CONTROLS** *window, and you will see the following image:*

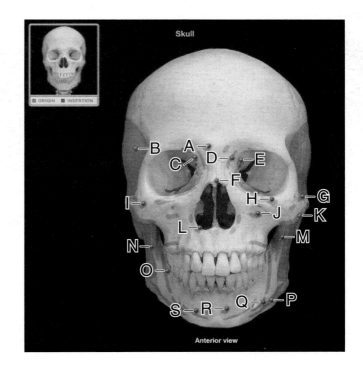

Anterior view

• *Mouse-over the pins on the screen to find the information necessary to identify the following structures:*

A. _____

B. _____

C. _____

D. _____

E. _____

F. _____

G. _____

H. _____

I. _____

J. _____

K. _____

L. _____

M. _____

N. _____

O. _____

P. _____

Q. _____

R. _____

S. _____

• *Click **LAYER 2** in the **LAYER CONTROLS** window, and you will see the following image:*

• *Mouse-over the pins on the screen to find the information necessary to identify the following structures:*

A. _____

B. _____

C. _____

D. _____

E. _____

F. _____

G. _____

H. _____

I. _____

J. _____

K. _____

L. _____

M. _____

N. _____

O. _____

P. _____

Q. _____

R. _____

S. _____

T. _____

U. _____

V. _____

CHECK POINT:

Skull, Anterior View

1. Name the structures transmitted through the mental foramen.
2. Name a feature of the skull that can be either a notch or a foramen.
3. What two bones contain teeth?
4. What is the name for the sockets of the teeth? (You may have to revisit earlier parts of this chapter to answer this one.)
5. The nasal septum consists of what specific bones or structures of bones?

EXERCISE 3.12:

Skull, Superior View

SELECT TOPIC	SELECT VIEW	
Skull	**Superior**	GO

• *Click* **LAYER 1** *in the* **LAYER CONTROLS** *window, and you will see the following image:*

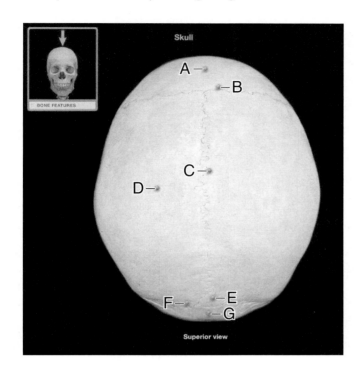

• *Mouse-over the pins on the screen to find the information necessary to identify the following structures:*

A. _____

B. _____

C. _____

D. _____

E. _____

F. _____

G. _____

CHECK POINT:

Skull, Superior View

1. Name the joint between the parietal bones.
2. Name the joint between the frontal and the parietal bones.
3. Name the bone type found in or near the sutures of the skull.
4. Where are these bones most often found?
5. Name the skull bone that articulates with the vertebral column.

EXERCISE 3.13:

Skull, Lateral View

SELECT TOPIC	SELECT VIEW	
Skull	**Lateral**	GO

• *Click* **LAYER 1** *in the* **LAYER CONTROLS** *window, and you will see the following image:*

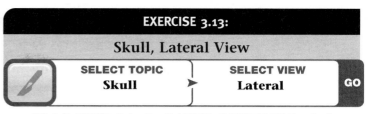
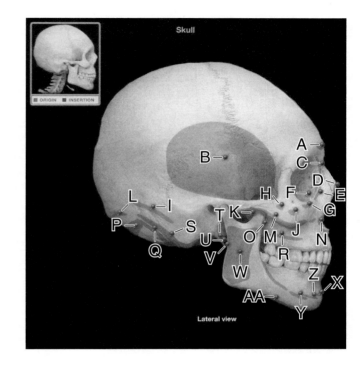

• *Mouse-over the pins on the screen to find the information necessary to identify the following structures:*

A. _____

B. _____

C. _____

D. _____

E. _____

F. _____

G. _____

H. _____

I. _____

J. _____

K. _____

L. _____

M. _____

N. _____

O. _____

P. _____

Q. _____

R. _____

S. _____

T. _____

U. _____

V. _____

W. _____

X. _____

Y. _____

Z. _____

AA. _____

- *Click* **LAYER 2** *in the* **LAYER CONTROLS** *window, and you will see the following image:*

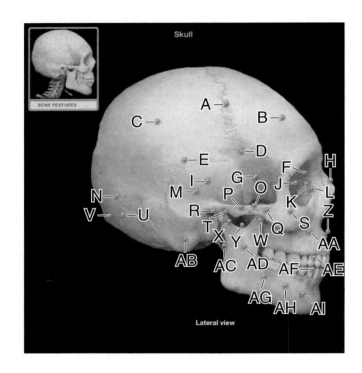

- *Mouse-over the pins on the screen to find the information necessary to identify the following structures:*

A. _____

B. _____

C. _____

D. _____

E. _____

F. _____

G. _____

H. _____

I. _____

J. _____

K. _____

L. _____

M. _____

N. _____

O. _____

P. _____

Q. _____

R. _____

S. _____

T. _____

U. _____

V. _____

W. _____

X. _____

Y. _____

Z. _____

AA. _____

AB. _____

AC. _____

AD. _____

AE. _____

AF. _____

AG. _____

AH. _____

AI. _____

C H E C K   P O I N T :

Skull, Lateral View

1. Name the specific bony structures that form the temporomandibular joint.
2. What two bones make up most of the lateral skull (one on each side)?
3. What suture is their point of articulation?

EXERCISE 3.14:

Skull, Posterior View

SELECT TOPIC	SELECT VIEW	
Skull	**Posterior**	**GO**

- *Click* **LAYER 1** *in the* **LAYER CONTROLS** *window, and you will see the following image:*

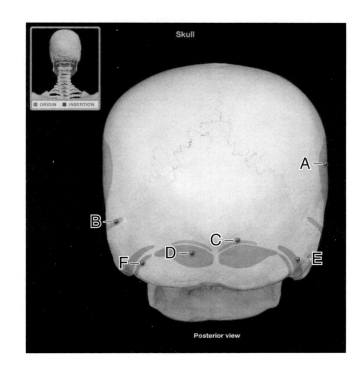

- *Mouse-over the pins on the screen to find the information necessary to identify the following structures:*

A. _____

B. _____

C. _____

D. _____

E. _____

F. _____

- *Click* **LAYER 2** *in the* **LAYER CONTROLS** *window, and you will see the following image:*

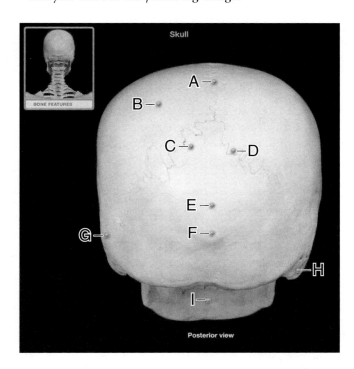

- *Mouse-over the pins on the screen to find the information necessary to identify the following structures:*

A. _____

B. _____

C. _____

D. _____

E. _____

F. _____

G. _____

H. _____

I. _____

CHECK POINT:

Skull, Posterior View

1. What suture forms the joint between the parietal and occipital bones?
2. What bone forms most of the posterior skull?
3. What is the attachment site on the skull for the nuchal ligament?

EXERCISE 3.15:

Skull, Mid-sagittal View

SELECT TOPIC	SELECT VIEW	
Skull	**Mid-sagittal**	GO

- Click the flashing **GO** button.

- Click **LAYER 1** in the **LAYER CONTROLS** window, and you will see the following image:

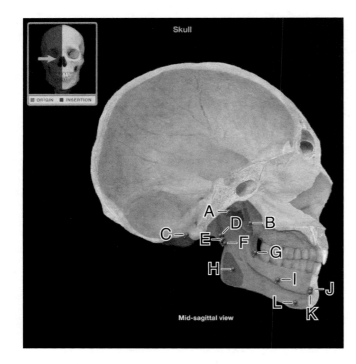

- Mouse-over the pins on the screen to find the information necessary to identify the following structures:

A. _____

B. _____

C. _____

D. _____

E. _____

F. _____

G. _____

H. _____

I. _____

J. _____

K. _____

L. _____

- Click **LAYER 2** in the **LAYER CONTROLS** window, and you will see the following image:

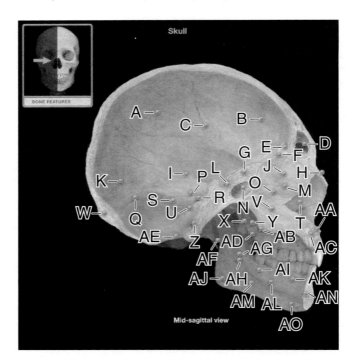

- Mouse-over the pins on the screen to find the information necessary to identify the following structures:

A. _____

B. _____

C. _____

D. _____

E. _____

F. _____

G. _____

H. _____

I. _____

J. _____

K. _____

L. _____

M. _____

N. _____

O. _____

P. _____

Q. _____

R. _____

S. _____

T. _____

U. _____

V. _____

W. _____

X. _____

Y. _____

Z. _____

AA. _____

AB. _____

AC. _____

AD. _____

AE. _____

AF. _____

AG. _____

AH. _____

AI. _____

AJ. _____

AK. _____

AL. _____

AM. _____

AN. _____

AO. _____

CHECK POINT:

Skull, Mid-sagittal View

1. When your dentist wants to numb your lower jaw by anesthetizing the nerves that serve the teeth and skin, what "hole" in what bone would be used to access those nerves?
2. Name the paranasal sinuses.
3. Name the structure that increases the surface area of the nasal cavity and plays an important role in warming inhaled air.
4. Name the structure that contains the sublingual salivary gland.
5. Name the muscles that attach to the styloid process of the temporal bone.

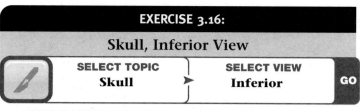

EXERCISE 3.16:

Skull, Inferior View

SELECT TOPIC	SELECT VIEW	
Skull	**Inferior**	GO

• *Click* **LAYER 1** *in the* **LAYER CONTROLS** *window, and you will see the following image:*

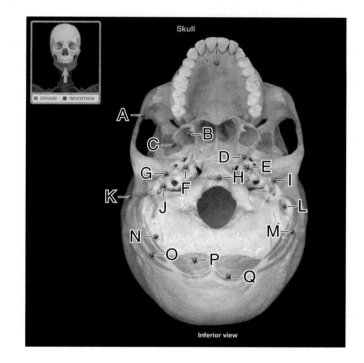

• *Mouse-over the pins on the screen to find the information necessary to identify the following structures:*

A. _____

B. _____

C. _____

D. _____

E. _____

F. _____

G. _____

H. _____

I. _____

J. _____

K. _____

L. _____

M. _____

N. _____

O. _____

P. _____

Q. _____

- *Click* **LAYER 2** *in the* **LAYER CONTROLS** *window, and you will see the following image:*

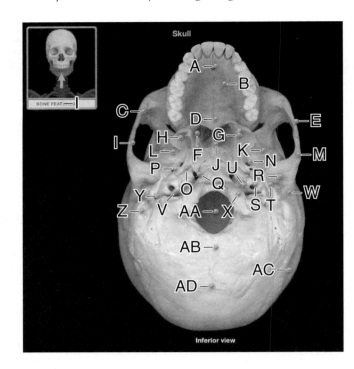

- *Mouse-over the pins on the screen to find the information necessary to identify the following structures:*

A. _____

B. _____

C. _____

D. _____

E. _____

F. _____

G. _____

H. _____

I. _____

J. _____

K. _____

L. _____

M. _____

N. _____

O. _____

P. _____

Q. _____

R. _____

S. _____

T. _____

U. _____

V. _____

W. _____

X. _____

Y. _____

Z. _____

AA. _____

AB. _____

AC. _____

AD. _____

CHECK POINT:

Skull, Inferior View

1. The zygomatic arch is made up of what two specific structures of what two bones?
2. What is the name for the large foramen on the inferior side of the occipital bone?
3. Name the irregular-shaped opening formed by the sphenoid, temporal, and occipital bones.

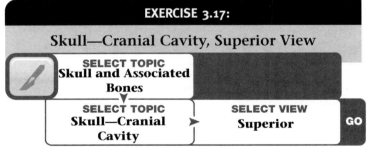

- *Click* **LAYER 1** *in the* **LAYER CONTROLS** *window, and you will see the following image:*

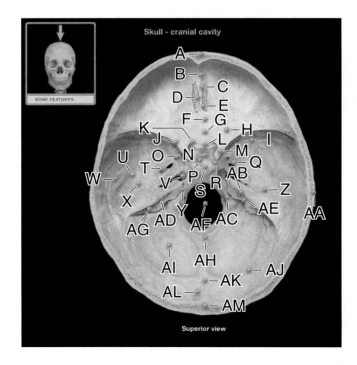

• *Mouse-over the pins on the screen to find the information necessary to identify the following structures:*

A. _____

B. _____

C. _____

D. _____

E. _____

F. _____

G. _____

H. _____

I. _____

J. _____

K. _____

L. _____

M. _____

N. _____

O. _____

P. _____

Q. _____

R. _____

S. _____

T. _____

U. _____

V. _____

W. _____

X. _____

Y. _____

Z. _____

AA. _____

AB. _____

AC. _____

AD. _____

AE. _____

AF. _____

AG. _____

AH. _____

AI. _____

AJ. _____

AK. _____

AL. _____

AM. _____

CHECK POINT:

Skull—Cranial Cavity, Superior View

1. Describe the cribriform plate.
2. Name the structures that traverse through the jugular foramen.
3. Name the structure that passes through the foramen ovale.
4. Name the structure contained in the hypophyseal fossa.
5. What structure is transmitted by the foramen rotundum?

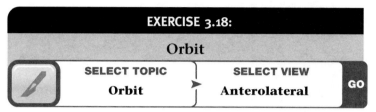

EXERCISE 3.18:

Orbit

SELECT TOPIC — **Orbit**   SELECT VIEW — **Anterolateral**   GO

• *Click **LAYER 1** in the **LAYER CONTROLS** window, and you will see the following image:*

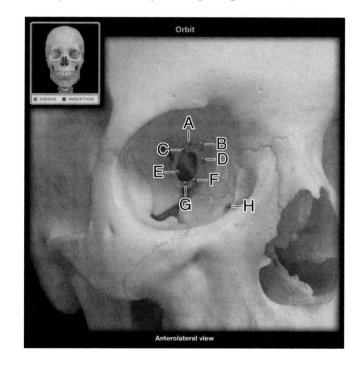

Anterolateral view

• *Mouse-over the pins on the screen to find the information necessary to identify the following structures:*

A. _____

B. _____

C. _____

D. _____

E. _____

F. _____

G. _____

H. _____

- *Click* **LAYER 2** *in the* **LAYER CONTROLS** *window, and you will see the following image:*

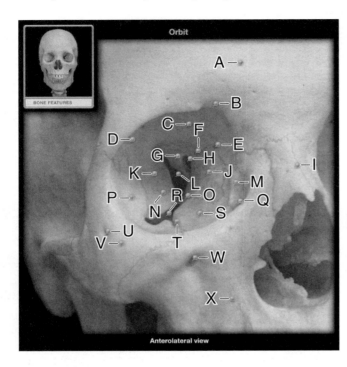

Orbit

Anterolateral view

- *Mouse-over the pins on the screen to find the information necessary to identify the following structures:*

A. _____

B. _____

C. _____

D. _____

E. _____

F. _____

G. _____

H. _____

I. _____

J. _____

K. _____

L. _____

M. _____

N. _____

O. _____

P. _____

Q. _____

R. _____

S. _____

T. _____

U. _____

V. _____

W. _____

X. _____

CHECK POINT:

Orbit

1. Name the seven bones that constitute the orbit.
2. Name the structure that transmits the optic nerve (CN II) and the ophthalmic artery.
3. Name the nerves and blood vessels transmitted by the superior orbital fissure.
4. Name the nerves and blood vessels transmitted by the inferior orbital fissure.
5. Name the nerves and blood vessels transmitted by the anterior ethmoidal foramen.

EXERCISE 3.19:

Ethmoid Bone

SELECT TOPIC	SELECT VIEW	
Ethmoid Bone ▶	**Superior-Posterior**	GO

- *Click* **LAYER 1** *in the* **LAYER CONTROLS** *window, and you will see the following image:*

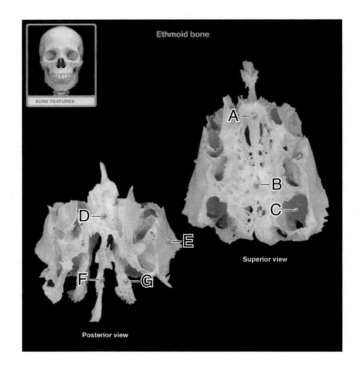

Ethmoid bone

Superior view

Posterior view

- *Mouse-over the pins on the screen to find the information necessary to identify the following structures:*

A. _____

B. _____

C. _____

D. _____

E. _____

F. _____

G. _____

CHECK POINT:

Ethmoid Bone

1. Name the characteristic features of the ethmoid bone.
2. Describe the cribriform plate of the ethmoid bone.
3. Name the structure that forms the superior part of the nasal septum.

CHECK POINT:

Frontal Bone

1. Name the structures that traverse the supraorbital notch.
2. Describe the supraorbital margin.
3. Name the sutures associated with the frontal bone. Name the bones that articulate the frontal bone at each suture.

EXERCISE 3.20:

Frontal Bone

| SELECT TOPIC Frontal Bone | ▶ | SELECT VIEW Anterior | GO |

- *Click* **LAYER 1** *in the* **LAYER CONTROLS** *window, and you will see the following image:*

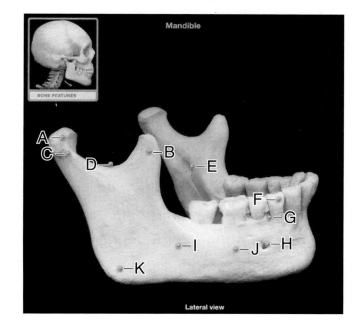

- *Mouse-over the pins on the screen to find the information necessary to identify the following structures:*

A. _____

B. _____

C. _____

D. _____

EXERCISE 3.21:

Mandible

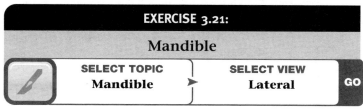

| SELECT TOPIC Mandible | ▶ | SELECT VIEW Lateral | GO |

- *Click* **LAYER 1** *in the* **LAYER CONTROLS** *window, and you will see the following image:*

- *Mouse-over the pins on the screen to find the information necessary to identify the following structures:*

A. _____

B. _____

C. _____

D. _____

E. _____

F. _____

G. _____

H. _____

I. _____

J. _____

K. _____

CHECK POINT:
_____

Mandible

1. Name the bone structures that constitute the temporomandibular joint.
2. What structure is also known as the mandibular incisure?

CHECK POINT:
_____

Maxilla

1. Name a structure of the maxilla bone that contributes to voice resonance.
2. Name the function of the incisive canal.
3. Name the structure that forms the anterior three-quarters of the hard palate.

EXERCISE 3.22:

Maxilla

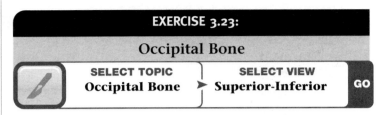

SELECT TOPIC	SELECT VIEW	
Maxilla ▶	Medial-Lateral	GO

• Click **LAYER 1** in the **LAYER CONTROLS** window, and you will see the following image:

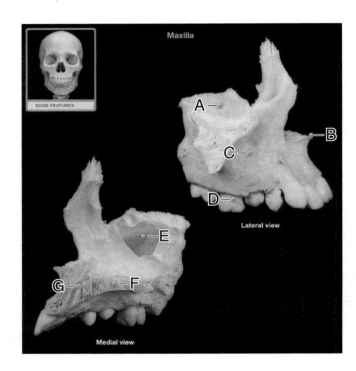

• Mouse-over the pins on the screen to find the information necessary to identify the following structures:

A. _____

B. _____

C. _____

D. _____

E. _____

F. _____

G. _____

EXERCISE 3.23:

Occipital Bone

SELECT TOPIC	SELECT VIEW	
Occipital Bone ▶	Superior-Inferior	GO

• Click **LAYER 1** in the **LAYER CONTROLS** window, and you will see the following image:

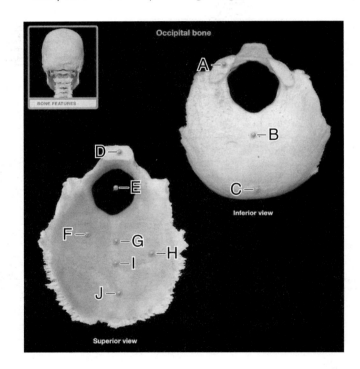

• Mouse-over the pins on the screen to find the information necessary to identify the following structures:

A. _____

B. _____

C. _____

D. _____

E. _____

F. _____

G. _____

H. _____

I. _____

J. _____

CHECK POINT:

Occipital Bone

1. Name the structures that pass through the foramen magnum.
2. What is the function of the occipital condyles?
3. What structures are contained within the cerebellar fossa?

EXERCISE 3.24:

Parietal Bone

SELECT TOPIC	SELECT VIEW	
Parietal Bone ▸	**Lateral**	**GO**

- *There are no* **TAGS** *for this bone, but be sure to read the information in the* **STRUCTURE INFORMATION** *window.*

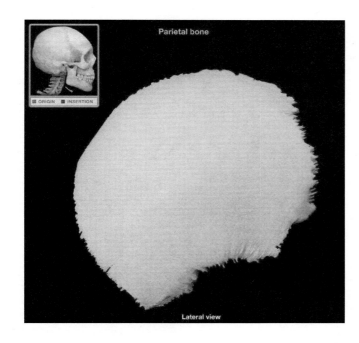

CHECK POINT:

Parietal Bone

1. Name the bones that articulate with the parietal bone.
2. Name the sutures involved with each articulation listed in question 1.

EXERCISE 3.25:

Sphenoid Bone

SELECT TOPIC	SELECT VIEW	
Sphenoid Bone ▸	**Superior-Posterior**	**GO**

- *Click* **LAYER 1** *in the* **LAYER CONTROLS** *window, and you will see the following image:*

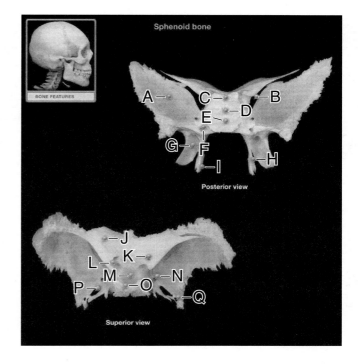

- *Mouse-over the pins on the screen to find the information necessary to identify the following structures:*

A. _____

B. _____

C. _____

D. _____

E. _____

F. _____

G. _____

H. _____

I. _____

J. _____

K. _____

L. _____

M. _____

N. _____

O. _____

P. _____

Q. _____

CHECK POINT:

Sphenoid Bone

1. What structure is contained in the sella turcica?
2. Name the two structures transmitted through the optic canal.
3. Name the paired lateral projections of the sphenoid bone.

CHECK POINT:

Temporal Bone

1. Name the bony canal that ends at the tympanic membrane.
2. Name the bony canal traversed by the facial and vestibulocochlear nerves.
3. Name the bony structure containing the organs of hearing and equilibrium.

EXERCISE 3.26:

Temporal Bone

| SELECT TOPIC | SELECT VIEW | |
| Temporal Bone ▶ | Medial-Lateral | GO |

- *Click the flashing* **GO** *button.*

- *Click* **LAYER 1** *in the* **LAYER CONTROLS** *window, and you will see the following image:*

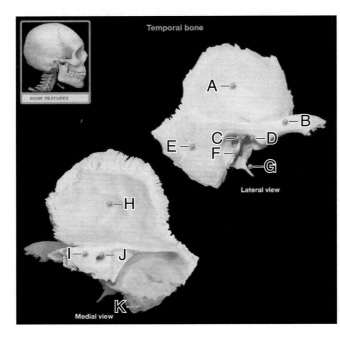

- *Mouse-over the pins on the screen to find the information necessary to identify the following structures:*

A. _____

B. _____

C. _____

D. _____

E. _____

F. _____

G. _____

H. _____

I. _____

J. _____

K. _____

EXERCISE 3.27:

Teeth

| SELECT TOPIC | SELECT VIEW | |
| Teeth ▶ | Superior-Inferior | GO |

- *Click* **LAYER 1** *in the* **LAYER CONTROLS** *window, and you will see the following image:*

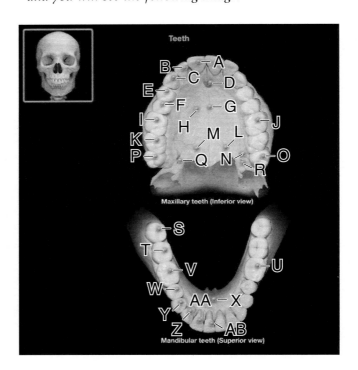

- *Mouse-over the pins on the screen to find the information necessary to identify the following structures:*

A. _____

B. _____

C. _____

D. _____

E. _____

F. _____

G. _____

H. _____

I. _____

J. _____

K. _____

L. _____

M. _____

N. _____

O. _____

P. _____

Q. _____

R. _____

S. _____

T. _____

U. _____

V. _____

W. _____

X. _____

Y. _____

Z. _____

AA. _____

AB. _____

CHECK POINT:

Teeth

1. What four teeth are also known as "wisdom teeth"?
2. What structures are transmitted through the lesser palatine foramen?
3. What structures are transmitted through the greater palatine foramen?

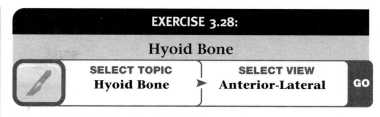

EXERCISE 3.28:

Hyoid Bone

SELECT TOPIC	SELECT VIEW	
Hyoid Bone	**Anterior-Lateral**	**GO**

- *Click* **LAYER 1** *in the* **LAYER CONTROLS** *window, and you will see the following image:*

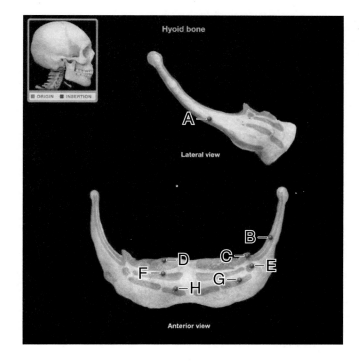

- *Mouse-over the pins on the screen to find the information necessary to identify the following structures:*

A. _____

B. _____

C. _____

D. _____

E. _____

F. _____

G. _____

H. _____

- *Click* **LAYER 2** *in the* **LAYER CONTROLS** *window, and you will see the following image:*

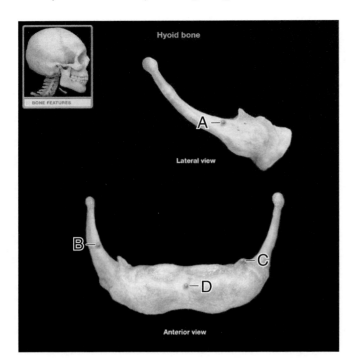

- *Mouse-over the pins on the screen to find the information necessary to identify the following structures:*

A. _____

B. _____

C. _____

D. _____

CHECK POINT:

Hyoid Bone

1. Name the function of the greater horn of the hyoid bone.
2. Name the function of the lesser horn of the hyoid bone.
3. What muscles attach to the body of the hyoid bone?

EXERCISE 3.29a:

Imaging—Head and Neck

SELECT TOPIC	SELECT VIEW	
Head and Neck	▶ Anterior-Posterior	GO

- *Click the* **TURN TAGS ON** *button, and you will see the following image:*

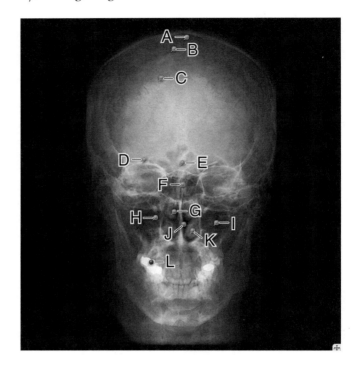

- *Mouse-over the green pins on the screen to find the information necessary to identify the following structures:*

A. _____

B. _____

C. _____

D. _____

E. _____

F. _____

G. _____

H. _____

I. _____

J. _____

K. _____

Non-skeletal System Structure (blue pin)

L. _____

EXERCISE 3.29b:

Imaging—Head and Neck

SELECT TOPIC	SELECT VIEW	
Head and Neck ▶	**Lateral**	**GO**

- *Click the* **TURN TAGS ON** *button and you will see the following image:*

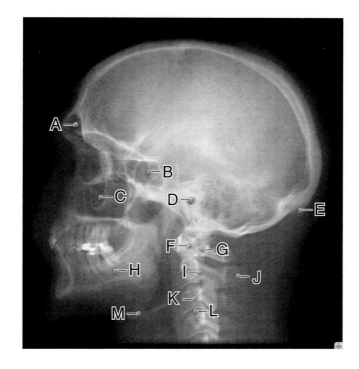

- *Mouse-over the pins on the screen to find the information necessary to identify the following structures:*

A. _____

B. _____

C. _____

D. _____

E. _____

F. _____

G. _____

H. _____

I. _____

J. _____

K. _____

L. _____

M. _____

Self Test

Take this opportunity to quiz yourself by taking the **SELF TEST.** See page 15 for a reminder on how to access the self-test for this section.

IN REVIEW

What Have I Learned?

The following questions cover the material that you just read, the introduction to the skeleton and the skeletal structures of the head and neck. Apply what you have learned in answering these questions on a separate piece of paper.

1. The hard palate consists of which bones?

2. What is the name for the large foramen on the inferior side of the occipital bone?

3. What passes through this foramen?

4. Name the two bones that form the bridge of the nose.

5. At the junction of the temporal and occipital bones, the internal jugular vein passes through which foramen?

6. The greater and lesser wings are parts of what bone?

7. Name the bony canal of the external ear. What bone is it a part of?

8. What is the attachment site for the sternocleidomastoid muscle on the skull?

9. What bony landmark is defined as "any bony prominence"?

10. What bony landmark is defined as "a pit or socket"?

11. What bony landmark is defined as "a hole through a bone, usually round"?

12. What bony landmark is defined as "a smooth, rounded knob"?

Thoracic Cage and Vertebral Column

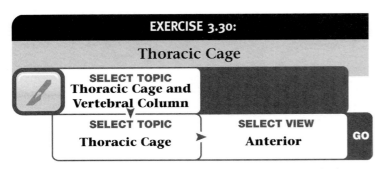

- *Click **LAYER 1** in the **LAYER CONTROLS** window, and you will see the following image:*

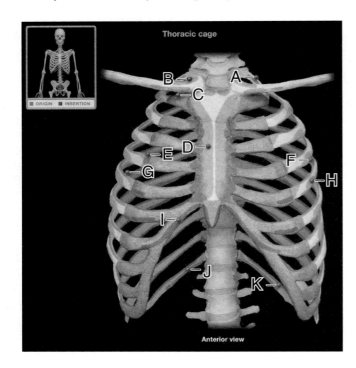

- *Mouse-over the pins on the screen to find the information necessary to identify the following structures:*

A. _____

B. _____

C. _____

D. _____

E. _____

F. _____

G. _____

H. _____

I. _____

J. _____

K. _____

- *Click **LAYER 2** in the **LAYER CONTROLS** window, and you will see the following image:*

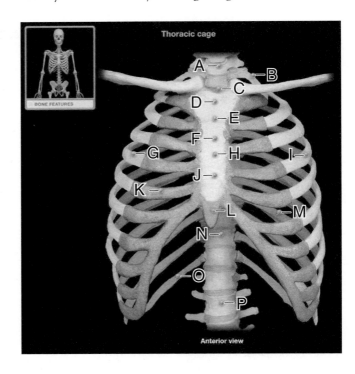

- *Mouse-over the pins on the screen to find the information necessary to identify the following structures:*

A. _____

B. _____

C. _____

D. _____

E. _____

F. _____

G. _____

H. _____

I. _____

J. _____

K. _____

L. _____

M. _____

N. _____

O. _____

P. _____

CHECK POINT:

Thoracic Cage

1. Describe the thoracic cage.
2. Define a true rib. How many are there?
3. Define a false rib. How many are there?
4. Define a floating rib. How many are there?

EXERCISE 3.31:

Imaging—Thorax

SELECT TOPIC	SELECT VIEW	
Thorax	▶ **Posterior-Anterior**	GO

- *Click the* **TURN TAGS ON** *button, and you will see the following image:*

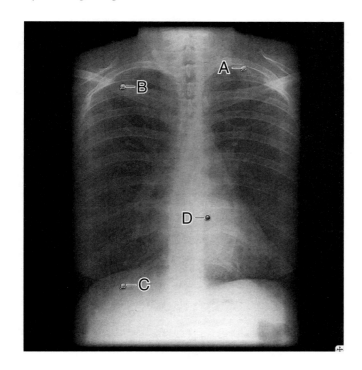

- *Mouse-over the green pins on the screen to find the information necessary to identify the following structures:*

A. _____

B. _____

C. _____

Non-skeletal System Structure (blue pin)

D. _____

EXERCISE 3.32:

Vertebral Column

SELECT TOPIC	SELECT VIEW	
Vertebral Column	▶ **Anterior-Posterior-Lateral**	GO

- *Click* **LAYER 1** *in the* **LAYER CONTROLS** *window, and you will see the following image:*

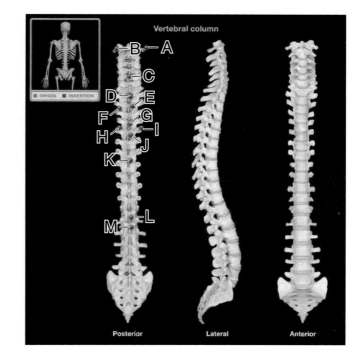

- *Mouse-over the pins on the screen to find the information necessary to identify the following structures:*

A. _____

B. _____

C. _____

D. _____

E. _____

F. _____

G. _____

H. _____

I. _____

J. _____

K. _____

L. _____

M. _____

- Click **LAYER 2** in the **LAYER CONTROLS** window, and you will see the following image:

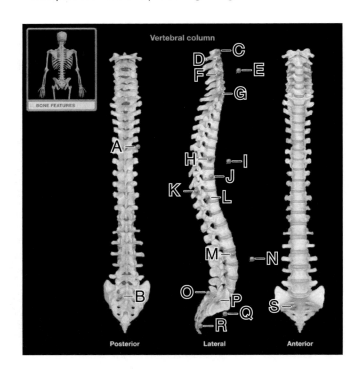

- Mouse-over the pins on the screen to find the information necessary to identify the following structures:

A. _____

B. _____

C. _____

D. _____

E. _____

F. _____

G. _____

H. _____

I. _____

J. _____

K. _____

L. _____

M. _____

N. _____

O. _____

P. _____

Q. _____

R. _____

S. _____

CHECK POINT:

Vertebral Column

1. Name the different groups of vertebrae. How many of each in a typical skeleton?
2. What structure passes through each intervertebral foramen?
3. Which vertebra has a dens?
4. Which vertebra has no vertebral body?

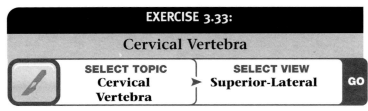

EXERCISE 3.33:

Cervical Vertebra

| SELECT TOPIC Cervical Vertebra | ▶ | SELECT VIEW Superior-Lateral | GO |

- Click **LAYER 1** in the **LAYER CONTROLS** window, and you will see the following image:

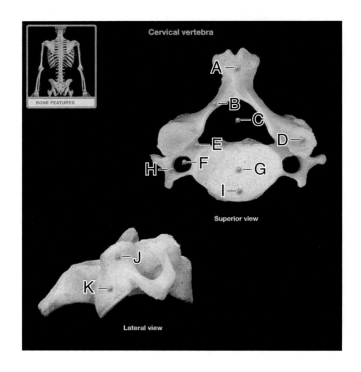

- Mouse-over the pins on the screen to find the information necessary to identify the following structures:

A. _____

B. _____

C. _____

D. _____

E. _____

F. _____

G. _____

H. _____

I. _____

J. _____

K. _____

H. _____

I. _____

J. _____

CHECK POINT:

Cervical Vertebra

1. Name a characteristic unique to the transverse processes of the cervical vertebrae.
2. What structure passes through this opening in C1–C7?
3. What is unique about the spinous process of most cervical vertebrae?
4. Which cervical vertebra does not have a spinous process?

CHECK POINT:

Atlas (C1 Vertebra)

1. What is the most posterior structure of the Atlas (C1 vertebra)?
2. What is its function?
3. What structure is lacking on the atlas and coccygeal vertebrae that is present on all other vertebrae?

EXERCISE 3.34:

Atlas (C1 Vertebra)

SELECT TOPIC	SELECT VIEW	
Atlas (C1 Vertebra) ▸	Superior	GO

- *Click* **LAYER 1** *in the* **LAYER CONTROLS** *window, and you will see the following image:*

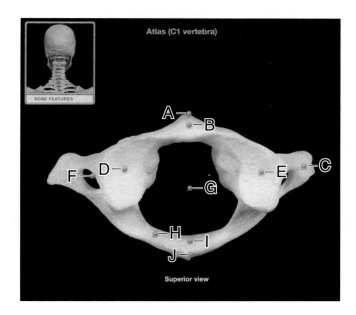

- *Mouse-over the pins on the screen to find the information necessary to identify the following structures:*

A. _____

B. _____

C. _____

D. _____

E. _____

F. _____

G. _____

EXERCISE 3.35:

Axis (C2 Vertebra)

SELECT TOPIC	SELECT VIEW	
Axis (C2 Vertebra) ▸	Posterior-Superior	GO

- *Click* **LAYER 1** *in the* **LAYER CONTROLS** *window, and you will see the following image:*

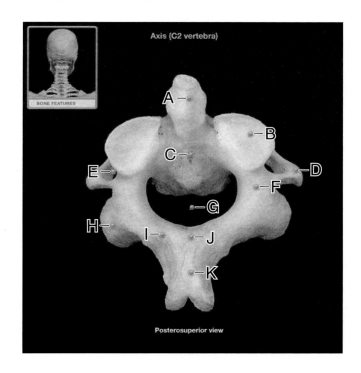

- *Mouse-over the pins on the screen to find the information necessary to identify the following structures:*

A. _____

B. _____

C. _____

D. _____

E. _____

F. _____

G. _____

H. _____

I. _____

J. _____

K. _____

F. _____

G. _____

H. _____

I. _____

J. _____

K. _____

L. _____

CHECK POINT:

Axis (C2 Vertebra)

1. Name a structure unique to the axis.
2. What is another name for this structure?
3. What does this structure represent?

CHECK POINT:

Thoracic Vertebra

1. List the characteristic features of thoracic vertebrae.
2. How many thoracic vertebrae do you have?
3. Where are they located?

EXERCISE 3.36:

Thoracic Vertebra

SELECT TOPIC	SELECT VIEW	
Thoracic Vertebra ▶	**Superior-Lateral**	GO

- *Click* **LAYER 1** *in the* **LAYER CONTROLS** *window, and you will see the following image:*

EXERCISE 3.37:

Lumbar Vertebra

SELECT TOPIC	SELECT VIEW	
Lumbar Vertebra ▶	**Superior-Lateral**	GO

- *Click* **LAYER 1** *in the* **LAYER CONTROLS** *window, and you will see the following image:*

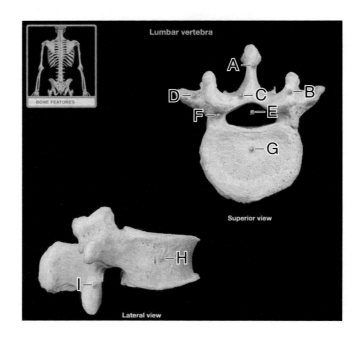

- *Mouse-over the pins on the screen to find the information necessary to identify the following structures:*

A. _____

B. _____

C. _____

D. _____

E. _____

- *Mouse-over the pins on the screen to find the information necessary to identify the following structures:*

A. _____

B. _____

C. _____

D. _____

E. _____

F. _____

G. _____

H. _____

I. _____

CHECK POINT:

Lumbar Vertebra

1. List the characteristic features of lumbar vertebrae.
2. How many lumbar vertebrae do you have?
3. Where are they located?

EXERCISE 3.38:

Sacrum and Coccyx

SELECT TOPIC	SELECT VIEW	
Sacrum and Coccyx	▶ **Anterior-Posterior**	**GO**

- *Click* **LAYER 1** *in the* **LAYER CONTROLS** *window, and you will see the following image:*

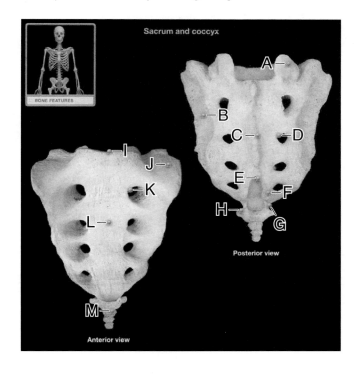

- *Mouse-over the pins on the screen to find the information necessary to identify the following structures:*

A. _____

B. _____

C. _____

D. _____

E. _____

F. _____

G. _____

H. _____

I. _____

J. _____

K. _____

L. _____

M. _____

- *Click* **LAYER 2** *in the* **LAYER CONTROLS** *window, and you will see the following image:*

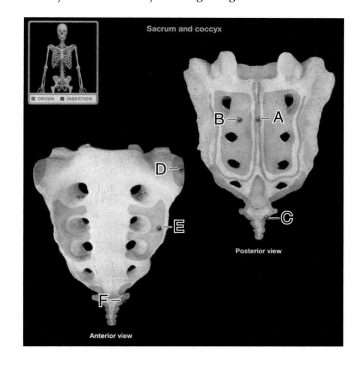

- *Mouse-over the pins on the screen to find the information necessary to identify the following structures:*

A. _____

B. _____

C. _____

D. _____

E. _____

F. _____

CHECK POINT:

Sacrum and Coccyx

1. Name the landmark for establishing female pelvic dimensions.
2. Name the route for injection in caudal epidural anesthesia.
3. Describe the coccyx.

Self Test

Take this opportunity to quiz yourself by taking the **SELF TEST.** See page 15 for a reminder on how to access the self-test for this section.

I N   R E V I E W

What Have I Learned?

The following questions cover the material that you just read, the thoracic cage and vertebral column. Apply what you have learned in answering these questions on a separate piece of paper.

1. Describe the thoracic cage.

2. Define a true rib. How many are there?

3. Define a false rib. How many are there?

4. Define a floating rib. How many are there?

5. Name the different groups of vertebrae. How many of each in a typical skeleton?

6. What structure passes through each intervertebral foramen?

7. Which vertebra has a dens?

8. Which vertebra has no vertebral body?

9. Name a characteristic unique to the transverse processes of the cervical vertebrae.

10. What structure passes through this opening in C1–C7?

11. What is unique about the spinous process of most cervical vertebrae?

12. Which vertebra does not have a spinous process?

13. What is the most posterior structure of the Atlas (C1 vertebra)?

14. What is its function?

15. What structure is lacking on the atlas and coccygeal vertebrae that is present on all other vertebrae?

16. Name a structure unique to the axis.

17. What is another name for this structure?

18. What does this structure represent?

19. List the characteristic features of thoracic vertebrae.

20. How many thoracic vertebrae do you have?

21. Where are they located?

22. List the characteristic features of lumbar vertebrae.

23. How many lumbar vertebrae do you have?

24. Where are they located?

25. Name the landmark for establishing female pelvic dimensions.

26. Name the route for injection in caudal epidural anesthesia.

27. Describe the coccyx.

Pectoral Girdle and Upper Limb

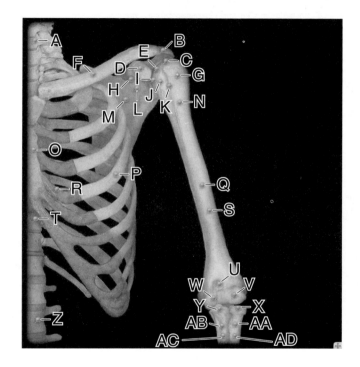

EXERCISE 3.39:

Shoulder and Arm, Anterior View

SELECT TOPIC
Pectoral Girdle and Upper Limb
▼
SELECT TOPIC
Shoulder and Arm ▶
SELECT VIEW
Anterior
GO

- *Click **LAYER 1** in the **LAYER CONTROLS** window, and you will see the following image:*

- *Mouse-over the pins on the screen to find the information necessary to identify the following structures:*

A. _____

B. _____

C. _____

D. _____

E. _____

F. _____

G. _____

H. _____

I. _____

J. _____

K. _____

L. _____

M. _____

N. _____

O. _____

P. _____

Q. _____

R. _____

S. _____

T. _____

U. _____

V. _____

W. _____

X. _____

Y. _____

Z. _____

AA. _____

AB. _____

AC. _____

AD. _____

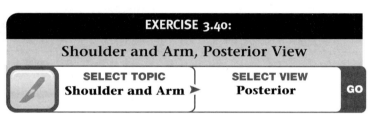

EXERCISE 3.40:

Shoulder and Arm, Posterior View

SELECT TOPIC
Shoulder and Arm ▶
SELECT VIEW
Posterior
GO

- *Click **LAYER 1** in the **LAYER CONTROLS** window, and you will see the following image:*

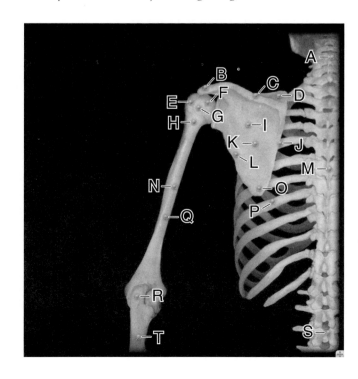

- *Mouse-over the pins on the screen to find the information necessary to identify the following structures:*

A. _____

B. _____

C. _____

D. _____

E. _____

F. _____

G. _____

H. _____

I. _____

J. _____

K. _____

L. _____

M. _____

N. _____

O. _____

P. _____

Q. _____

R. _____

S. _____

T. _____

CHECK POINT:

Shoulder and Arm

1. Which bony structures form the glenohumeral joint?
2. How many ribs do you have?
3. Both true and false ribs articulate with _____.
4. Name the characteristic features of the scapula.

EXERCISE 3.41:

Imaging—Shoulder and Arm

 | SELECT TOPIC **Shoulder and Arm** | ► | SELECT VIEW **Anterior-Posterior** | GO

- *Click the* **TURN TAGS ON** *button, and you will see the following image:*

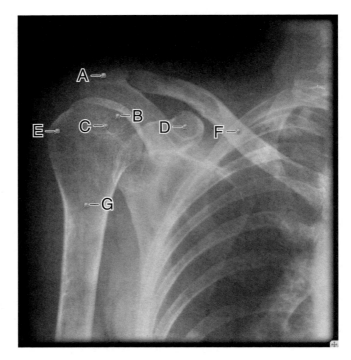

- *Mouse-over the pins on the screen to find the information necessary to identify the following structures:*

A. _____

B. _____

C. _____

D. _____

E. _____

F. _____

G. _____

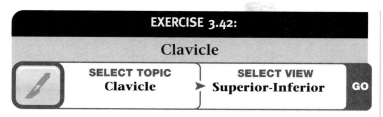

• *Click* **LAYER 1** *in the* **LAYER CONTROLS** *window, and you will see the following image:*

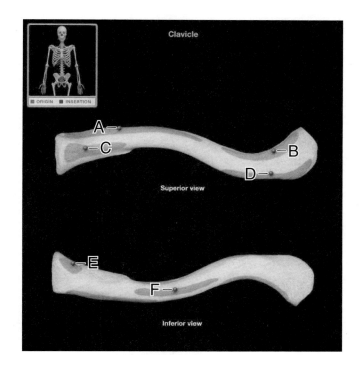

• *Mouse-over the pins on the screen to find the information necessary to identify the following structures:*

A. _____

B. _____

C. _____

D. _____

E. _____

F. _____

• *Click* **LAYER 2** *in the* **LAYER CONTROLS** *window, and you will see the following image:*

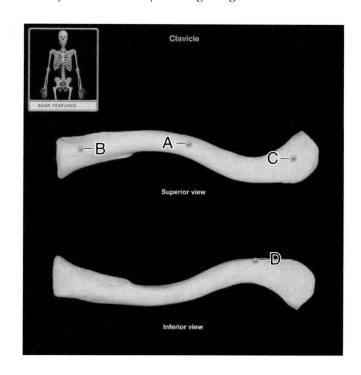

• *Mouse-over the pins on the screen to find the information necessary to identify the following structures:*

A. _____

B. _____

C. _____

D. _____

CHECK POINT:

Clavicle

1. The medial end of the clavicle articulates. . . .
2. The lateral end of the clavicle articulates. . . .
3. Another name for the clavicle is. . . .

EXERCISE 3.43:

Scapula

SELECT TOPIC	SELECT VIEW	
Scapula	▶ **Anterior-Posterior-Lateral**	**GO**

• *Click* **LAYER 1** *in the* **LAYER CONTROLS** *window, and you will see the following image:*

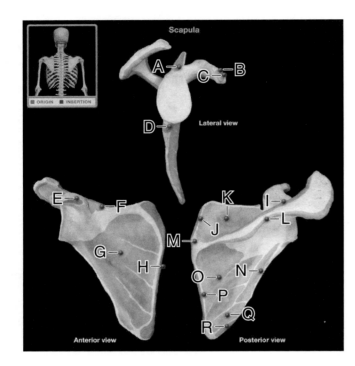

• *Mouse-over the pins on the screen to find the information necessary to identify the following structures:*

A. _____

B. _____

C. _____

D. _____

E. _____

F. _____

G. _____

H. _____

I. _____

J. _____

K. _____

L. _____

M. _____

N. _____

O. _____

P. _____

Q. _____

R. _____

• *Click* **LAYER 2** *in the* **LAYER CONTROLS** *window, and you will see the following image:*

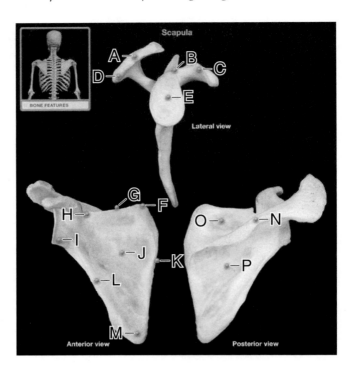

• *Mouse-over the pins on the screen to find the information necessary to identify the following structures:*

A. _____

B. _____

C. _____

D. _____

E. _____

F. _____

G. _____

H. _____

I. _____

J. _____

K. _____

L. _____

M. _____

N. _____

O. _____

P. _____

CHECK POINT:

Scapula

1. Name a landmark for intramuscular injections.
2. Name another visible subcutaneous landmark.

EXERCISE 3.44:

Humerus

SELECT TOPIC
Humerus

SELECT VIEW
Anterior-Posterior

GO

• *Click* **LAYER 1** *in the* **LAYER CONTROLS** *window, and you will see the following image:*

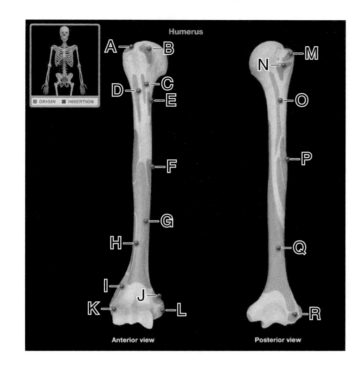

• *Mouse-over the pins on the screen to find the information necessary to identify the following structures:*

A. _____

B. _____

C. _____

D. _____

E. _____

F. _____

G. _____

H. _____

I. _____

J. _____

K. _____

L. _____

M. _____

N. _____

O. _____

P. _____

Q. _____

R. _____

• *Click* **LAYER 2** *in the* **LAYER CONTROLS** *window, and you will see the following image:*

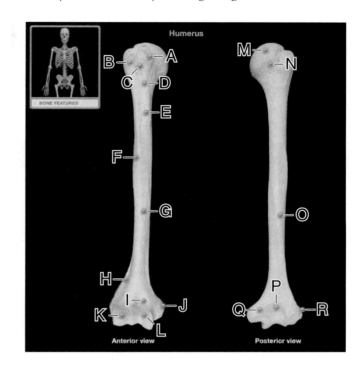

• *Mouse-over the pins on the screen to find the information necessary to identify the following structures:*

A. _____

B. _____

C. _____

D. _____

E. _____

F. _____

G. _____

H. _____

I. _____

J. _____

K. _____

L. _____

M. _____

N. _____

O. _____

P. _____

Q. _____

R. _____

CHECK POINT:

Humerus

1. How many necks are on the proximal end of the humerus? Where are they located?
2. Which neck is a common site for fractures?
3. Why is your elbow referred to as your "funny bone"?

EXERCISE 3.45:

Forearm and Hand, Anterior View

SELECT TOPIC	SELECT VIEW	
Forearm and Hand	▶ Anterior	GO

- *Click **LAYER 1** in the **LAYER CONTROLS** window, and you will see the following image:*

- *Mouse-over the pins on the screen to find the information necessary to identify the following structures:*

A. _____

B. _____

C. _____

D. _____

E. _____

F. _____

G. _____

H. _____

I. _____

J. _____

K. _____

L. _____

M. _____

N. _____

O. _____

P. _____

Q. _____

R. _____

S. _____

T. _____

U. _____

V. _____

W. _____

X. _____

Y. _____

Z. _____

AA. _____

AB. _____

AC. _____

AD. _____

AE. _____

AF. _____

EXERCISE 3.46:

Forearm and Hand, Posterior View

| | SELECT TOPIC **Forearm and Hand** | ▶ | SELECT VIEW **Posterior** | GO |

- *Click* **LAYER 1** *in the* **LAYER CONTROLS** *window, and you will see the following image:*

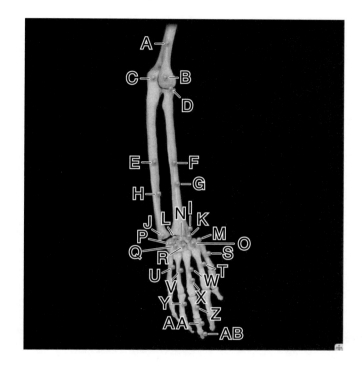

- *Mouse-over the pins on the screen to find the information necessary to identify the following structures:*

A. _____

B. _____

C. _____

D. _____

E. _____

F. _____

G. _____

H. _____

I. _____

J. _____

K. _____

L. _____

M. _____

N. _____

O. _____

P. _____

Q. _____

R. _____

S. _____

T. _____

U. _____

V. _____

W. _____

X. _____

Y. _____

Z. _____

AA. _____

AB. _____

CHECK POINT:

Forearm and Hand

1. List the proximal carpal bones from lateral to medial.
2. List the distal carpal bones from lateral to medial.
3. List the phalanges of fingers II through V from proximal to distal.
4. List the phalanges of finger I from proximal to distal.

EXERCISE 3.47:

Radius and Ulna

| | SELECT TOPIC **Radius and Ulna** | ▶ | SELECT VIEW **Anterior-Posterior** | GO |

- *Click* **LAYER 1** *in the* **LAYER CONTROLS** *window, and you will see the following image:*

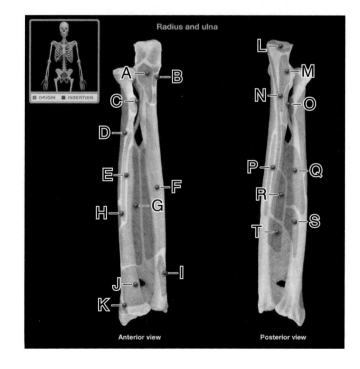

- *Mouse-over the pins on the screen to find the information necessary to identify the following structures:*

A. _____

B. _____

C. _____

D. _____

E. _____

F. _____

G. _____

H. _____

I. _____

J. _____

K. _____

L. _____

M. _____

N. _____

O. _____

P. _____

Q. _____

R. _____

S. _____

T. _____

- *Click **LAYER 2** in the **LAYER CONTROLS** window, and you will see the following image:*

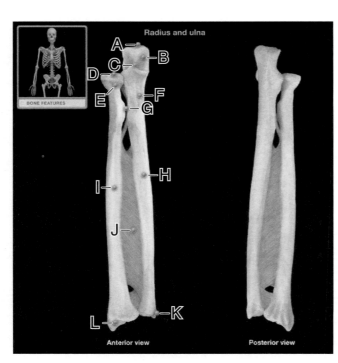

- *Mouse-over the pins on the screen to find the information necessary to identify the following structures:*

A. _____

B. _____

C. _____

D. _____

E. _____

F. _____

G. _____

H. _____

I. _____

J. _____

K. _____

L. _____

CHECK POINT:

Radius and Ulna

1. Describe the interosseous membrane of the forearm.
2. What structure holds the radio-ulnar joint in place?
3. Name the distal pointed projection of the radius. Of the ulna?

EXERCISE 3.48:

Wrist and Hand, Anterior View

SELECT TOPIC	SELECT VIEW	
Wrist and Hand ▸	**Anterior**	**GO**

- *Click **LAYER 1** in the **LAYER CONTROLS** window and you will see the following image:*

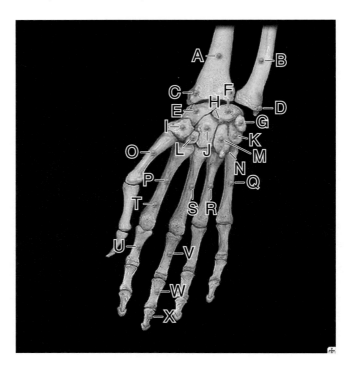

- *Mouse-over the pins on the screen to find the information necessary to identify the following structures:*

A. _____

B. _____

C. _____

D. _____

E. _____

F. _____

G. _____

H. _____

I. _____

J. _____

K. _____

L. _____

M. _____

N. _____

O. _____

P. _____

Q. _____

R. _____

S. _____

T. _____

U. _____

V. _____

W. _____

X. _____

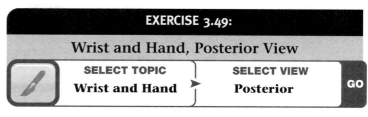

- *Click **LAYER 1** in the **LAYER CONTROLS** window, and you will see the following image:*

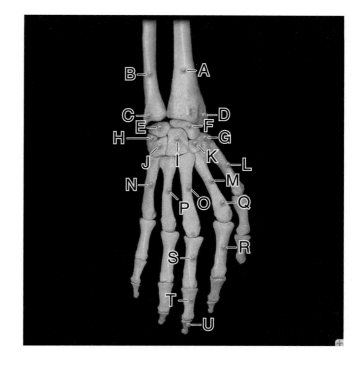

- *Mouse-over the pins on the screen to find the information necessary to identify the following structures:*

A. _____

B. _____

C. _____

D. _____

E. _____

F. _____

G. _____

H. _____

I. _____

J. _____

K. _____

L. _____

M. _____

N. _____

O. _____

P. _____

Q. _____

R. _____

S. _____

T. _____

U. _____

EXERCISE 3.50:

Wrist and Hand, Anterior and Posterior View

| SELECT TOPIC Wrist and Hand | ▸ SELECT VIEW Anterior -Posterior | GO |

- *Click **LAYER 1** in the **LAYER CONTROLS** window, and you will see the following image:*

- *Mouse-over the pins on the screen to find the information necessary to identify the following structures:*

A. _____

B. _____

C. _____

D. _____

E. _____

F. _____

G. _____

H. _____

I. _____

J. _____

K. _____

L. _____

M. _____

N. _____

O. _____

P. _____

Q. _____

R. _____

S. _____

T. _____

U. _____

V. _____

W. _____

X. _____

Y. _____

Z. _____

AA. _____

AB. _____

AC. _____

AD. _____

AE. _____

AF. _____

- *Click **LAYER 2** in the **LAYER CONTROLS** window, and you will see the following image:*

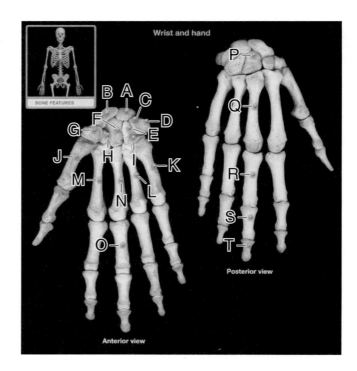

• *Mouse-over the pins on the screen to find the information necessary to identify the following structures:*

A. _____

B. _____

C. _____

D. _____

E. _____

F. _____

G. _____

H. _____

I. _____

J. _____

K. _____

L. _____

M. _____

N. _____

O. _____

P. _____

Q. _____

R. _____

S. _____

T. _____

CHECK POINT:

Wrist and Hand

1. Name the lateral bone of the forearm.
2. Name the medial bone of the forearm.
3. How are the fingers numbered?

EXERCISE 3.51a:

Imaging—Wrist and Hand

SELECT TOPIC	SELECT VIEW	
Wrist and Hand ▶	**Anterior-Posterior**	GO

• *Click the* **TURN TAGS ON** *button, and you will see the following image:*

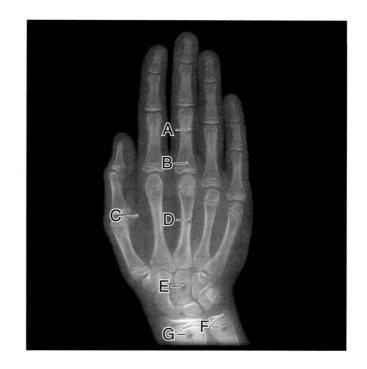

• *Mouse-over the pins on the screen to find the information necessary to identify the following structures:*

A. _____

B. _____

C. _____

D. _____

E. _____

F. _____

G. _____

EXERCISE 3.51b:

Imaging—Wrist and Hand

SELECT TOPIC	SELECT VIEW	
Wrist and Hand	➤ **Posterior-Anterior**	GO

- *Click the* **TURN TAGS ON** *button, and you will see the following image:*

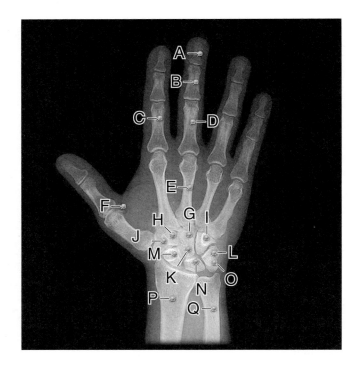

- *Mouse-over the pins on the screen to find the information necessary to identify the following structures:*

A. _____

B. _____

C. _____

D. _____

E. _____

F. _____

G. _____

H. _____

I. _____

J. _____

K. _____

L. _____

M. _____

N. _____

O. _____

P. _____

Q. _____

Self Test
Take this opportunity to quiz yourself by taking the **SELF TEST.** See page 15 for a reminder on how to access the self-test for this section.

I N   R E V I E W

What Have I Learned?

The following questions cover the material that you just learned: the pectoral girdle and upper limb. Apply what you have learned in answering these questions on a separate piece of paper.

1. Which bony structures form the glenohumeral joint?

2. Name the characteristic features of the scapula.

3. The lateral end of the clavicle articulates . . . .

4. Name a landmark of the scapula for intramuscular injections.

5. Name another visible subcutaneous landmark of the scapula.

6. How many necks are on the proximal end of the humerus? Where are they located?

7. Which neck is a common site for fractures?

8. Why is your elbow referred to as your "funny bone"?

9. List the proximal carpal bones from lateral to medial.

10. List the distal carpal bones from lateral to medial.

11. List the phalanges of fingers II through V from proximal to distal.

12. List the phalanges of finger I from proximal to distal.

13. Describe the interosseous membrane of the forearm.

Continued

14. What structure holds the radio-ulnar joint in place?

15. Name the distal pointed projection of the radius. Of the ulna.

16. Name the lateral bone of the forearm.

17. Name the medial bone of the forearm.

18. Name the bones and the bony structures that form the shoulder joint.

19. Name the wrist bone with a prominent hook (hamulus).

20. What term describes a bone embedded in a tendon? Which wrist bone is embedded in a tendon? Which tendon?

21. Name the prominent ridge on the posterior scapula.

22. Name the eight wrist bones.

Pelvic Girdle and Lower Limb

EXERCISE 3.52:

Hip and Thigh, Anterior View

SELECT TOPIC
Pelvic Girdle and Lower Limb
▼
SELECT TOPIC
Hip and Thigh ▶ SELECT VIEW
Anterior   GO

• *Click* **LAYER 1** *in the* **LAYER CONTROLS** *window, and you will see the following image:*

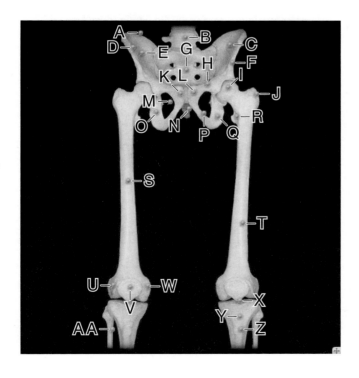

• *Mouse-over the pins on the screen to find the information necessary to identify the following structures:*

A. _____

B. _____

C. _____

D. _____

E. _____

F. _____

G. _____

H. _____

I. _____

J. _____

K. _____

L. _____

M. _____

N. _____

O. _____

P. _____

Q. _____

R. _____

S. _____

T. _____

U. _____

V. _____

W. _____

X. _____

Y. _____

Z. _____

AA. _____

EXERCISE 3.53:

Hip and Thigh, Posterior View

| SELECT TOPIC | SELECT VIEW | |
| Hip and Thigh ▶ | Posterior | GO |

• *Click* **LAYER 1** *in the* **LAYER CONTROLS** *window, and you will see the following image:*

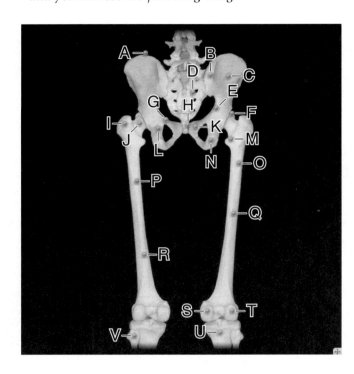

• *Mouse-over the pins on the screen to find the information necessary to identify the following structures:*

A. _____

B. _____

C. _____

D. _____

E. _____

F. _____

G. _____

H. _____

I. _____

J. _____

K. _____

L. _____

M. _____

N. _____

O. _____

P. _____

Q. _____

R. _____

S. _____

T. _____

U. _____

V. _____

CHECK POINT:

Hip and Thigh

1. Describe the patella.
2. Describe the sacrum.
3. Name the enlarged distal ends of the femur.

EXERCISE 3.54:

Pelvis

| SELECT TOPIC | SELECT VIEW | |
| Pelvis ▶ | Superior | GO |

• *Click* **LAYER 1** *in the* **LAYER CONTROLS** *window, and you will see the following image:*

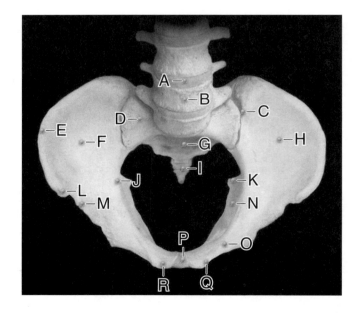

• *Mouse-over the pins on the screen to find the information necessary to identify the following structures:*

A. _____

B. _____

C. _____

D. _____

E. _____

F. _____

G. _____

H. _____

I. _____

J. _____

K. _____

L. _____

M. _____

N. _____

O. _____

P. _____

Q. _____

R. _____

CHECK POINT:

Pelvis

1. Name the synovial joint between the sacrum and the ilium.
2. How much movement does this joint allow? Why?
3. Name the landmark for administering anesthetic during childbirth.

EXERCISE 3.55:

Pelvic Girdle—Female

| SELECT TOPIC Pelvic Girdle—Female | ▸ | SELECT VIEW Anterior | GO |

- *Click* **LAYER 1** *in the* **LAYER CONTROLS** *window, and you will see the following image:*

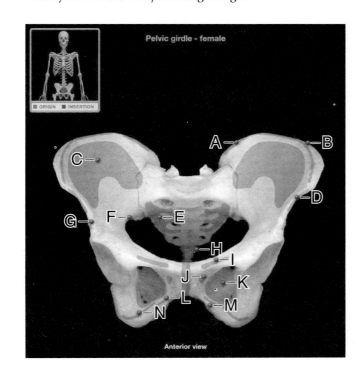

- *Mouse-over the pins on the screen to find the information necessary to identify the following structures:*

A. _____

B. _____

C. _____

D. _____

E. _____

F. _____

G. _____

H. _____

I. _____

J. _____

K. _____

L. _____

M. _____

N. _____

- *Click* **LAYER 2** *in the* **LAYER CONTROLS** *window, and you will see the following image:*

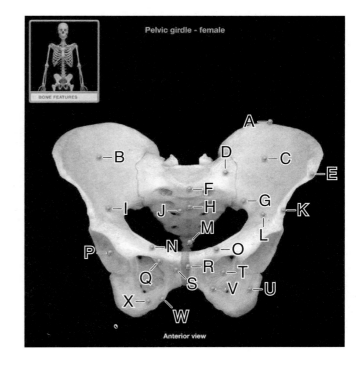

- *Mouse-over the pins on the screen to find the information necessary to identify the following structures:*

A. _____

B. _____

C. _____

D. _____

E. _____

F. _____

G. _____

H. _____

I. _____

J. _____

K. _____

L. _____

M. _____

N. _____

O. _____

P. _____

Q. _____

R. _____

S. _____

T. _____

U. _____

V. _____

W. _____

X. _____

C H E C K   P O I N T :

Pelvic Girdle—Female

1. In the pregnant female, what events occur concerning the pubic symphysis?
2. Describe the subpubic angle.
3. The subpubic angle in females is usually _____ degrees.

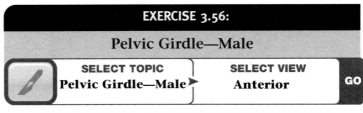

EXERCISE 3.56:

Pelvic Girdle—Male

| SELECT TOPIC | SELECT VIEW | |
| Pelvic Girdle—Male ▶ | Anterior | GO |

- *Click* **LAYER 1** *in the* **LAYER CONTROLS** *window, and you will see the following image:*

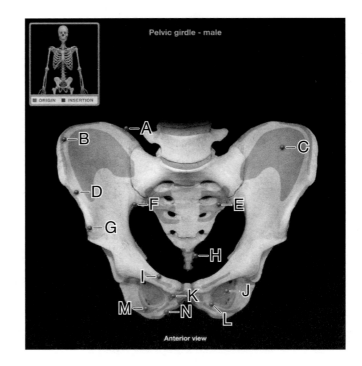

- *Mouse-over the pins on the screen to find the information necessary to identify the following structures:*

A. _____

B. _____

C. _____

D. _____

E. _____

F. _____

G. _____

H. _____

I. _____

J. _____

K. _____

L. _____

M. _____

N. _____

• Click **LAYER 2** in the **LAYER CONTROLS** window, and you will see the following image:

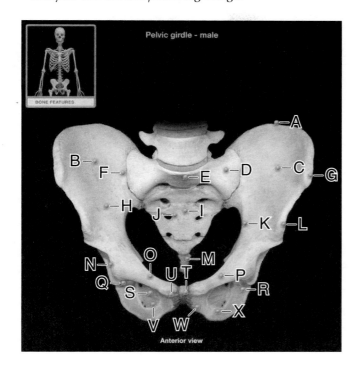

• Mouse-over the pins on the screen to find the information necessary to identify the following structures:

A. _____

B. _____

C. _____

D. _____

E. _____

F. _____

G. _____

H. _____

I. _____

J. _____

K. _____

L. _____

M. _____

N. _____

O. _____

P. _____

Q. _____

R. _____

S. _____

T. _____

U. _____

V. _____

W. _____

X. _____

CHECK POINT:

Pelvic Girdle—Male

1. Describe the difference between the male and female obturator foramena.
2. What structures are found in the right iliac fossa? The left iliac fossa?
3. The subpubic angle in males is usually _____ degrees.

EXERCISE 3.57:

Hip Bone

SELECT TOPIC	SELECT VIEW	
Hip Bone	**Medial-Lateral**	GO

• Click **LAYER 1** in the **LAYER CONTROLS** window, and you will see the following image:

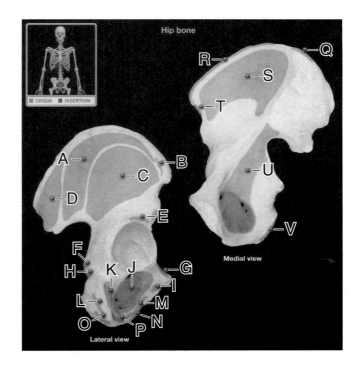

• Mouse-over the pins on the screen to find the information necessary to identify the following structures:

A. _____

B. _____

C. _____

D. _____

E. _____

F. _____

G. _____

H. _____

I. _____

J. _____

K. _____

L. _____

M. _____

N. _____

O. _____

P. _____

Q. _____

R. _____

S. _____

T. _____

U. _____

V. _____

- *Click* **LAYER 2** *in the* **LAYER CONTROLS** *window, and you will see the following image:*

- *Mouse-over the pins on the screen to find the information necessary to identify the following structures:*

A. _____

B. _____

C. _____

D. _____

E. _____

F. _____

G. _____

H. _____

I. _____

J. _____

K. _____

L. _____

M. _____

N. _____

O. _____

P. _____

Q. _____

R. _____

S. _____

T. _____

U. _____

V. _____

W. _____

X. _____

Y. _____

Z. _____

CHECK POINT:

Hip Bone

1. Describe the acetabulum.
2. Name a hip landmark for intramuscular injections.
3. Name a hip landmark for administering anesthetic during childbirth.

EXERCISE 3.58:

Femur

| | SELECT TOPIC
Femur | ➤ | SELECT VIEW
Anterior-Posterior | GO |

- *Click* **LAYER 1** *in the* **LAYER CONTROLS** *window, and you will see the following image:*

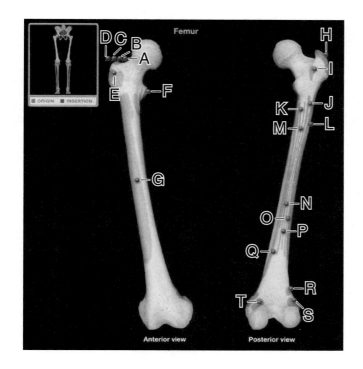

- *Mouse-over the pins on the screen to find the information necessary to identify the following structures:*

A. _____

B. _____

C. _____

D. _____

E. _____

F. _____

G. _____

H. _____

I. _____

J. _____

K. _____

L. _____

M. _____

N. _____

O. _____

P. _____

Q. _____

R. _____

S. _____

T. _____

- *Click* **LAYER 2** *in the* **LAYER CONTROLS** *window, and you will see the following image:*

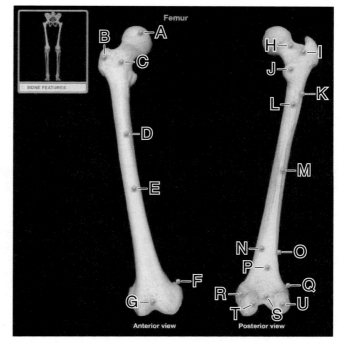

- *Mouse-over the pins on the screen to find the information necessary to identify the following structures:*

A. _____

B. _____

C. _____

D. _____

E. _____

F. _____

G. _____

H. _____

I. _____

J. _____

K. _____

L. _____

M. _____

N. _____

O. _____

P. _____

Q. _____

R. _____

S. _____

T. _____

U. _____

CHECK POINT:

Femur

1. Name a common site of femur fractures, especially in the elderly.
2. What correlation exists between the length of the femur and body height?
3. Name the central depression of the head of the femur.

EXERCISE 3.59:

Patella

SELECT TOPIC	SELECT VIEW	
Patella	► Anterior-Posterior	GO

- *Click **LAYER 1** in the **LAYER CONTROLS** window, and you will see the following image:*

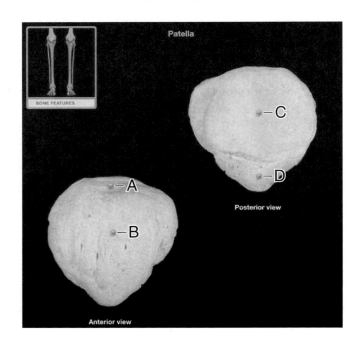

- *Mouse-over the pins on the screen to find the information necessary to identify the following structures:*

A. _____

B. _____

C. _____

D. _____

CHECK POINT:

Patella

1. Describe the location of the patella.
2. What structures form the knee joint?
3. What affect does the patella have on the tendon of the quadriceps femoris muscle?

EXERCISE 3.60:

Leg and Foot, Anterior View

SELECT TOPIC	SELECT VIEW	
Leg and Foot	► Anterior	GO

- *Click **LAYER 1** in the **LAYER CONTROLS** window, and you will see the following image:*

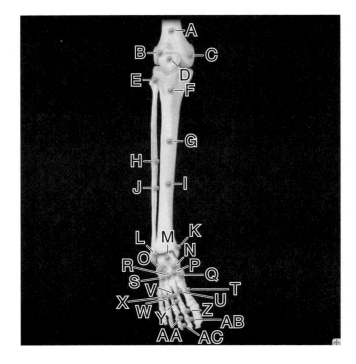

- *Mouse-over the pins on the screen to find the information necessary to identify the following structures:*

A. _____

B. _____

C. _____

D. _____

E. _____

F. _____

G. _____

H. _____

I. _____

J. _____

K. _____

L. _____

M. _____

N. _____

O. _____

P. _____

Q. _____

R. _____

S. _____

T. _____

U. _____

V. _____

W. _____

X. _____

Y. _____

Z. _____

AA. _____

AB. _____

AC. _____

EXERCISE 3.61:

Leg and Foot, Posterior View

SELECT TOPIC	SELECT VIEW	
Leg and Foot ▶	**Posterior**	GO

- *Click **LAYER 1** in the **LAYER CONTROLS** window, and you will see the following image:*

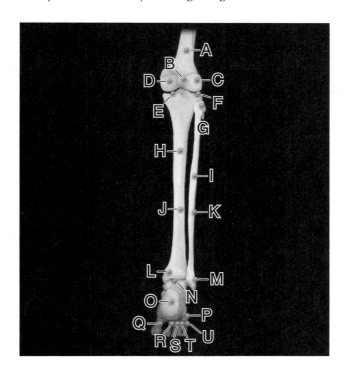

- *Mouse-over the pins on the screen to find the information necessary to identify the following structures:*

A. _____

B. _____

C. _____

D. _____

E. _____

F. _____

G. _____

H. _____

I. _____

J. _____

K. _____

L. _____

M. _____

N. _____

O. _____

P. _____

Q. _____

R. _____

S. _____

T. _____

U. _____

CHECK POINT:

Leg and Foot

1. What bone forms the connecting link between the foot and the leg?
2. What bone contributes to the knee and ankle joints?
3. Name the largest tarsal bone. Where is it located?

EXERCISE 3.62:

Tibia and Fibula

SELECT TOPIC	SELECT VIEW	
Tibia and Fibula	▶ **Anterior-Posterior**	**GO**

• *Click* **LAYER 1** *in the* **LAYER CONTROLS** *window, and you will see the following image:*

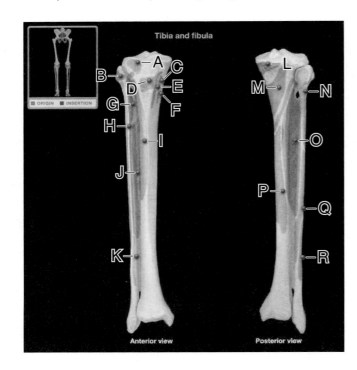

• *Mouse-over the pins on the screen to find the information necessary to identify the following structures:*

A. _____

B. _____

C. _____

D. _____

E. _____

F. _____

G. _____

H. _____

I. _____

J. _____

K. _____

L. _____

M. _____

N. _____

O. _____

P. _____

Q. _____

R. _____

• *Click* **LAYER 2** *in the* **LAYER CONTROLS** *window, and you will see the following image:*

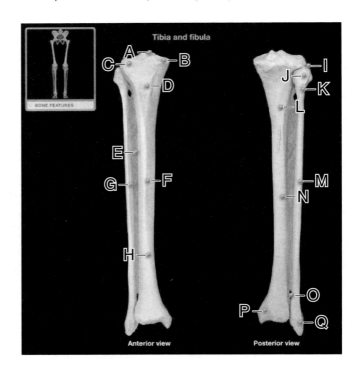

• *Mouse-over the pins on the screen to find the information necessary to identify the following structures:*

A. _____

B. _____

C. _____

D. _____

E. _____

F. _____

G. _____

H. _____

I. _____

J. _____

K. _____

L. _____

M. _____

N. _____

O. _____

P. _____

Q. _____

CHECK POINT:

Tibia and Fibula

1. What structure on the tibia contributes to the ankle joint?
2. What structure on the fibula contributes to the ankle joint?
3. What is unique about the anterior shaft of the tibia?

EXERCISE 3.63:

Ankle and Foot

| SELECT TOPIC **Ankle and Foot** | ▶ | SELECT VIEW **Superior-Inferior** | GO |

- *Click* **LAYER 1** *in the* **LAYER CONTROLS** *window, and you will see the following image:*

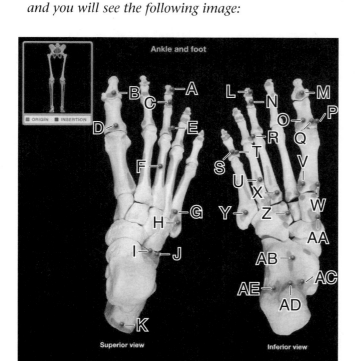

- *Mouse-over the pins on the screen to find the information necessary to identify the following structures:*

A. _____

B. _____

C. _____

D. _____

E. _____

F. _____

G. _____

H. _____

I. _____

J. _____

K. _____

L. _____

M. _____

N. _____

O. _____

P. _____

Q. _____

R. _____

S. _____

T. _____

U. _____

V. _____

W. _____

X. _____

Y. _____

Z. _____

AA. _____

AB. _____

AC. _____

AD. _____

AE. _____

- *Click* **LAYER 2** *in the* **LAYER CONTROLS** *window, and you will see the following image:*

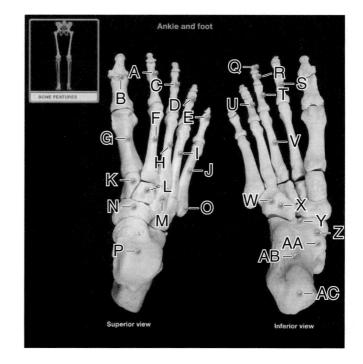

• *Mouse-over the pins on the screen to find the information necessary to identify the following structures:*

A. _____

B. _____

C. _____

D. _____

E. _____

F. _____

G. _____

H. _____

I. _____

J. _____

K. _____

L. _____

M. _____

N. _____

O. _____

P. _____

Q. _____

R. _____

S. _____

T. _____

U. _____

V. _____

W. _____

X. _____

Y. _____

Z. _____

AA. _____

AB. _____

AC. _____

CHECK POINT:

Ankle and Foot

1. Name the structure commonly called the "ball of the foot".
2. Name the structures that are the surface contact points on the plantar foot.
3. List three common names for digit I of the foot.

Self Test

Take this opportunity to quiz yourself by taking the **SELF TEST.** See page 15 for a reminder on how to access the self-test for this section.

I N   R E V I E W

What Have I Learned?

The following questions cover the material that you have just learned: the pelvic girdle and lower limb. Apply what you have learned in answering these questions on a separate piece of paper.

1. Where on the femur does the lateral collateral ligament attach?

2. How many tarsal bones are there on each foot? Name them.

3. The head of which bone is commonly referred to as the "ball of the foot"?

4. Name the small bones of the toes. How many of these bones make up each toe?

5. Name the two enlargements of the distal femur. What structures do they articulate with?

6. Describe the patella.

7. Describe the sacrum.

8. Name the synovial joint between the sacrum and the ilium.

9. How much movement does this joint allow? Why?

10. Name the landmark for administering anesthetic during childbirth.

11. In the pregnant female, what events occur concerning the pubic symphysis?

12. The subpubic angle in females is usually _____ degrees.

13. The subpubic angle in males is usually _____ degrees.

Continued

14. Describe the difference between the male and female obturator foramena.

15. Name a hip landmark for intramuscular injections.

16. Name a hip landmark for administering anesthetic during childbirth.

17. Name a common site of femur fractures, especially in the elderly.

18. What correlation exists between the length of the femur and body height?

19. Name the central depression of the head of the femur.

20. What structures form the knee joint?

21. What bone forms the connecting link between the foot and the leg?

22. What bone contributes to the knee and ankle joints?

23. Name the largest tarsal bone. Where is it located?

24. What structure on the tibia contributes to the ankle joint?

25. What structure on the fibula contributes to the ankle joint?

HEADS UP!

*With the study of the joints, or articulations, we can tie together the concepts that you have learned in the **Skeletal System** and integrate them with those you will learn in the next chapter – the **Muscular System**. With this in mind, you will want to refer back to this section of the **Skeletal System** while you are working on the **Muscular System**.*

Animation: Synovial Joint

SELECT ANIMATION
Synovial Joint   PLAY

• *After viewing the animation, answer these questions:*

1. A joint is . . .

2. What is the most common type of joint?

3. How many types of synovial joints are found in the body? What distinguishes each one from the others?

4. Which type of synovial joint allows the most mobility? Give an example.

5. Which type of synovial joint allows the least mobility? Give an example.

6. A typical synovial joint is characterized by . . .

7. Describe the joint capsule.

8. What is synovial fluid? Where is it produced?

9. What tissue covers the surface of adjoining bones? What is its function?

10. What is the function of the meniscus of the knee joint?

11. What is the function of bursae?

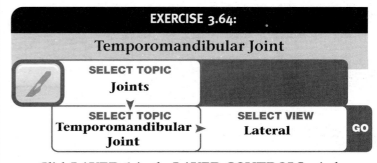

EXERCISE 3.64:

Temporomandibular Joint

SELECT TOPIC
Joints

SELECT TOPIC
Temporomandibular Joint

SELECT VIEW
Lateral   GO

• *Click **LAYER 1** in the **LAYER CONTROLS** window, and you will see the following image:*

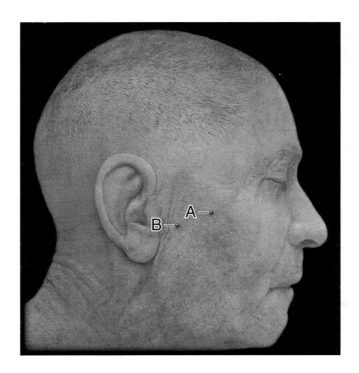

- *Mouse-over the pins on the screen to find the information necessary to identify the following structures:*

A. _____

B. _____

- *Click **LAYER 2** in the **LAYER CONTROLS** window, and you will see the following image:*

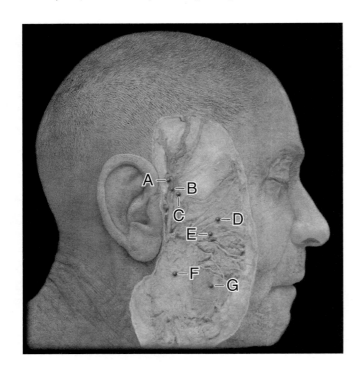

- *Mouse-over the pins on the screen to find the information necessary to identify the following structures:*

A. _____

B. _____

C. _____

D. _____

E. _____

F. _____

G. _____

- *Click **LAYER 3** in the **LAYER CONTROLS** window, and you will see the following image:*

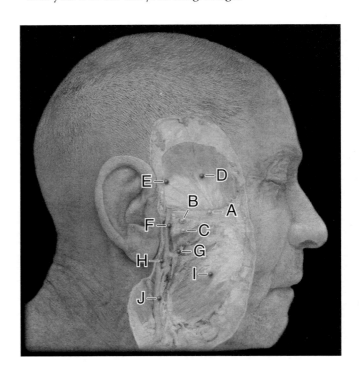

- *Mouse-over the green pins on the screen to find the information necessary to identify the following structures:*

A. _____

B. _____

C. _____

Non-skeletal System Structures (blue pins)

D. _____

E. _____

F. _____

G. _____

H. _____

I. _____

J. _____

• *Click* **LAYER 4** *in the* **LAYER CONTROLS** *window, and you will see the following image:*

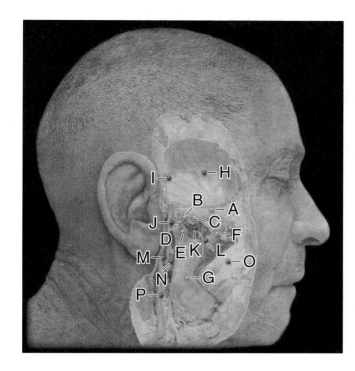

• *Mouse-over the green pins on the screen to find the information necessary to identify the following structures:*

A. _____

B. _____

C. _____

D. _____

E. _____

F. _____

G. _____

Non-skeletal System Structures (blue pins)

H. _____

I. _____

J. _____

K. _____

L. _____

M. _____

N. _____

O. _____

P. _____

• *Click* **LAYER 5** *in the* **LAYER CONTROLS** *window, and you will see the following image:*

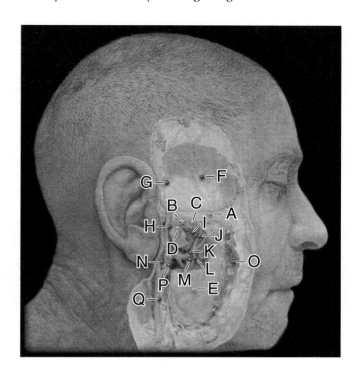

• *Mouse-over the green pins on the screen to find the information necessary to identify the following structures:*

A. _____

B. _____

C. _____

D. _____

E. _____

Non-skeletal System Structures (blue pins)

F. _____

G. _____

H. _____

I. _____

J. _____

K. _____

L. _____

M. _____

N. _____

O. _____

P. _____

Q. _____

CHECK POINT:

Temporomandibular Joint

1. Name the structure that encloses the temporomandibular joint.
2. What is the function of the synovial membrane?
3. What is the function of the lateral ligament of the temporomandibular joint?
4. What is the cause of "jaw clicking"?
5. What is the location of the articular disk of the temporomandibular joint?

EXERCISE 3.65:

Glenohumeral (Shoulder) Joint

SELECT TOPIC	SELECT VIEW	
Glenohumeral ➤ (Shoulder) Joint	Anterior	**GO**

- *Click* **LAYER 1** *in the* **LAYER CONTROLS** *window, and you will see the following image:*

- *Mouse-over the pins on the screen to find the information necessary to identify the following structures:*

A. _____

B. _____

C. _____

- *Click* **LAYER 2** *in the* **LAYER CONTROLS** *window, and you will see the following image:*

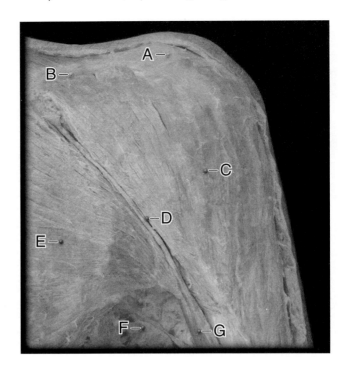

- *Mouse-over the green pins on the screen to find the information necessary to identify the following structures:*

A. _____

B. _____

Non-skeletal System Structures (blue pins)

C. _____

D. _____

E. _____

F. _____

G. _____

- *Click* **LAYER 3** *in the* **LAYER CONTROLS** *window, and you will see the following image:*

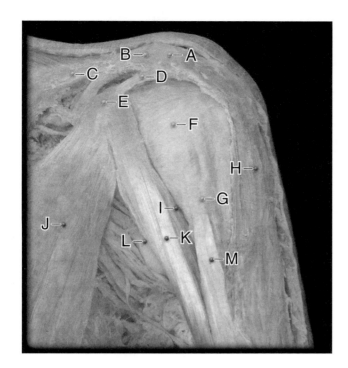

- *Mouse-over the green pins on the screen to find the information necessary to identify the following structures:*

A. _____

B. _____

C. _____

D. _____

E. _____

F. _____

G. _____

Non-skeletal System Structures (blue pins)

H. _____

I. _____

J. _____

K. _____

L. _____

M. _____

- *Click* **LAYER 4** *in the* **LAYER CONTROLS** *window, and you will see the following image:*

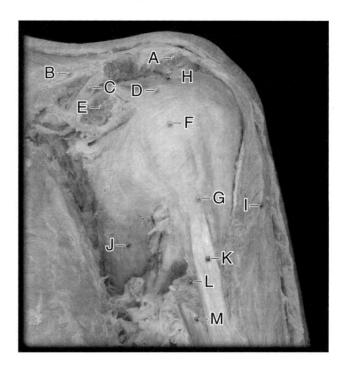

- *Mouse-over the green pins on the screen to find the information necessary to identify the following structures:*

A. _____

B. _____

C. _____

D. _____

E. _____

F. _____

G. _____

Non-skeletal System Structures (blue pins)

H. _____

I. _____

J. _____

K. _____

L. _____

M. _____

- *Click* **LAYER 5** *in the* **LAYER CONTROLS** *window, and you will see the following image:*

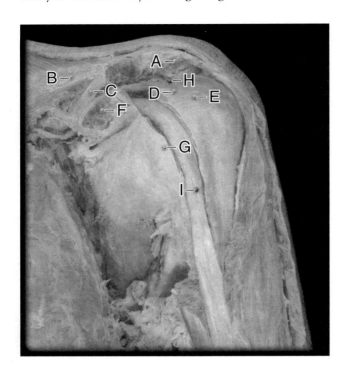

- *Mouse-over the green pins on the screen to find the information necessary to identify the following structures:*

A. _____

B. _____

C. _____

D. _____

E. _____

F. _____

G. _____

Non-skeletal System Structures (blue pins)

H. _____

I. _____

- *Click* **LAYER 6** *in the* **LAYER CONTROLS** *window, and you will see the following image:*

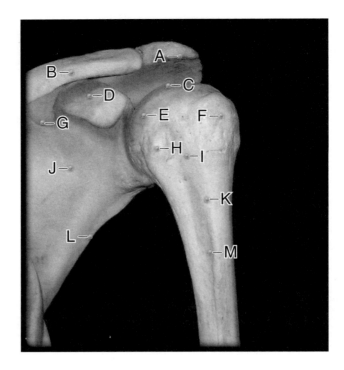

- *Mouse-over the pins on the screen to find the information necessary to identify the following structures:*

A. _____

B. _____

C. _____

D. _____

E. _____

F. _____

G. _____

H. _____

I. _____

J. _____

K. _____

L. _____

M. _____

C H E C K   P O I N T :

Glenohumeral (Shoulder) Joint

1. Name the structure enclosing the acromioclavicular joint.
2. What is the function of the coracohumeral ligament?
3. What is the function of the coracoclavicular ligament?
4. What is the result of a torn coracoclavicular ligament?
5. Name the tough, fibrous envelope of the glenohumeral joint.

EXERCISE 3.66:

Elbow Joint

SELECT TOPIC	SELECT VIEW	
Elbow Joint	▶ **Anterior**	**GO**

• *Click* **LAYER 1** *in the* **LAYER CONTROLS** *window, and you will see the following image:*

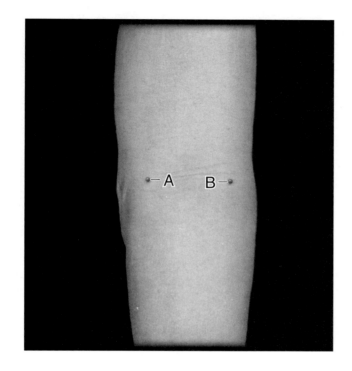

• *Mouse-over the pins on the screen to find the information necessary to identify the following structures:*

A. _____

B. _____

• *Click* **LAYER 2** *in the* **LAYER CONTROLS** *window, and you will see the following image:*

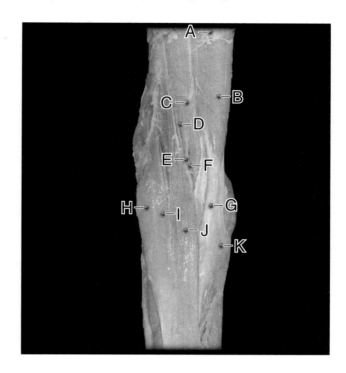

• *Mouse-over the pins on the screen to find the information necessary to identify the following structures:*

A. _____

B. _____

C. _____

D. _____

E. _____

F. _____

G. _____

H. _____

I. _____

J. _____

K. _____

- *Click* **LAYER 3** *in the* **LAYER CONTROLS** *window, and you will see the following image:*

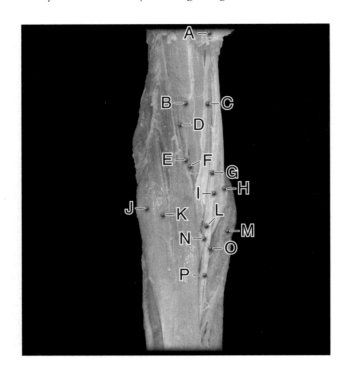

- *Mouse-over the pins on the screen to find the information necessary to identify the following structures:*

A. _____

B. _____

C. _____

D. _____

E. _____

F. _____

G. _____

H. _____

I. _____

J. _____

K. _____

L. _____

M. _____

N. _____

O. _____

P. _____

- *Click* **LAYER 4** *in the* **LAYER CONTROLS** *window, and you will see the following image:*

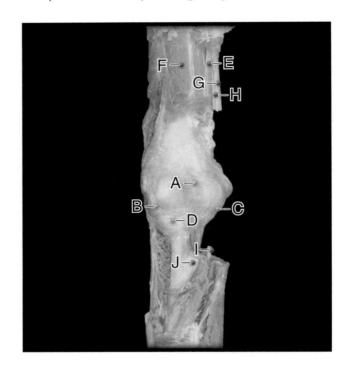

- *Mouse-over the green pins on the screen to find the information necessary to identify the following structures:*

A. _____

B. _____

C. _____

D. _____

Non-skeletal System Structures (blue pins)

E. _____

F. _____

G. _____

H. _____

I. _____

J. _____

• *Click* **LAYER 5** *in the* **LAYER CONTROLS** *window, and you will see the following image:*

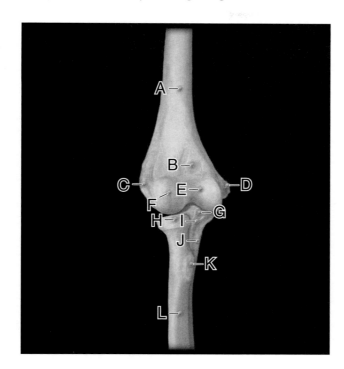

• *Mouse-over the pins on the screen to find the information necessary to identify the following structures:*

A. _____

B. _____

C. _____

D. _____

E. _____

F. _____

G. _____

H. _____

I. _____

J. _____

K. _____

L. _____

CHECK POINT:

Elbow Joint

1. Name the structure that encloses the elbow joint.
2. What is the function of the radial collateral ligament?
3. What is the function of the ulnar collateral ligament?
4. What is the function of the anular ligament of the radius?
5. Name the bones associated with the elbow joint.

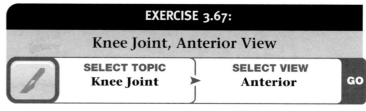

EXERCISE 3.67:

Knee Joint, Anterior View

SELECT TOPIC	SELECT VIEW	
Knee Joint	**Anterior**	**GO**

• *Click* **LAYER 1** *in the* **LAYER CONTROLS** *window, and you will see the following image:*

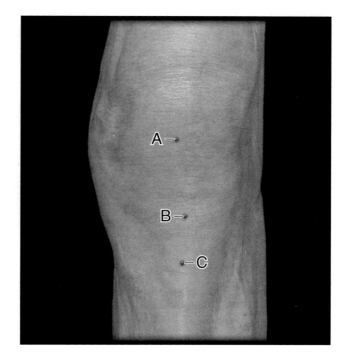

• *Mouse-over the pins on the screen to find the information necessary to identify the following structures:*

A. _____

B. _____

C. _____

- *Click* **LAYER 2** *in the* **LAYER CONTROLS** *window, and you will see the following image:*

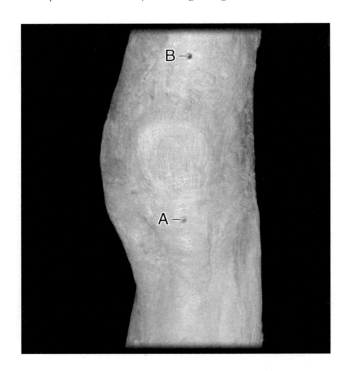

- *Mouse-over the green pin on the screen to find the information necessary to identify the following structure:*

A. _____

Non-skeletal System Structure (blue pin)

B. _____

- *Click* **LAYER 3** *in the* **LAYER CONTROLS** *window, and you will see the following image:*

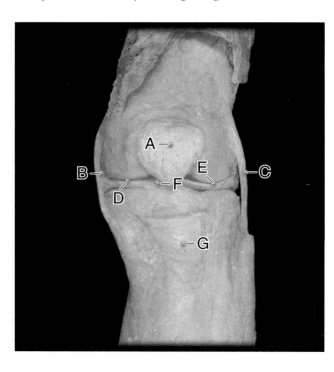

- *Mouse-over the pins on the screen to find the information necessary to identify the following structures:*

A. _____

B. _____

C. _____

D. _____

E. _____

F. _____

G. _____

- *Click* **LAYER 4** *in the* **LAYER CONTROLS** *window, and you will see the following image:*

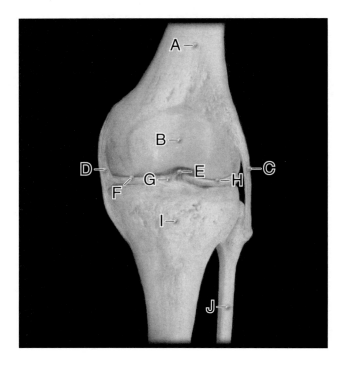

- *Mouse-over the pins on the screen to find the information necessary to identify the following structures:*

A. _____

B. _____

C. _____

D. _____

E. _____

F. _____

G. _____

H. _____

I. _____

J. _____

- *Click* **LAYER 5** *in the* **LAYER CONTROLS** *window, and you will see the following image:*

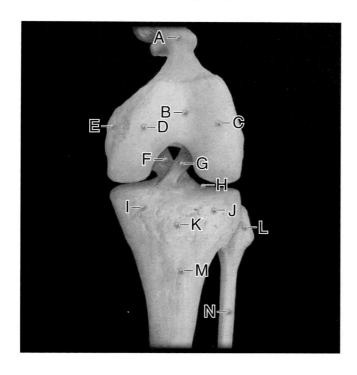

- *Mouse-over the pins on the screen to find the information necessary to identify the following structures:*

A. _____

B. _____

C. _____

D. _____

E. _____

F. _____

G. _____

H. _____

I. _____

J. _____

K. _____

L. _____

M. _____

N. _____

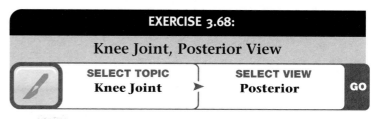

EXERCISE 3.68:

Knee Joint, Posterior View

| SELECT TOPIC | SELECT VIEW | |
| Knee Joint | Posterior | GO |

- *Click* **LAYER 1** *in the* **LAYER CONTROLS** *window, and you will see the following image:*

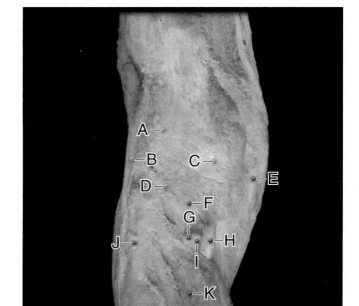

- *Mouse-over the pin on the screen to find the information necessary to identify the following structure:*

A. _____

- *Click* **LAYER 2** *in the* **LAYER CONTROLS** *window, and you will see the following image:*

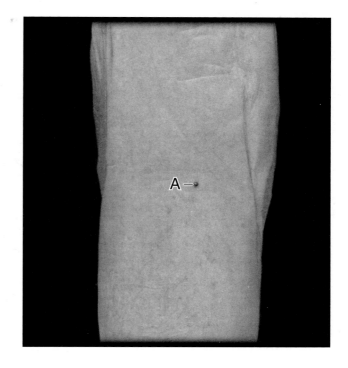

- *Mouse-over the green pins on the screen to find the information necessary to identify the following structures:*

A. _____

B. _____

C. _____

D. _____

Non-skeletal System Structures (blue pins)

E. _____

F. _____

G. _____

H. _____

I. _____

J. _____

K. _____

- *Click **LAYER 3** in the **LAYER CONTROLS** window, and you will see the following image:*

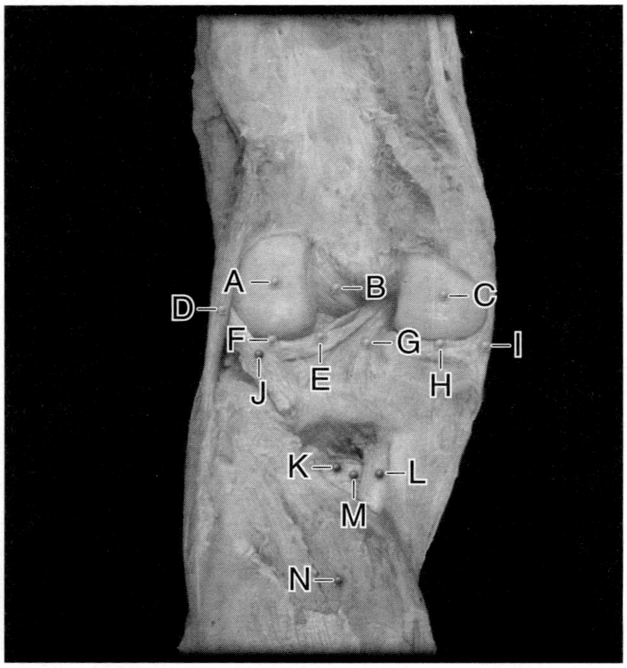

- *Mouse-over the green pins on the screen to find the information necessary to identify the following structures:*

A. _____

B. _____

C. _____

D. _____

E. _____

F. _____

G. _____

H. _____

I. _____

Non-skeletal System Structures (blue pins)

J. _____

K. _____

L. _____

M. _____

N. _____

- *Click **LAYER 4** in the **LAYER CONTROLS** window, and you will see the following image:*

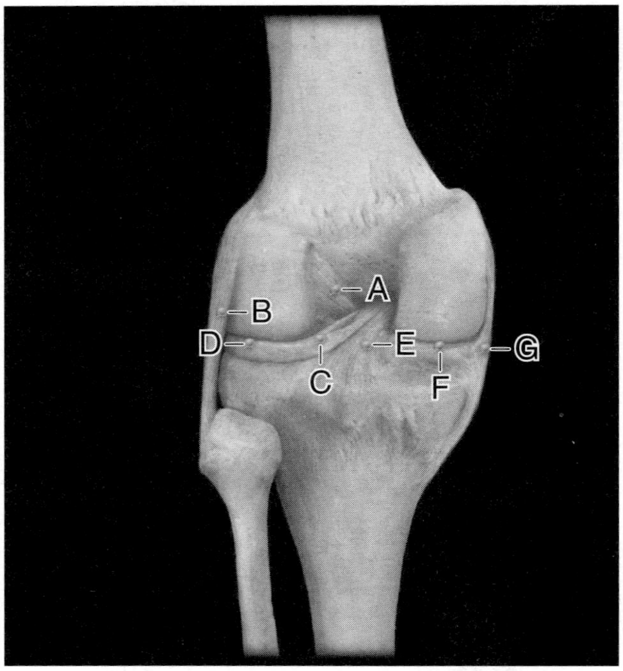

- *Mouse-over the pins on the screen to find the information necessary to identify the following structures:*

A. _____

B. _____

C. _____

D. _____

E. _____

F. _____

G. _____

- *Click* **LAYER 5** *in the* **LAYER CONTROLS** *window and you will see the following image:*

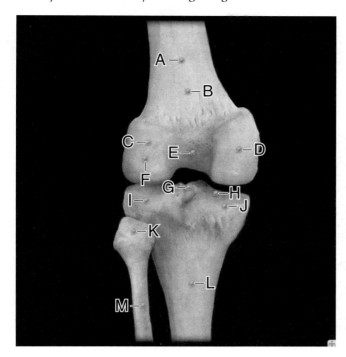

- *Mouse-over the pins on the screen to find the information necessary to identify the following structures:*

A. _____

B. _____

C. _____

D. _____

E. _____

F. _____

G. _____

H. _____

I. _____

J. _____

K. _____

L. _____

M. _____

CHECK POINT:

Knee Joint

1. Name the structures that make up the "unhappy triad."
2. Why are these structures referred to as the "unhappy triad"?
3. Name a ligament absent in approximately 40 percent of knees.
4. What is the function of the anterior cruciate ligament?
5. What is the function of the posterior cruciate ligament?

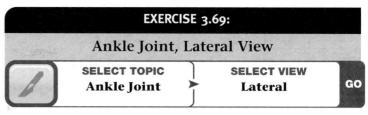

EXERCISE 3.69:

Ankle Joint, Lateral View

SELECT TOPIC	SELECT VIEW	
Ankle Joint	**Lateral**	GO

- *Click* **LAYER 1** *in the* **LAYER CONTROLS** *window, and you will see the following image:*

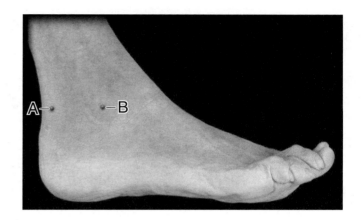

- *Mouse-over the pins on the screen to find the information necessary to identify the following structures:*

A. _____

B. _____

- *Click* **LAYER 2** *in the* **LAYER CONTROLS** *window, and you will see the following image:*

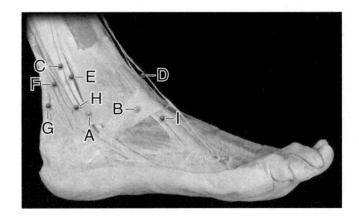

- *Mouse-over the green pins on the screen to find the information necessary to identify the following structures:*

A. _____

B. _____

Non-skeletal System Structures (blue pins)

C. _____

D. _____

E. _____

F. _____

G. _____

H. _____

I. _____

- *Click* **LAYER 3** *in the* **LAYER CONTROLS** *window, and you will see the following image:*

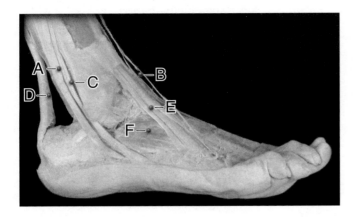

- *Mouse-over the pins on the screen to find the information necessary to identify the following structures:*

A. _____

B. _____

C. _____

D. _____

E. _____

F. _____

- *Click* **LAYER 4** *in the* **LAYER CONTROLS** *window, and you will see the following image:*

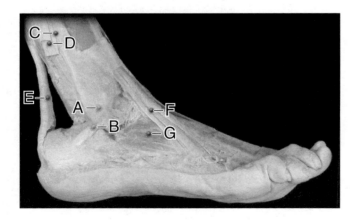

- *Mouse-over the green pins on the screen to find the information necessary to identify the following structures:*

A. _____

B. _____

Non-skeletal System Structures (blue pins)

C. _____

D. _____

E. _____

F. _____

G. _____

- *Click* **LAYER 5** *in the* **LAYER CONTROLS** *window, and you will see the following image:*

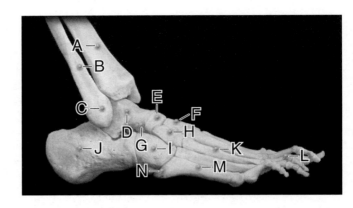

- *Mouse-over the pins on the screen to find the information necessary to identify the following structures:*

A. _____

B. _____

C. _____

D. _____

E. _____

F. _____

G. _____

H. _____

I. _____

J. _____

K. _____

L. _____

M. _____

N. _____

EXERCISE 3.70:

Ankle Joint, Medial View

SELECT TOPIC	SELECT VIEW	
Ankle Joint ▶	**Medial**	GO

- *Click* **LAYER 1** *in the* **LAYER CONTROLS** *window, and you will see the following image:*

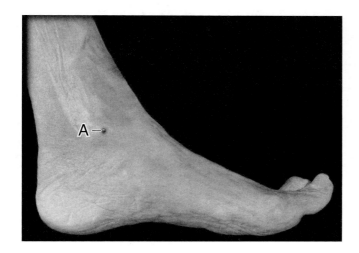

- *Mouse-over the pin on the screen to find the information necessary to identify the following structure:*

A. _____

- *Click* **LAYER 2** *in the* **LAYER CONTROLS** *window, and you will see the following image:*

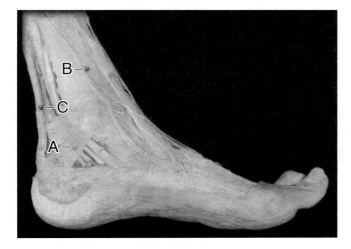

- *Mouse-over the green pin on the screen to find the information necessary to identify the following structure:*

A. _____

Non-skeletal System Structures (blue pins)

B. _____

C. _____

- *Click* **LAYER 3** *in the* **LAYER CONTROLS** *window, and you will see the following image:*

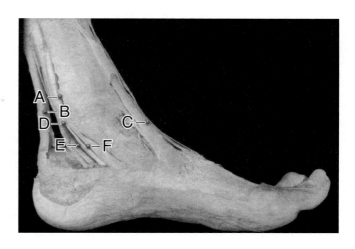

- *Mouse-over the pins on the screen to find the information necessary to identify the following structures:*

A. _____

B. _____

C. _____

D. _____

E. _____

F. _____

- *Click* **LAYER 4** *in the* **LAYER CONTROLS** *window, and you will see the following image:*

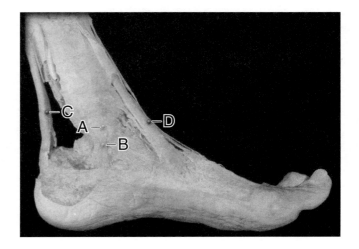

- *Mouse-over the green pins on the screen to find the information necessary to identify the following structures:*

A. _____

B. _____

Non-skeletal System Structures (blue pins)

C. _____

D. _____

- *Click **LAYER 5** in the **LAYER CONTROLS** window, and you will see the following image:*

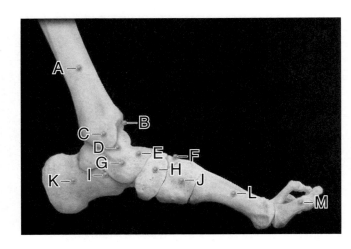

- *Mouse-over the pins on the screen to find the information necessary to identify the following structures:*

A. _____

B. _____

C. _____

D. _____

E. _____

F. _____

G. _____

H. _____

I. _____

J. _____

K. _____

L. _____

M. _____

CHECK POINT: _____

Ankle Joint

1. Name the strongest tendon of the body. Why is it considered the strongest?
2. What is the extensor retinaculum of the foot? What is its function?
3. What is the function of the lateral ligament of the ankle?
4. What is the flexor retinaculum of the foot? What is its function?
5. What is the function of the medial (deltoid) ligament of the ankle?

Self Test

Take this opportunity to quiz yourself by taking the **SELF TEST.** See page 15 for a reminder on how to access the self-test for this section.

IN REVIEW

What Have I Learned?

The following questions cover the material that you have just learned: joints. Apply what you have learned in answering these questions on a separate piece of paper.

1. Name the structure that encloses the temporomandibular joint.

2. What is the function of the synovial membrane?

3. What is the cause of "jaw clicking"?

4. Name the structure enclosing the acromioclavicular joint.

5. What is the result of a torn coracoclavicular ligament?

6. Name the tough, fibrous envelope of the glenohumeral joint.

7. Name the structure that encloses the elbow joint.

8. Name the bones associated with the elbow joint.

9. Name the structures that make up the "unhappy triad."

10. Name a ligament absent in approximately 40 percent of knees.

11. What is the function of the anterior cruciate ligament?

12. What is the function of the posterior cruciate ligament?

13. Name the strongest tendon of the body. Why is it considered the strongest?

14. What is the extensor retinaculum of the foot? What is its function?

EXERCISE 3.71:

Elastic Cartilage, Histology

SELECT TOPIC	SELECT VIEW	
Elastic Cartilage ➤		**GO**

- *Click the* **TURN TAGS ON** *button, and you will see the following image:*

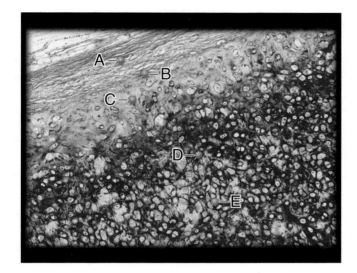

- *Mouse-over the pins on the screen to find the information necessary to identify the following structures:*

A. _____

B. _____

C. _____

D. _____

E. _____

EXERCISE 3.72:

Fibrous Cartilage, Histology

SELECT TOPIC	SELECT VIEW	
Fibrous Cartilage ➤		**GO**

- *Click the* **TURN TAGS ON** *button, and you will see the following image:*

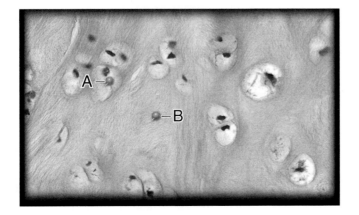

- *Mouse-over the pins on the screen to find the information necessary to identify the following structures:*

A. _____

B. _____

EXERCISE 3.73:

Hyaline Cartilage, Histology

SELECT TOPIC	SELECT VIEW	
Hyaline Cartilage ➤		**GO**

- *Click the* **TURN TAGS ON** *button, and you will see the following image:*

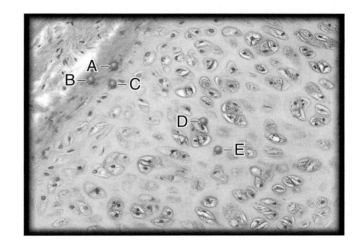

- *Mouse-over the pins on the screen to find the information necessary to identify the following structures:*

A. _____

B. _____

C. _____

D. _____

E. _____

EXERCISE 3.74a:

Compact Bone, Histology

SELECT TOPIC	SELECT VIEW	
Compact Bone	▶ Low Magnification	GO

- *Click the* **TURN TAGS ON** *button, and you will see the following image:*

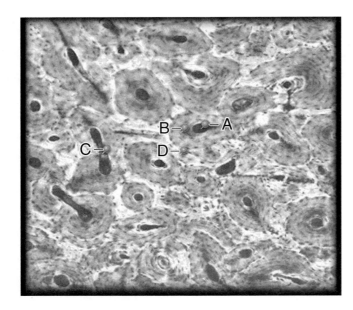

- *Mouse-over the pins on the screen to find the information necessary to identify the following structures:*

A. _____

B. _____

C. _____

D. _____

EXERCISE 3.74b:

Compact Bone, Histology

SELECT TOPIC	SELECT VIEW	
Compact Bone	▶ High Magnification	GO

- *Click the* **TURN TAGS ON** *button, and you will see the following image:*

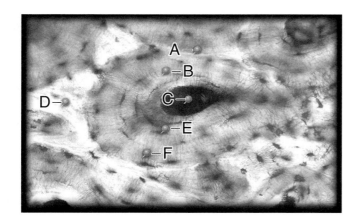

- *Mouse-over the pins on the screen to find the information necessary to identify the following structures:*

A. _____

B. _____

C. _____

D. _____

E. _____

F. _____

EXERCISE 3.75:

Cancellous Bone, Histology

SELECT TOPIC	SELECT VIEW	
Cancellous Bone	▶	GO

- *Click the* **TURN TAGS ON** *button, and you will see the following image:*

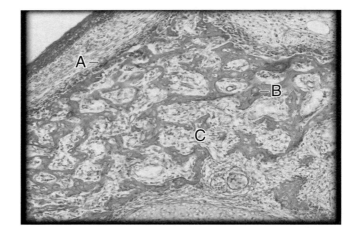

- *Mouse-over the pins on the screen to find the information necessary to identify the following structures:*

A. _____

B. _____

C. _____

Self Test

Take this opportunity to quiz yourself by taking the **SELF TEST.** See page 15 for a reminder on how to access the self-test for this section.

I N   R E V I E W

What Have I Learned?

The following questions cover the material that you have just learned: histology. Apply what you have learned in answering these questions on a separate piece of paper.

1. Name the outer layer of most cartilage.

2. Give four examples of elastic cartilage.

3. What are lacunae? What are contained within the lacunae?

4. Give four examples of fibrous cartilage.

5. What is the function of the chondrogenic layer of the perichondrium?

6. Describe an osteon.

7. Describe a perforating canal.

8. Interstitial lamellae result from . . .

9. What component of bone tissue is essential to bone nutrition, growth, and repair?

10. What are trabeculae?

EXERCISE 3.76:

Coloring Exercise

Look up the bones of the skull and then color them in with colored pens or pencils. Write the names of important bones or bone features in the blank label lines.

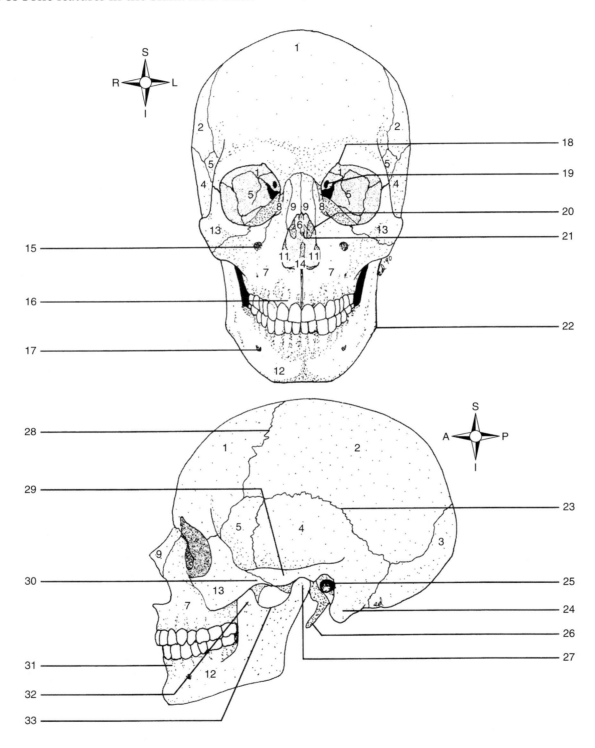

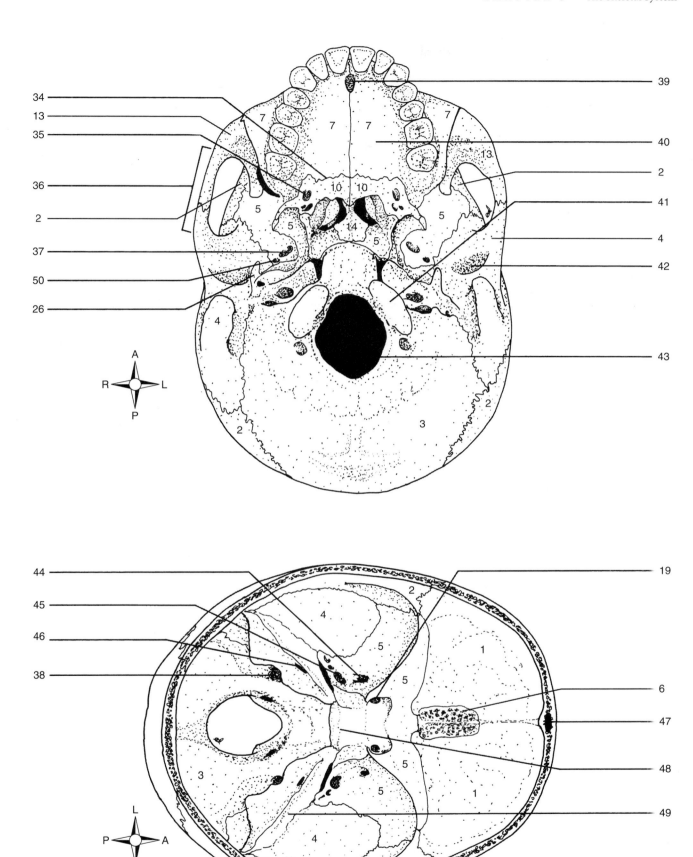

34
13
35
36
2
37
50
26

39
40
2
41
4
42
43

44
45
46
38

19
6
47
48
49

Horizontal
section

CHAPTER 4

The Muscular System

Overview: Muscular System

Without the skeletal muscular system, the bones that we learned in the previous chapter would be unable to move. Our bodies are equipped with some 600 skeletal muscles to not only put those 206 bones into motion, but also to generate as much as 85% of our body heat, maintain our posture, control the openings involved with the entrance and exit of materials, and to express our emotions and thoughts through movements of our facial muscles.

Three important structural terms to understand as you begin your study of the skeletal muscular system are a muscle's **origin, insertion,** and **belly.** Most muscles are attached to different bones at each end. This assures that each muscle or its tendon will span at least one joint. When a muscle contracts, it causes movement where one bone remains relatively stationary and the other bone will move. As we have previously stated, the end of the muscle attached to the relatively stationary bone is called the **origin,** while the end of the muscle attached to the moving bone is called the **insertion.** Again, one way to remember the difference is to think of your birthplace, your **origin.** No matter where you may move throughout your life, your **origin** remains the same—it doesn't move! You may **insert** yourself at several locations throughout your life—away to college, a job in a different town, and so on. These require moving! Also, many muscles are narrow at each end, their origin and insertion, and thick in the middle. This thicker middle region is called the **belly.**

- *From the **Home Screen,** click the drop-down box on the **Select system** menu.*

- *From the systems listed, click on **Muscular,** and you will see the image to the right:*

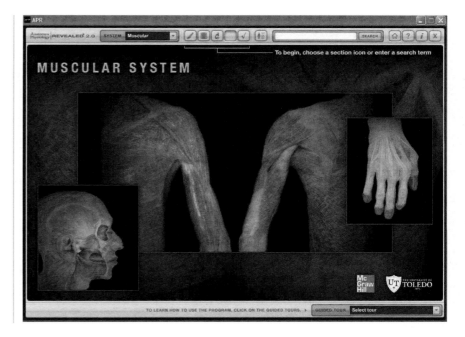

- This is the opening screen for the **Muscular System.**

Animations: Anatomy and Physiology

Before we begin our study of the muscular system, it is important that we first have an understanding of how the muscles function—not only their anatomy, but also their physiology. The five animations presented here are designed to help you comprehend the anatomy and physiology of muscles and muscle contractions. After we lay this foundation, we will be better prepared to build on this foundation by learning their names, origins, and insertions.

The first animation will introduce you to the structure of muscle fibers, and then, through the series of the second through the fifth animations, will walk you through the process of muscle contraction. If after viewing these animations the process is not clear to you, repeat the series of animations in sequence until you understand how this important process unfolds. You can pause the animations and repeat sections as needed to answer the questions following each animation.

Animation: Skeletal Muscle

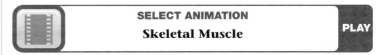

- *After viewing the animation, answer these questions:*

1. _____ is responsible for voluntary movement of the human body.

2. What is the structure of these muscles?

3. What is the epimysium?

4. At the ends of the muscle, the epimysium is continuous with. . . .

5. What is the perimysium?

6. What is a fascicle?

7. What is the endomysium?

8. A muscle fiber consists of. . . .

9. What is unique about the nuclei of a muscle fiber? Where are they located?

10. What structures dominate the interior of the muscle fiber?

11. What are the two types of protein filaments that compose the myofibril?

12. Name the orderly contractile unit of the muscle fiber.

13. The shortened sarcomeres result in _____, and ultimately _____.

Animation: Neuromuscular Junction

SELECT ANIMATION
Neuromuscular Junction
PLAY

- *After viewing the animation, answer these questions:*

1. What sequence of events occurs when an action potential arrives at the presynaptic terminal?

2. What sequence of events occurs when the calcium ions enter the presynaptic terminal?

3. What is the function of acetylcholine?

4. The movement of _____ ions into the muscle cell results in _____.

5. Once threshold has been reached, ...

6. What happens to the acetylcholine after the generation of an action potential on the muscle cell membrane?

Animation: Sliding Filament

SELECT ANIMATION
Sliding Filament
PLAY

- *After viewing the animation, answer these questions:*

1. Describe the muscle myofilaments in a relaxed muscle.

2. What interaction do these myofilaments have during contraction?

3. What is the result?

4. Describe the muscle myofilaments in a fully contracted muscle.

Animation: Excitation-Contraction Coupling

SELECT ANIMATION
Excitation-Contraction Coupling
PLAY

- *After viewing the animation, answer these questions:*

1. Action potentials are propagated over which structure of the skeletal muscle fiber?

2. How does the action potential arrive into the interior of the muscle fiber?

3. What does the entry of the action potential cause to occur inside the muscle fiber?

4. This causes _____ ions to diffuse from the _____ into the _____.

5. Where is the tropomyosin located? What is it covering?

6. Describe the molecules attached to the tropomyosin.

7. What events occur when the calcium ions bind to the troponin molecule?

8. How are cross-bridges formed?

Animation: Cross-Bridge Cycle

SELECT ANIMATION
Cross-Bridge Cycle
PLAY

- *After viewing the animation, answer these questions:*

1. During contraction of a muscle, _____ bind to _____. This moves _____ out of the way and uncovers _____ for _____ on the _____.

2. What molecules are attached to the myosin head from the previous cycle of movement?

3. How are cross-bridges formed? What is released in the process?

4. How is the myosin head moved? What does this cause? What is released from the myosin head when this occurs?

5. How is the bond between the actin and myosin filaments broken? How is energy released? What happens to this energy?

6. After the myosin head returns to its upright position, what occurs if calcium ions are still present?

Muscular System: Head and Neck

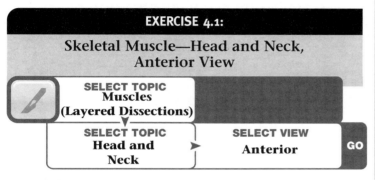

EXERCISE 4.1:

Skeletal Muscle—Head and Neck, Anterior View

SELECT TOPIC
**Muscles
(Layered Dissections)**

SELECT TOPIC
Head and Neck ▶ SELECT VIEW
Anterior   **GO**

• *Click* **LAYER 1** *in the* **LAYER CONTROLS** *window, and you will see the following image:*

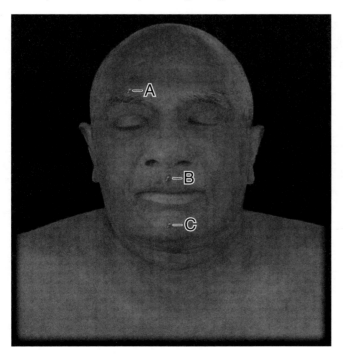

• *Mouse-over the blue pins on the screen to find the information necessary to identify the following non-muscular system structures:*

A. _____

B. _____

C. _____

C H E C K   P O I N T :

Head and Neck, Anterior View

1. Name the ridge superior to each orbit on the anterior side.
2. What bone is that ridge part of?
3. What is the name of the shallow midline groove of the upper lip?

• *Click* **LAYER 2** *in the* **LAYER CONTROLS** *window, and you will see the following image:*

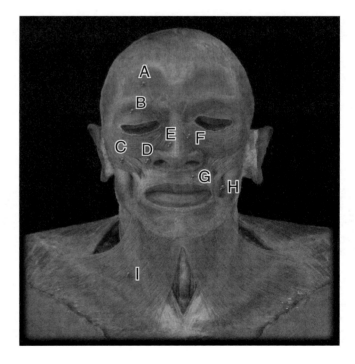

• *Mouse-over the pins on the screen to find the information necessary to identify the following structures:*

A. _____

B. _____

C. _____

D. _____

E. _____

F. _____

G. _____

H. _____

I. _____

• *Click **LAYER 4** in the **LAYER CONTROLS** window, and you will see the following image:*

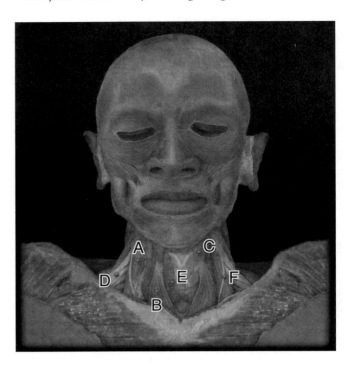

• *Mouse-over the green pins on the screen to find the information necessary to identify the following structures:*

A. _____

B. _____

C. _____

Non-muscular System Structures (blue pins)

D. _____

E. _____

F. _____

CHECK POINT:

Head and Neck, Anterior View, cont'd

10. The roots of the brachial plexus are located between the anterior and middle _____ muscles.
11. Name a muscle responsible for elevation of the larynx and depression of the hyoid bone.
12. Name the four infrahyoid muscles.

CHECK POINT:

Head and Neck, Anterior View, cont'd

4. Name the muscle that closes the eye when winking or blinking.
5. Name the muscle responsible for compression of the cheek as in inflating a balloon or playing a wind instrument.
6. Name the muscle that closes and protrudes the lips.

• *Click **LAYER 3** in the **LAYER CONTROLS** window, and you will see the following image:*

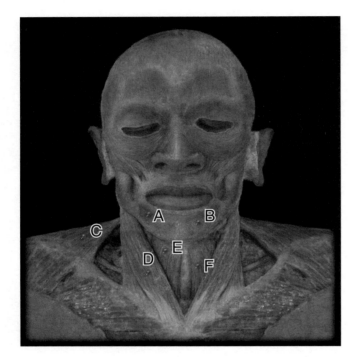

• *Mouse-over the pins on the screen to find the information necessary to identify the following structures:*

A. _____

B. _____

C. _____

D. _____

E. _____

F. _____

CHECK POINT:

Head and Neck, Anterior View, cont'd

7. Name two muscles whose insertion is the hyoid bone.
8. Name the muscle responsible for depression of the angle of the mouth to grimace.
9. Name the muscle responsible for depression of the lower lip while pouting.

- *Click* **LAYER 5** *in the* **LAYER CONTROLS** *window, and you will see the following image:*

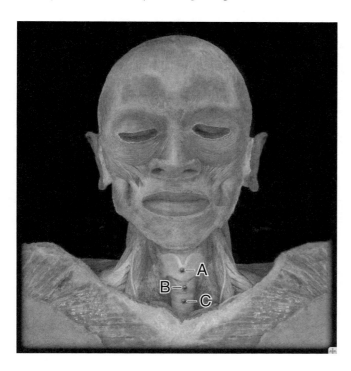

- *Mouse-over the blue pins on the screen to find the information necessary to identify the following non-muscular system structures:*

A. _____

B. _____

C. _____

Animations: Muscle Actions

SELECT ANIMATION
Buccinator Muscle   **PLAY**

- *After viewing this animation, select and view these additional animations:*
  - **Frontalis muscle**
  - **Levator labii superioris alaeque nasi muscle**
  - **Orbicularis oculi muscle**
  - **Orbicularis oris muscle**
  - **Trapezius muscle**

EXERCISE 4.2:

Skeletal Muscle—Head and Neck, Lateral View

SELECT TOPIC	**SELECT VIEW**	
Head and Neck ▸	**Lateral**	**GO**

- *Click* **LAYER 1** *in the* **LAYER CONTROLS** *window and you will see the following image:*

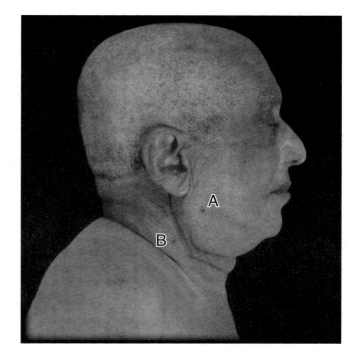

- *Mouse-over the blue pins on the screen to find the information necessary to identify the following non-muscular system structures:*

A. _____

B. _____

CHECK POINT:

Head and Neck, Lateral View

1. What location on the mandible provides an attachment site for the masseter muscle?
2. What other muscle attaches at this point?

- *Click **LAYER 2** in the **LAYER CONTROLS** window, and you will see the following image:*

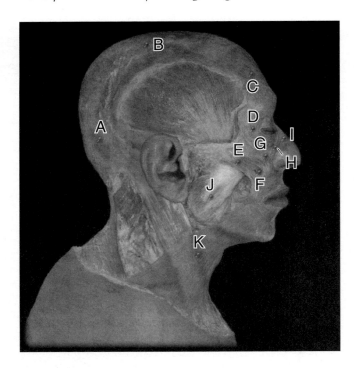

- *Mouse-over the pins on the screen to find the information necessary to identify the following structures:*

A. _____

B. _____

C. _____

D. _____

E. _____

F. _____

G. _____

H. _____

I. _____

J. _____

K. _____

CHECK POINT:

Head and Neck, Lateral View, cont'd

3. Name the muscle responsible for elevation of the upper lip in a sneer.
4. Name the two muscles responsible for elevation of the upper lip in a smile.
5. Name the muscle that elevates and creases the skin of the neck as well as depresses the lower lip and the angle of the mouth.

- *Click **LAYER 3** in the **LAYER CONTROLS** window, and you will see the following image:*

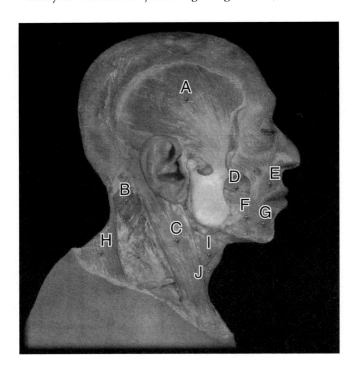

- *Mouse-over the pins on the screen to find the information necessary to identify the following structures:*

A. _____

B. _____

C. _____

D. _____

E. _____

F. _____

G. _____

H. _____

I. _____

J. _____

CHECK POINT:

Head and Neck, Lateral View, cont'd

6. Name a muscle with two bellies (superior and inferior) joined by an intermediate tendon.
7. What is the "kissing muscle"?

• *Click* **LAYER 4** *in the* **LAYER CONTROLS** *window, and you will see the following image:*

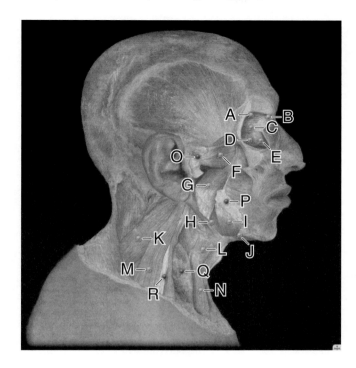

• *Mouse-over the green pins on the screen to find the information necessary to identify the following structures:*

A. _____

B. _____

C. _____

D. _____

E. _____

F. _____

G. _____

H. _____

I. _____

J. _____

K. _____

L. _____

M. _____

N. _____

Non-muscular System Structures (blue pins)

O. _____

P. _____

Q. _____

R. _____

CHECK POINT:

Head and Neck, Lateral View, cont'd

8. Name a muscle responsible for the protrusion of the mandible.
9. Name a muscle responsible for the elevation of the scapula, as in shrugging the shoulders.
10. Name the muscle involved in abduction of the eyeball.

• *Click* **LAYER 5** *in the* **LAYER CONTROLS** *window, and you will see the following image:*

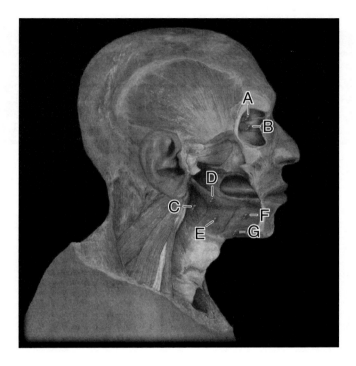

• *Mouse-over the pins on the screen to find the information necessary to identify the following structures:*

A. _____

B. _____

C. _____

D. _____

E. _____

F. _____

G. _____

CHECK POINT:

Head and Neck, Lateral View, cont'd

11. Name the muscle involved with adduction of the eyeball.
12. Name the muscle whose tendon passes through a trochlea.
13. Which muscle allows you to stick out your tongue?

Animations: Muscle Actions

SELECT ANIMATION
Buccinator Muscle

PLAY

- *After viewing this animation, select and view these additional animations:*
  - **Frontalis muscle**
  - **Levator labii superioris alaeque nasi muscle**
  - **Masseter muscle**
  - **Platysma muscle**
  - **Temporalis muscle**
  - **Trapezius muscle**

EXERCISE 4.3:

Skeletal Muscles—Head and Neck, Mid-sagittal View

SELECT TOPIC	SELECT VIEW	
Head and Neck	**Mid-sagittal**	**GO**

- *Click* **LAYER 1** *in the* **LAYER CONTROLS** *window, and you will see the following image:*

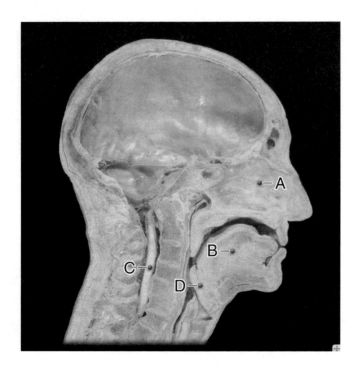

- *Mouse-over the blue pins on the screen to find the information necessary to identify the following non-muscular system structures:*

A. _____

B. _____

C. _____

D. _____

- *Click* **LAYER 2** *in the* **LAYER CONTROLS** *window, and you will see the following image:*

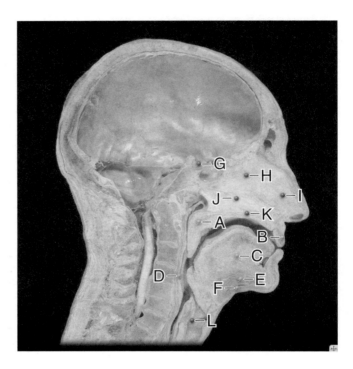

- *Mouse-over the green pins on the screen to find the information necessary to identify the following structures:*

A. _____

B. _____

C. _____

D. _____

E. _____

F. _____

Non-muscular System Structures (blue pins)

G. _____

H. _____

I. _____

J. _____

K. _____

L. _____

CHECK POINT:

Head and Neck, Mid-sagittal View

1. Name the muscular structure that separates the oropharynx from the nasopharynx.
2. Name a muscle that blends with the musculature of the tongue.

- *Click **LAYER 3** in the **LAYER CONTROLS** window, and you will see the following image:*

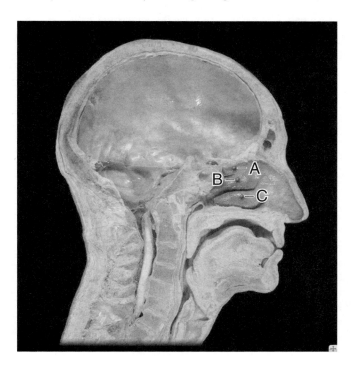

- *Mouse-over the blue pins on the screen to find the information necessary to identify the following non-muscular system structures:*

A. _____

B. _____

C. _____

EXERCISE 4.4:

Skeletal Muscle—Head and Neck, Posterior View

SELECT TOPIC	SELECT VIEW	
Head and Neck >	**Posterior**	GO

- *Click **LAYER 1** in the **LAYER CONTROLS** window, and you will see the following image:*

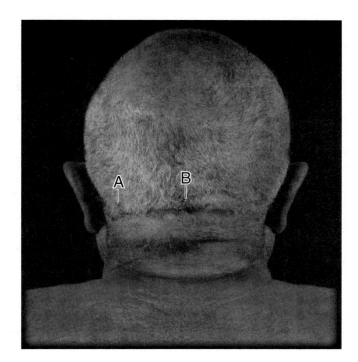

- *Mouse-over the blue pins on the screen to find the information necessary to identify the following non-muscular system structures:*

A. _____

B. _____

CHECK POINT:

Head and Neck, Posterior View

1. Name the three muscles that attach to the mastoid process.
2. What is the attachment point to the skull for the nuchal ligament?

- *Click **LAYER 2** in the **LAYER CONTROLS** window, and you will see the following image:*

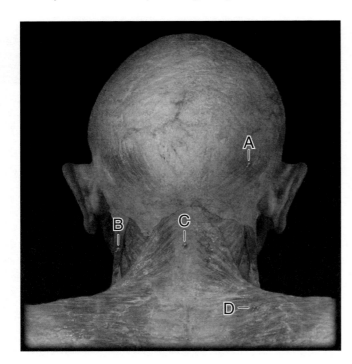

- *Mouse-over the pins on the screen to find the information necessary to identify the following structures:*

A. _____

B. _____

C. _____

D. _____

CHECK POINT:

Head and Neck, Posterior View, cont'd

3. Name two muscles attached to the nuchal ligament.
4. Name the two origins and the one insertion for the sternocleidomastoid muscle.
5. Name the large superficial muscle located from the posterior neck to the shoulders and the posterior midline.

- *Click **LAYER 3** in the **LAYER CONTROLS** window, and you will see the following image:*

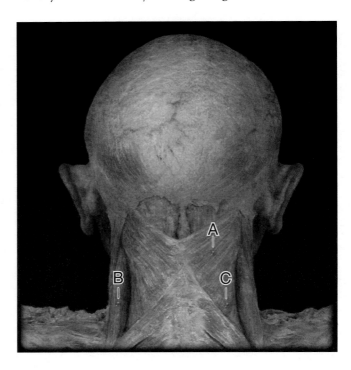

- *Mouse-over the pins on the screen to find the information necessary to identify the following structures:*

A. _____

B. _____

C. _____

- *Click **LAYER 4** in the **LAYER CONTROLS** window, and you will see the following image:*

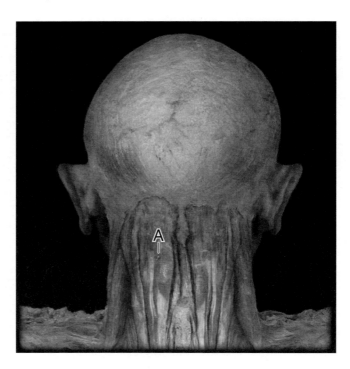

• *Mouse-over the pin on the screen to find the information necessary to identify the following structure:*

A. _____

• *Click **LAYER 6** in the **LAYER CONTROLS** window, and you will see the following image:*

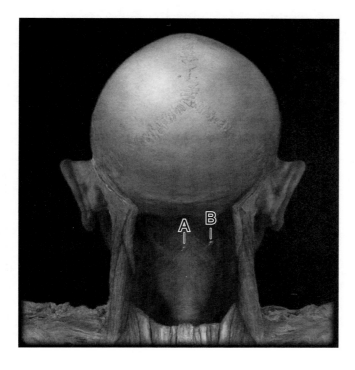

• *Mouse-over the pins on the screen to find the information necessary to identify the following structures:*

A. _____

B. _____

CHECK POINT:

Head and Neck, Posterior View, cont'd

6. Name a muscle responsible for elevation of the pharynx during swallowing.

Self Test

Take this opportunity to quiz yourself by taking the **SELF TEST.** See page 15 for a reminder on how to access the self-test for this section.

IN REVIEW

What Have I Learned?

The following questions cover the material that you have just learned, the muscles of the head and neck. Apply what you have learned to answer these questions on a separate piece of paper.

1. Name a muscle responsible for elevation of the larynx.

2. Name the muscle that flares the nostrils.

3. The scapula is elevated by which muscles?

4. When this muscle contracts, the head rotates so that the face turns downward and to the opposite side.

5. Name three muscles responsible for closing the mouth.

6. Name three muscles responsible for depression of the hyoid bone.

7. What muscle is responsible for flexion of the head to look downward?

8. Name the group of muscles responsible for the peristaltic waves of swallowing.

9. Name three muscles involved in moving the tongue.

10. Name the muscle involved in elevating the eyebrow and creasing the skin of the forehead.

11. Name a muscle responsible for depression of the larynx.

12. There is a muscle complex that lies deep to the scalp from the forehead to the posterior skull. What is the name of that complex and the two muscles that it consists of?

13. List all of the muscles involved with eye movement, and describe the movement involved with each muscle.

14. Name the anatomical structure commonly called the "chin."

Muscular System: Trunk, Shoulder Girdle, and Upper Limb

EXERCISE 4.5:

Skeletal Muscle—Thorax, Anterior View

SELECT TOPIC	SELECT VIEW	
Thorax	**Anterior**	**GO**

- *Click **LAYER 1** in the **LAYER CONTROLS** window, and you will see the following image:*

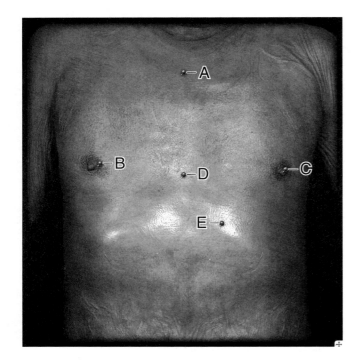

- *Mouse-over the blue pins on the screen to find the information necessary to identify the following non-muscular system structures:*

A. _____

B. _____

C. _____

D. _____

E. _____

CHECK POINT:

Thorax, Anterior View

1. What is the name for the superficially visible inferior border of costal cartilages 7–10?
2. What structures attach to this location?
3. What are the two names for the shallow notch in the superficially visible superior border of the manubrium?

- *Click **LAYER 2** in the **LAYER CONTROLS** window, and you will see the following image:*

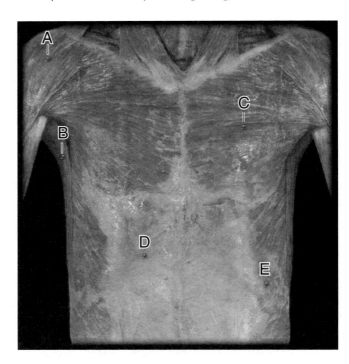

- *Mouse-over the pins on the screen to find the information necessary to identify the following structures:*

A. _____

B. _____

C. _____

D. _____

E. _____

CHECK POINT:

Thorax, Anterior View, cont'd

4. Name the muscle involved with adduction, extension, and medial rotation of the arm.
5. Name the muscle involved with abduction, flexion, extension, and lateral and medial rotation of the arm.
6. What is the name for the fibrous compartment enclosing the rectus abdominis muscle?
7. What is an aponeurosis?

• *Click* **LAYER 3** *in the* **LAYER CONTROLS** *window, and you will see the following image:*

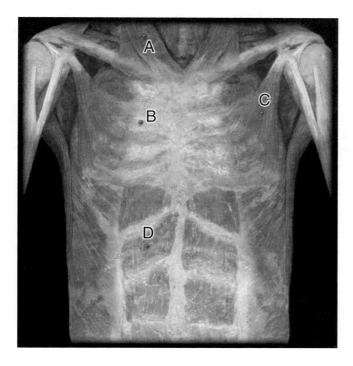

• *Mouse-over the pins on the screen to find the information necessary to identify the following structures:*

A. _____

B. _____

C. _____

D. _____

CHECK POINT:

Thorax, Anterior View, cont'd

8. Name the muscle that consists of three to four bellies, separated by tendinous intersections.
9. Name the muscle with its origin at the medial clavicle and the manubrium of the sternum and its insertion at the mastoid process.
10. Name the muscle that stabilizes the scapula and is involved in its lateral rotation.

• *Click* **LAYER 4** *in the* **LAYER CONTROLS** *window, and you will see the following image:*

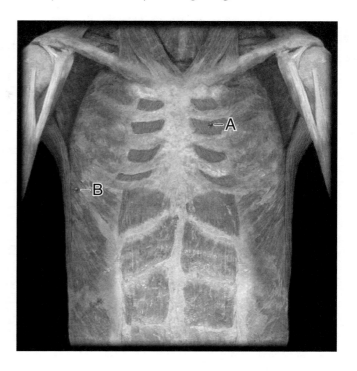

• *Mouse-over the pins on the screen to find the information necessary to identify the following structures:*

A. _____

B. _____

• *Click* **LAYER 5** *in the* **LAYER CONTROLS** *window, and you will see the following image:*

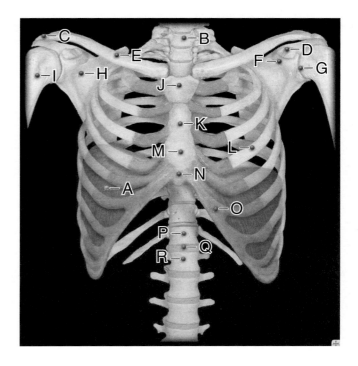

• *Mouse-over the green pin on the screen to find the information necessary to identify the following structure:*

A. _____

Non-muscular System Structures (blue pins)

B. _____

C. _____

D. _____

E. _____

F. _____

G. _____

H. _____

I. _____

J. _____

K. _____

L. _____

M. _____

N. _____

O. _____

P. _____

Q. _____

R. _____

Animations: Muscle Actions

SELECT ANIMATION	
Deltoid Muscle	PLAY

• *After viewing this animation, select and view these additional animations:*

  – **External abdominal oblique muscle**
  – **Latissimus dorsi muscle**
  – **Pectoralis major muscle**
  – **Rectus abdominus muscle**
  – **Serratus anterior muscle**

Self Test

Take this opportunity to quiz yourself by taking the **SELF TEST.** See page 15 for a reminder on how to access the self-test for this section.

I N   R E V I E W

What Have I Learned?

The following questions cover the material that you have just learned, the muscles of the thorax. Apply what you have learned to answer these questions on a separate piece of paper.

1. Name the structure formed by the tendons of three abdominal muscles.

2. Name the three primary muscles of respiration.

3. Name the muscle responsible for the adduction, extension, and medial rotation of the humerus.

4. Name the two muscles that stabilize the scapula.

5. Name the muscle that is the site of intramuscular injections of the arm.

EXERCISE 4.6:

Skeletal Muscle—Abdomen, Anterior View

SELECT TOPIC	SELECT VIEW	
Abdomen >	**Anterior**	GO

• *Click* **LAYER 1** *in the* **LAYER CONTROLS** *window, and you will see the following image:*

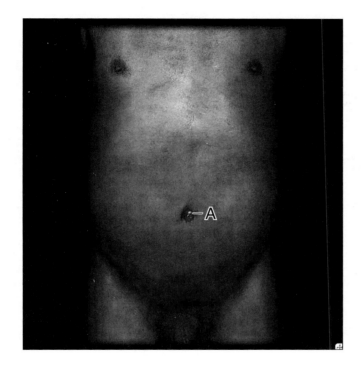

• *Mouse-over the blue pin on the screen to find the information necessary to identify the following non-muscular system structure:*

A. _____

CHECK POINT:

Abdomen, Anterior View

1. Describe umbilicus variability.
2. What is the umbilicus a landmark for?
3. Where is it located on lean individuals?

• *Click* **LAYER 2** *in the* **LAYER CONTROLS** *window, and you will see the following image:*

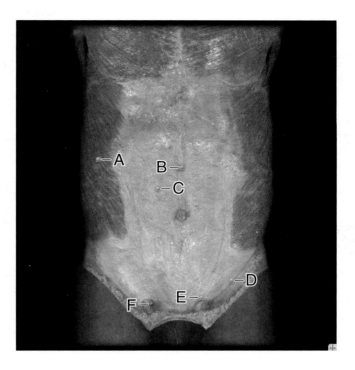

• *Mouse-over the green pins on the screen to find the information necessary to identify the following structures:*

A. _____

B. _____

C. _____

D. _____

E. _____

Non-Muscular System Structure (blue pin)

F. _____

CHECK POINT:

Abdomen, Anterior View, cont'd

4. Name the common site for male inguinal hernias.
5. Opening the abdominal wall by incision through the _____ avoids cutting muscle fibers.
6. What abdominal muscle has its fibers running at right angles to the internal abdominal oblique?

- *Click* **LAYER 3** *in the* **LAYER CONTROLS** *window, and you will see the following image:*

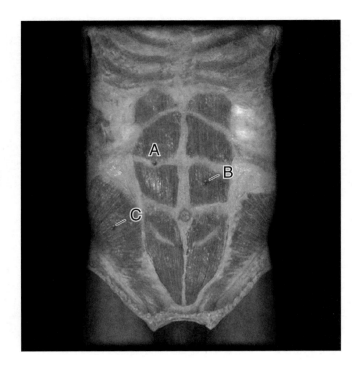

- *Mouse-over the pins on the screen to find the information necessary to identify the following structures:*

A. _____

B. _____

C. _____

CHECK POINT:

Abdomen, Anterior View, cont'd

7. Name the structures that subdivide the rectus abdominis muscle into three to four bellies.
8. What abdominal muscle has its fibers running at right angles to the external abdominal oblique?
9. Name the abdominal muscles in this view important in straining and abdominal breathing.

- *Click* **LAYER 4** *in the* **LAYER CONTROLS** *window, and you will see the following image:*

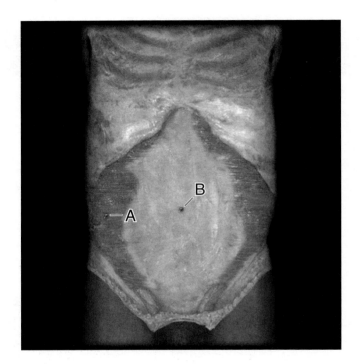

- *Mouse-over the pins on the screen to find the information necessary to identify the following structures:*

A. _____

B. _____

CHECK POINT:

Abdomen, Anterior View, cont'd

10. Name the abdominal muscle whose fibers run in a transverse plane.
11. What is the anatomical term for "flat tendons"?
12. What two structures come together to form the posterior rectus sheath?

• *Click* **LAYER 5** *in the* **LAYER CONTROLS** *window, and you will see the following image:*

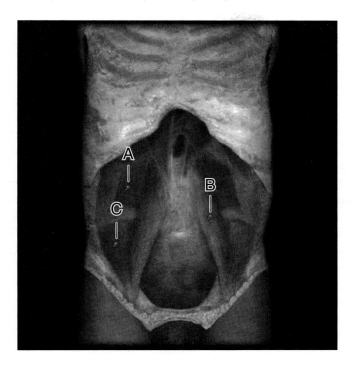

• *Mouse-over the pins on the screen to find the information necessary to identify the following structures:*

A. _____

B. _____

C. _____

CHECK POINT:

Abdomen, Anterior View, cont'd

13. Name a muscle of the posterior abdominal wall involved in respiration.

Animations: Muscle Actions

 SELECT ANIMATION
External Abdominal Oblique Muscle PLAY

• *After viewing this animation, select and view these additional animations:*

– **Iliacus muscle**
– **Rectus abdominis muscle**

Self Test
Take this opportunity to quiz yourself by taking the **SELF TEST.** See page 15 for a reminder on how to access the self-test for this section.

IN REVIEW

What Have I Learned?

The following questions cover the material that you have just learned, the muscles of the thorax. Apply what you have learned to answer these questions on a separate piece of paper.

1. Name the abdominal wall muscles responsible for abdominal breathing.

2. What is the term for a "seam" where two structures meet?

3. Two individual muscles of the abdomen unite to form a single muscle, the most powerful flexor of the hip. Name those two individual muscles and the muscle they unite to form.

4. Two pairs of abdominal wall muscles have their structures running at right angles to each other. What are those two pairs of muscles?

5. Name the abdominal wall muscles important in straining, such as while lifting.

EXERCISE 4.7:

Skeletal Muscle—Pelvis, Superior View

SELECT TOPIC **Pelvis**	▶	SELECT VIEW **Superior**	GO

- *Click* **LAYER 1** *in the* **LAYER CONTROLS** *window, and you will see the following image:*

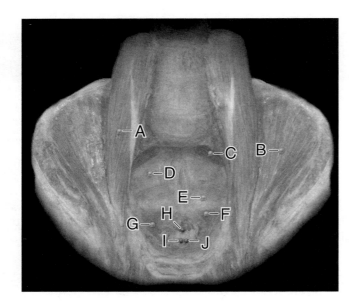

- *Mouse-over the green pins on the screen to find the information necessary to identify the following structures:*

A. _____

B. _____

C. _____

D. _____

E. _____

F. _____

G. _____

Non-Muscular System Structures (blue pins)

H. _____

I. _____

J. _____

Self Test

Take this opportunity to quiz yourself by taking the **SELF TEST.** See page 15 for a reminder on how to access the self-test for this section.

IN REVIEW

What Have I Learned?

The following questions cover the material that you have just learned, the muscles of the pelvis. Apply what you have learned to answer these questions on a separate piece of paper.

1. Name the pelvic muscle involved with lateral rotation of the femur and that exits the pelvis through the greater sciatic foramen.

2. Name the two muscles that make up the pelvic diaphragm. What are their functions?

3. Name the structure that serves as the origin for part of the levator ani muscle.

EXERCISE 4.8:

Skeletal Muscle—Back, Posterior View

SELECT TOPIC	SELECT VIEW	
Back	▶ **Posterior**	GO

- *Click **LAYER 1** in the **LAYER CONTROLS** window, and you will see the following image:*

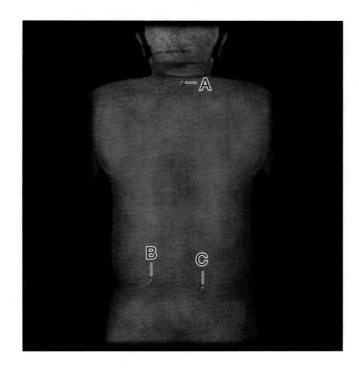

- *Mouse-over the blue pins on the screen to find the information necessary to identify the following non-muscular system structures:*

A. _____

B. _____

C. _____

CHECK POINT:

Back, Posterior View

1. Name the landmark for intramuscular injections of the hip.
2. The shallow skin depression (dimple) in the lower back marks what point?
3. What is the name for the prominent surface projection produced by the spinous process of vertebra C7?

- *Click **LAYER 2** in the **LAYER CONTROLS** window, and you will see the following image:*

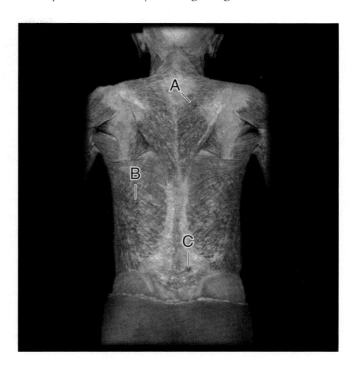

- *Mouse-over the pins on the screen to find the information necessary to identify the following structures:*

A. _____

B. _____

C. _____

CHECK POINT:

Back, Posterior View, cont'd

4. Name the superficial "kite-shaped" muscle of the back that spans from the nuchal line of the occipital bone to vertebra T12.
5. Name the deep fascia whose attached structures include the latissimus dorsi muscle.
6. Name the superficial muscle whose name describes its location as it spans from the back to the side of the body.

- *Click* **LAYER 3** *in the* **LAYER CONTROLS** *window, and you will see the following image:*

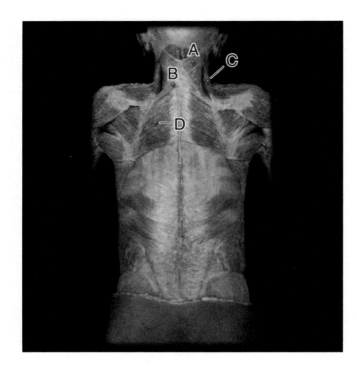

- *Mouse-over the pins on the screen to find the information necessary to identify the following structures:*

A. _____

B. _____

C. _____

D. _____

CHECK POINT:

Back, Posterior View, cont'd

7. Name two muscles involved in the retraction and elevation of the scapula.
8. Name a muscle that allows the shrugging of the shoulders.

- *Click* **LAYER 4** *in the* **LAYER CONTROLS** *window, and you will see the following image:*

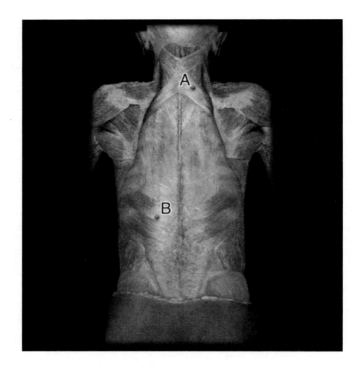

- *Mouse-over the pins on the screen to find the information necessary to identify the following structures:*

A. _____

B. _____

- *Click* **LAYER 5** *in the* **LAYER CONTROLS** *window, and you will see the following image:*

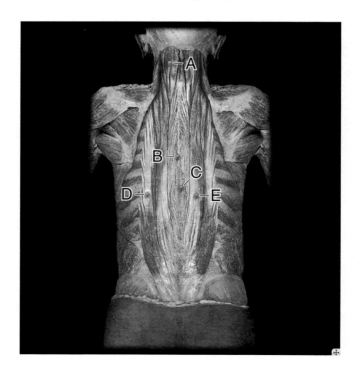

- *Mouse-over the pins on the screen to find the information necessary to identify the following structures:*

A. _____

B. _____

C. _____

D. _____

E. _____

CHECK POINT:

Back, Posterior View, cont'd

9. Name the muscle known as the "antigravity muscle."
10. This muscle consists of three separate muscles. What are they?

Animations: Muscle Actions

SELECT ANIMATION
Erector Spinae Muscle (extension)   **PLAY**

- *After viewing this animation, select and view these additional animations:*
  - **Erector spinae muscle (lateral flexion)**
  - **Infraspinatus muscle**
  - **Latissimus dorsi muscle**
  - **Rhomboid major and minor muscles**
  - **Subscapularis muscle**
  - **Supraspinatus muscle**
  - **Trapezius muscle**

Self Test

Take this opportunity to quiz yourself by taking the **SELF TEST.** See page 15 for a reminder on how to access the self-test for this section.

EXERCISE 4.9:

Skeletal Muscle—Shoulder and Arm, Anterior View

SELECT TOPIC	SELECT VIEW	
Shoulder and Arm ▶	**Anterior**	**GO**

- *Click **LAYER 1** in the **LAYER CONTROLS** window, and you will see the following image:*

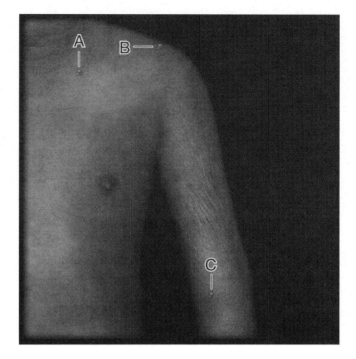

- *Mouse-over the blue pins on the screen to find the information necessary to identify the following non-muscular system structures:*

A. _____

B. _____

C. _____

CHECK POINT:

Shoulder and Arm, Anterior View

1. Name the structure referred to as the collar bone.
2. Name the structure that is the flattened, lateral part of the scapular spine.
3. What is the name for the triangular concavity of the anterior elbow?

- *Click* **LAYER 2** *in the* **LAYER CONTROLS** *window, and you will see the following image:*

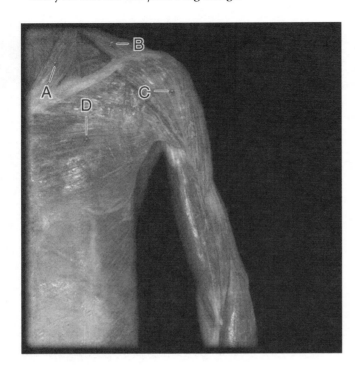

- *Mouse-over the pins on the screen to find the information necessary to identify the following structures:*

A. _____

B. _____

C. _____

D. _____

CHECK POINT:

Shoulder and Arm, Anterior View, cont'd

4. Name the superficial muscle of the chest.
5. Name the muscle that contributes to the roundness of the shoulder.
6. Name the mostly posterior muscle that has its insertion at the clavicle and scapula.

- *Click* **LAYER 3** *in the* **LAYER CONTROLS** *window, and you will see the following image:*

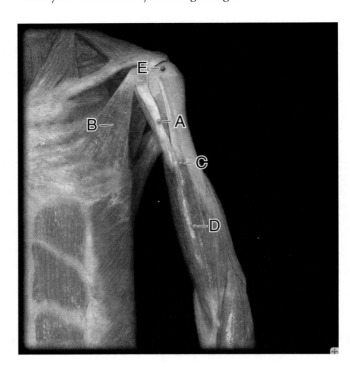

- *Mouse-over the green pins on the screen to find the information necessary to identify the following structures:*

A. _____

B. _____

C. _____

D. _____

Non-muscular System Structure (blue pin)

E. _____

CHECK POINT:

Shoulder and Arm, Anterior View, cont'd

7. Name the muscle of the arm that has two heads.
8. Name the tough fibrous envelope that surrounds the joint where the arm attaches to the pectoral girdle.
9. Name the two muscles referred to as the "pecs."

- *Click* **LAYER 4** *in the* **LAYER CONTROLS** *window, and you will see the following image:*

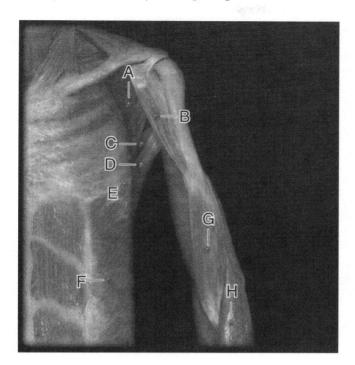

- *Mouse-over the pins on the screen to find the information necessary to identify the following structures:*

A. _____

B. _____

C. _____

D. _____

E. _____

F. _____

G. _____

H. _____

CHECK POINT:

Shoulder and Arm, Anterior View, cont'd

10. Name the four rotator cuff muscles.
11. What is the function of the rotator cuff muscles?
12. Name the muscle deep to the biceps brachii.

- *Click* **LAYER 5** *in the* **LAYER CONTROLS** *window, and you will see the following image:*

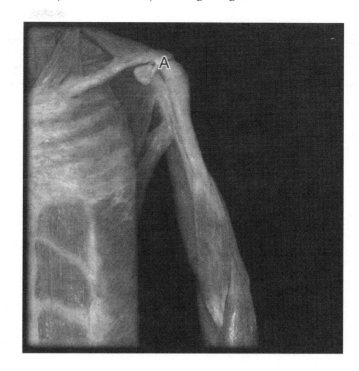

- *Mouse-over the blue pin on the screen to find the information necessary to identify the following non-muscular system structure:*

A. _____

Animations: Muscle Actions

SELECT ANIMATION	PLAY
Biceps Brachii Muscle	

- *After viewing this animation, select and view these additional animations:*

  – **Brachialis muscle**
  – **Deltoid muscle**
  – **External abdominal oblique muscle**
  – **Latissimus dorsi muscle**
  – **Pectoralis major muscle**
  – **Serratus anterior muscle**
  – **Subscapularis muscle**
  – **Teres major muscle**
  – **Trapezius muscle**
  – **Triceps brachii muscle**

EXERCISE 4.10:

Skeletal Muscle—Shoulder and Arm, Posterior View

SELECT TOPIC	SELECT VIEW	
Shoulder and Arm ▶	**Posterior**	**GO**

- *Click* **LAYER 1** *in the* **LAYER CONTROLS** *window, and you will see the following image:*

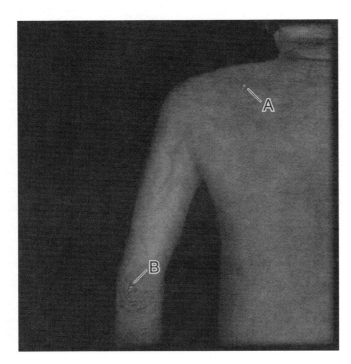

- *Mouse-over the blue pins on the screen to find the information necessary to identify the following non-muscular system structures:*

A. _____

B. _____

CHECK POINT:

Shoulder and Arm, Posterior View

1. What is the name for the point of the elbow?
2. What specific structure of what bone constitutes the point of the elbow?
3. Name the prominent ridge on the posterior surface of the scapula.

- *Click* **LAYER 2** *in the* **LAYER CONTROLS** *window, and you will see the following image:*

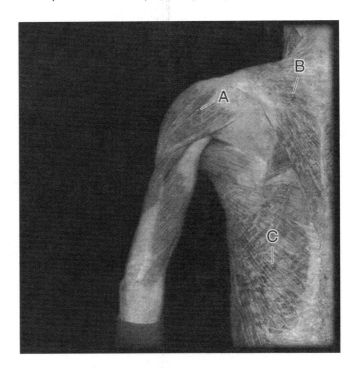

- *Mouse-over the pins on the screen to find the information necessary to identify the following structures:*

A. _____

B. _____

C. _____

CHECK POINT:

Shoulder and Arm, Posterior View, cont'd

4. Name the triangle-shaped muscle of the shoulder.
5. Name the large lateral muscle responsible for adduction, extension, and medial rotation of the arm.
6. Name the muscle responsible for the elevation, medial rotation, adduction, and depression of the scapula.

• *Click* **LAYER 3** *in the* **LAYER CONTROLS** *window, and you will see the following image:*

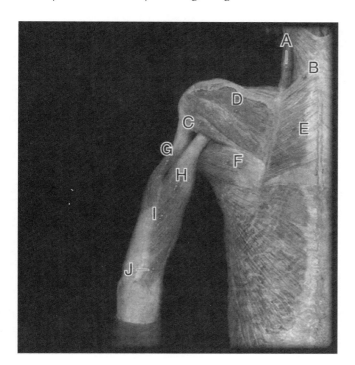

• *Mouse-over the pins on the screen to find the information necessary to identify the following structures:*

A. _____

B. _____

C. _____

D. _____

E. _____

F. _____

G. _____

H. _____

I. _____

J. _____

CHECK POINT:

Shoulder and Arm, Posterior View, cont'd

7. Name the muscle of the arm with three heads.
8. Name the muscle found in the infraspinous fossa of the scapula.
9. Name two muscles with their insertions on the medial border of the scapula.

• *Click* **LAYER 4** *in the* **LAYER CONTROLS** *window, and you will see the following image:*

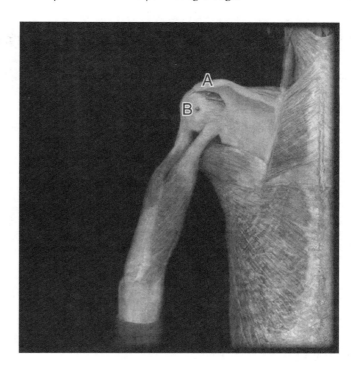

• *Mouse-over the green pin on the screen to find the information necessary to identify the following structure:*

A. _____

Non-muscular System Structure (blue pin)

B. _____

CHECK POINT:

Shoulder and Arm, Posterior View, cont'd

10. Name the muscle located in the supraspinous fossa of the scapula.
11. Name a muscle that holds the head of the humerus in the glenoid cavity.

Self Test
Take this opportunity to quiz yourself by taking the **SELF TEST.** See page 15 for a reminder on how to access the self-test for this section.

IN REVIEW

What Have I Learned?

The following questions cover the material that you have just learned, the muscles of the shoulder and arm. Apply what you have learned to answer these questions on a separate piece of paper.

1. Name the landmark for intramuscular injections of the shoulder.

2. Name the superficial "kite-shaped" muscle of the back that spans from the nuchal line of the occipital bone to vertebra T12.

3. Name the superficial muscle whose name describes its location as it spans from the back to the side of the body.

4. Name two muscles involved in the retraction and elevation of the scapula.

5. Name a muscle that allows the shrugging of the shoulders.

6. Name the muscle known as the "antigravity muscle." What three muscles combine to form this muscle?

7. Name the superficial muscle of the chest.

8. Name the muscle of the arm that has two heads.

9. Name the muscle of the arm that has three heads.

10. Name the four rotator cuff muscles. What is their function?

11. Name the large lateral muscle responsible for adduction, extension, and medial rotation of the humerus.

12. Name the muscle responsible for the elevation, medial rotation, adduction, and depression of the scapula.

13. Name a muscle that holds the head of the humerus in the glenoid cavity.

EXERCISE 4.11:

Skeletal Muscles—Forearm and Hand, Anterior View

SELECT TOPIC	SELECT VIEW	
Forearm and Hand ▶	**Anterior**	**GO**

- *Click* **LAYER 1** *in the* **LAYER CONTROLS** *window, and you will see the following image:*

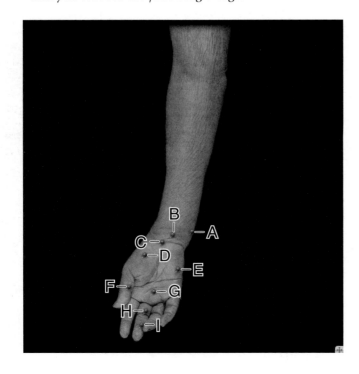

- *Mouse-over the blue pins on the screen to find the information necessary to identify the following non-muscular system structures:*

A. _____

B. _____

C. _____

D. _____

E. _____

F. _____

G. _____

H. _____

I. _____

- *Click* **LAYER 2** *in the* **LAYER CONTROLS** *window, and you will see the following image:*

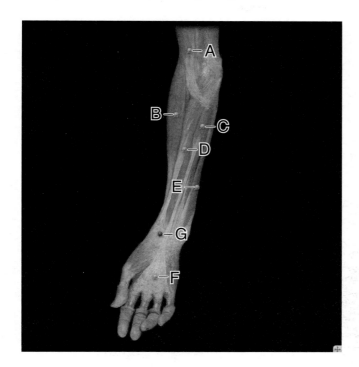

- *Mouse-over the green pins on the screen to find the information necessary to identify the following structures:*

A. _____

B. _____

C. _____

D. _____

E. _____

F. _____

Non-muscular System Structures (blue pin)

G. _____

CHECK POINT:

Forearm and Hand, Anterior View

1. Name three muscles that flex the wrist/hand.
2. Name two muscles that flex the forearm.
3. Name the structure that forms the carpal tunnel.

- *Click* **LAYER 3** *in the* **LAYER CONTROLS** *window, and you will see the following image:*

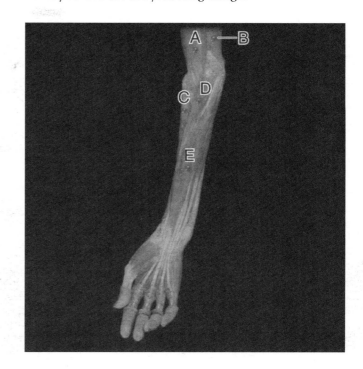

- *Mouse-over the pins on the screen to find the information necessary to identify the following structures:*

A. _____

B. _____

C. _____

D. _____

E. _____

CHECK POINT:

Forearm and Hand, Anterior View, cont'd

4. Name a muscle involved in the pronation of the forearm.
5. Name a muscle involved in the supination of the forearm.
6. Name a muscle involved with the extension of the forearm.

• *Click* **LAYER 4** *in the* **LAYER CONTROLS** *window, and you will see the following image:*

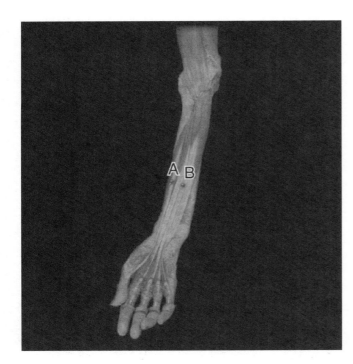

• *Mouse-over the pins on the screen to find the information necessary to identify the following structures:*

A. _____

B. _____

CHECK POINT:

Forearm and Hand, Anterior View, cont'd

7. Name the only muscle that flexes the distal interphalangeal joint.
8. Name the only muscle that flexes the distal phalanx of the thumb.
9. What is the anatomical name for the thumb?

• *Click* **LAYER 5** *in the* **LAYER CONTROLS** *window, and you will see the following image:*

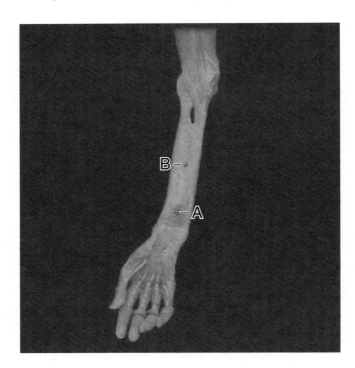

• *Mouse-over the green pin on the screen to find the information necessary to identify the following structure:*

A. _____

Non-muscular System Structure (blue pin)

B. _____

CHECK POINT:

Forearm and Hand, Anterior View, cont'd

10. Name the thick sheet of connective tissue between the ulna and the radius.
11. What is its function?

EXERCISE 4.12:

Skeletal Muscles—Forearm and Hand, Posterior View

SELECT TOPIC	SELECT VIEW	
Forearm and Hand ▶	Posterior	GO

- *Click* **LAYER 1** *in the* **LAYER CONTROLS** *window, and you will see the following image:*

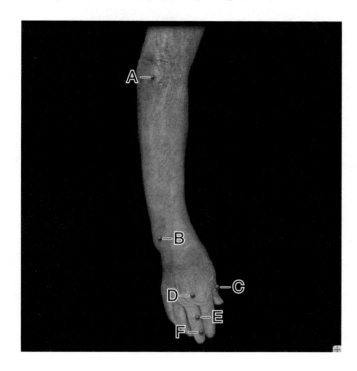

- *Mouse-over the blue pins on the screen to find the information necessary to identify the following non-muscular structures:*

A. _____

B. _____

C. _____

D. _____

E. _____

F. _____

- *Click* **LAYER 2** *in the* **LAYER CONTROLS** *window, and you will see the following image:*

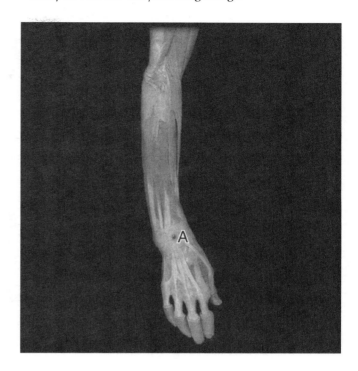

- *Mouse-over the blue pin on the screen to find the information necessary to identify the following non-muscular system structure:*

A. _____

CHECK POINT:

Forearm and Hand, Posterior View

1. What is the relationship between the retinaculum and the extensor tendons of the forearm?

- *Click **LAYER 3** in the **LAYER CONTROLS** window, and you will see the following image:*

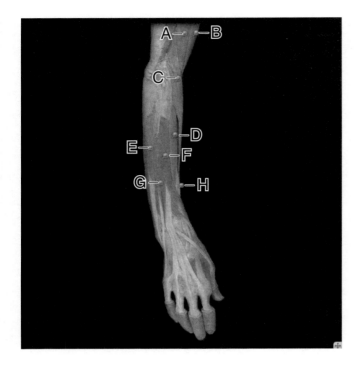

- *Mouse-over the pins on the screen to find the information necessary to identify the following structures:*

A. _____

B. _____

C. _____

D. _____

E. _____

F. _____

G. _____

H. _____

CHECK POINT:

Forearm and Hand, Posterior View, cont'd

2. Name three muscles that extend the hand.
3. Name a muscle that assists in both pronation and supination of the forearm.
4. Name a muscle that extends the fifth finger.

- *Click **LAYER 4** in the **LAYER CONTROLS** window, and you will see the following image:*

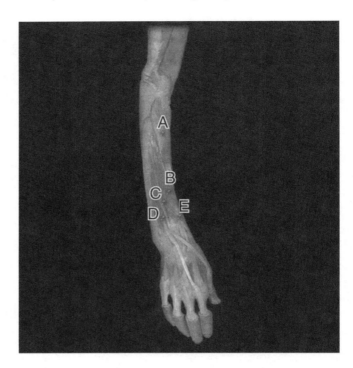

- *Mouse-over the pins on the screen to find the information necessary to identify the following structures:*

A. _____

B. _____

C. _____

D. _____

E. _____

CHECK POINT:

Forearm and Hand, Posterior View, cont'd

5. Name a muscle that both extends and abducts the thumb.
6. Name two muscles that only extend the thumb.
7. Name a muscle that extends the index finger.

Self Test

Take this opportunity to quiz yourself by taking the **SELF TEST.** See page 15 for a reminder on how to access the self-test for this section.

I N   R E V I E W

What Have I Learned?

The following questions cover the material that you have just learned, the muscles of the forearm and hand. Apply what you have learned to answer these questions on a separate piece of paper.

1. Name three muscles that flex the wrist.

2. Name two muscles that flex the forearm.

3. Name the structure that forms the carpal tunnel.

4. Name a muscle involved in the pronation of the forearm.

5. Name a muscle involved in the supination of the forearm.

6. What is the anatomical name for the thumb?

7. Name the thick sheet of connective tissue between the ulna and the radius. What is its function?

8. Name three muscles that extend the hand.

9. Name a muscle that both extends and abducts the thumb.

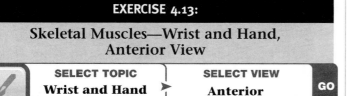

EXERCISE 4.13:

Skeletal Muscles—Wrist and Hand, Anterior View

SELECT TOPIC	SELECT VIEW	
Wrist and Hand ▶	**Anterior**	**GO**

• *Click* **LAYER 1** *in the* **LAYER CONTROLS** *window, and you will see the following image:*

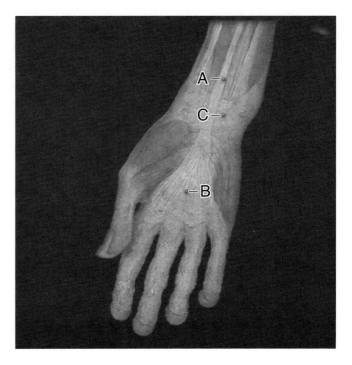

• *Mouse-over the blue pins on the screen to find the information necessary to identify the following non-muscular system structures:*

A. _____

B. _____

C H E C K   P O I N T :

Wrist and Hand, Anterior View

1. Name the thick, fleshy eminence at the base of the first digit.
2. Name the thick, fleshy eminence at the base of the fifth digit.

• *Click* **LAYER 2** *in the* **LAYER CONTROLS** *window, and you will see the following image:*

- *Mouse-over the green pins on the screen to find the information necessary to identify the following structures:*

A. _____

B. _____

Non-muscular System Structure (blue pin)

C. _____

C H E C K   P O I N T :

Wrist and Hand, Anterior View, cont'd

3. Name a muscle often missing on one or both forearms.

- *Click **LAYER 3** in the **LAYER CONTROLS** window, and you will see the following image:*

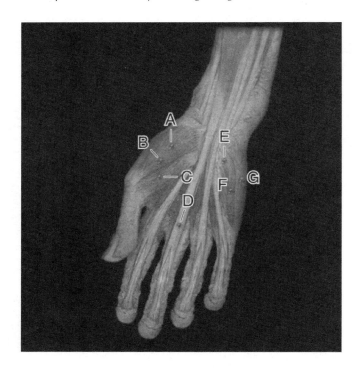

- *Mouse-over the pins on the screen to find the information necessary to identify the following structures:*

A. _____

B. _____

C. _____

D. _____

E. _____

F. _____

G. _____

C H E C K   P O I N T :

Wrist and Hand, Anterior View, cont'd

4. Name the three thenar muscles.
5. Name the three hypothenar muscles.
6. Which digits do each of the above six muscles act upon?

- *Click **LAYER 4** in the **LAYER CONTROLS** window, and you will see the following image:*

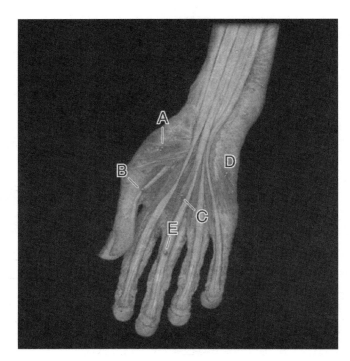

- *Mouse-over the pins on the screen to find the information necessary to identify the following structures:*

A. _____

B. _____

C. _____

D. _____

E. _____

C H E C K   P O I N T :

Wrist and Hand, Anterior View, cont'd

7. Name the muscles that both flex the metacarpophalangeal joint and extend the interphalangeal joints.
8. Name the muscle that allows the fifth finger to touch the tip of the first finger.
9. Name the muscle that allows the tip of the first finger to touch the tips of the other fingers.

• *Click* **LAYER 5** *in the* **LAYER CONTROLS** *window, and you will see the following image:*

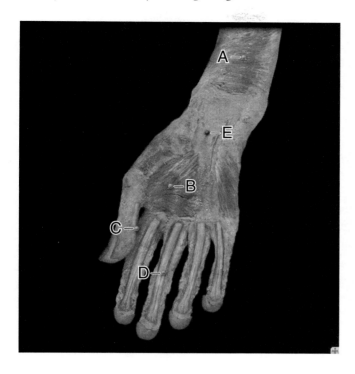

• *Mouse-over the green pins on the screen to find the information necessary to identify the following structures:*

A. _____

B. _____

C. _____

D. _____

Non-muscular System Structures (blue pin)

E. _____

CHECK POINT:

Wrist and Hand, Anterior View, cont'd

10. Name the only muscle that flexes the distal phalanx of the first digit.
11. Name the only muscle that flexes the distal interphalangeal joint of digits 2–5.
12. Name the distal pronator of the forearm.

EXERCISE 4.14:

Skeletal Muscles—Wrist and Hand, Posterior View

SELECT TOPIC	SELECT VIEW	
Wrist and Hand ▶	**Posterior**	GO

• *Click* **LAYER 1** *in the* **LAYER CONTROLS** *window, and you will see the following image:*

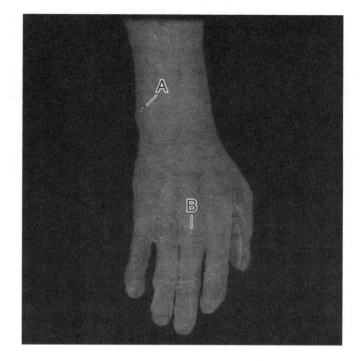

• *Mouse-over the blue pins on the screen to find the information necessary to identify the following non-muscular system structures:*

A. _____

B. _____

CHECK POINT:

Wrist and Hand, Posterior View

1. Flexion of which joint makes the knuckles prominent?
2. What structures are visible as the knuckles?

- *Click* **LAYER 2** *in the* **LAYER CONTROLS** *window, and you will see the following image:*

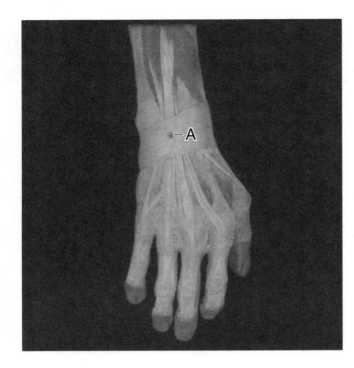

- *Mouse-over the blue pin on the screen to find the information necessary to identify the following non-muscular system structure:*

A. _____

CHECK POINT:

Wrist and Hand, Posterior View, cont'd

3. Name a muscle often missing on one or both forearms.

- *Click* **LAYER 3** *in the* **LAYER CONTROLS** *window, and you will see the following image:*

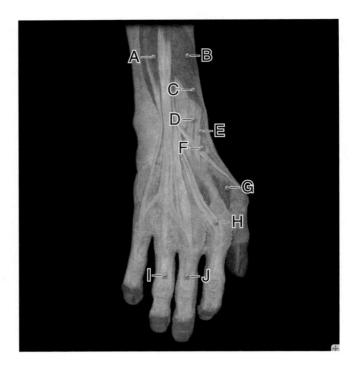

- *Mouse-over the pins on the screen to find the information necessary to identify the following structures:*

A. _____

B. _____

C. _____

D. _____

E. _____

F. _____

G. _____

H. _____

I. _____

J. _____

CHECK POINT:

Wrist and Hand, Posterior View, cont'd

4. Name a muscle that extends the index finger.
5. Name a muscle that extends fingers two through five.
6. Name a muscle that extends the thumb.
7. Name a muscle that extends the little finger.

Animations: Muscle Actions

SELECT ANIMATION	
Extensor Digitorum Muscle	PLAY

- *After viewing this animation, select and view this additional animation:*

  – Flexor digitorum superficialis and profundus muscles

Self Test

Take this opportunity to quiz yourself by taking the **SELF TEST.** See page 15 for a reminder on how to access the self-test for this section.

I N   R E V I E W

What Have I Learned?

The following questions cover the material that you have just learned, the muscles of the wrist and hand. Apply what you have learned to answer these questions on a separate piece of paper.

1. Name the thick, fleshy eminence at the base of the first digit.

2. Name the thick, fleshy eminence at the base of the fifth digit.

3. Name a muscle often missing on one or both forearms.

4. Name the only muscle that flexes the distal phalanx of the first digit.

5. Flexion of which joint makes the knuckles prominent?

6. What structures are visible as the knuckles?

EXERCISE 4.15:
Skeletal Muscles—Hip and Thigh, Anterior View

SELECT TOPIC	SELECT VIEW	
Hip and Thigh ▸	**Anterior**	GO

- *Click* **LAYER 1** *in the* **LAYER CONTROLS** *window, and you will see the following image:*

- *Mouse-over the blue pins on the screen to find the information necessary to identify the following non-muscular system structures:*

A. _____

B. _____

C. _____

D. _____

CHECK POINT:

Hip and Thigh, Anterior View

1. Name the superficially visible anterior subcutaneous end of the iliac crest.
2. Name the point of attachment for the quadriceps femoris muscles by way of the patellar ligament.
3. Name the ligament that connects the patella to the tuberosity of the tibia.

• *Click* **LAYER 2** *in the* **LAYER CONTROLS** *window, and you will see the following image:*

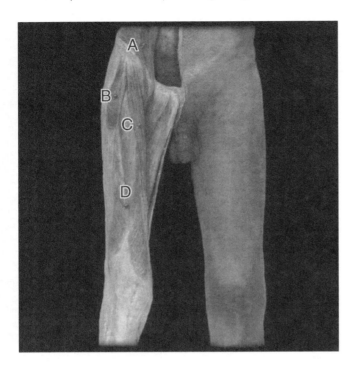

• *Mouse-over the pins on the screen to find the information necessary to identify the following structures:*

A. _____

B. _____

C. _____

D. _____

C H E C K   P O I N T :

Hip and Thigh, Anterior View, cont'd

4. Name the muscle whose origin is the anterior superior iliac spine of the ilium and whose insertion is the proximal medial shaft of the tibia.
5. Name the four muscles of the quadriceps femoris.
6. Name the most powerful flexor of the hip joint.

• *Click* **LAYER 3** *in the* **LAYER CONTROLS** *window, and you will see the following image:*

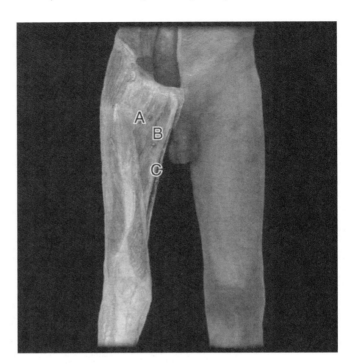

• *Mouse-over the pins on the screen to find the information necessary to identify the following structures:*

A. _____

B. _____

C. _____

C H E C K   P O I N T :

Hip and Thigh, Anterior View, cont'd

7. Name the muscle of the thigh that is weak in humans and used in muscle transplants.
8. Name the muscle often involved in a "pulled groin."

- *Click* **LAYER 4** *in the* **LAYER CONTROLS** *window, and you will see the following image:*

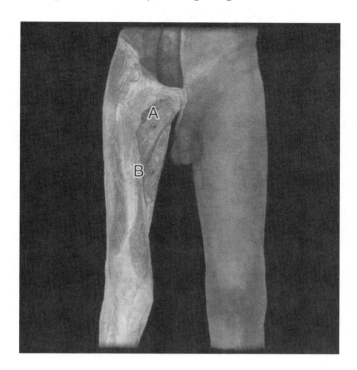

- *Mouse-over the pins on the screen to find the information necessary to identify the following structures:*

A. _____

B. _____

- *Click* **LAYER 5** *in the* **LAYER CONTROLS** *window, and you will see the following image:*

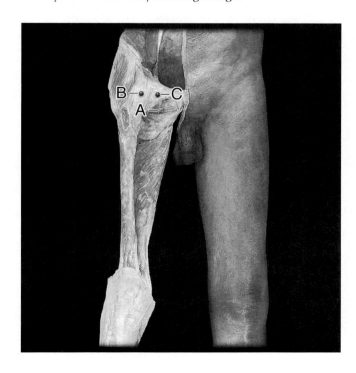

- *Mouse-over the green pin on the screen to find the information necessary to identify the following structure:*

A. _____

Non-muscular System Structures (blue pins)

B. _____

C. _____

CHECK POINT:

Hip and Thigh, Anterior View, cont'd

9. Name the strongest ligament around the hip joint.
10. Name the ligament that resists excessive abduction of the hip.
11. Name the ligament that resists hyperextension of the hip joint.

EXERCISE 4.16:

Skeletal Muscles—Hip and Thigh, Posterior View

SELECT TOPIC	SELECT VIEW	
Hip and Thigh	**Posterior**	GO

- *Click* **LAYER 1** *in the* **LAYER CONTROLS** *window, and you will see the following image:*

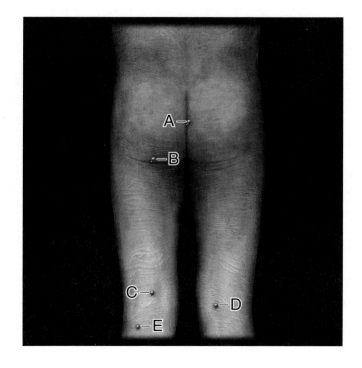

- *Mouse-over the blue pins on the screen to find the information necessary to identify the following non-muscular system structures:*

A. _____

B. _____

C. _____

D. _____

E. _____

CHECK POINT:

Hip and Thigh, Posterior View

1. Name the muscle whose tendon is the lateral hamstring.
2. Name the muscles whose tendons are the medial hamstring.
3. Name the structure that provides attachment for the fibular collateral ligament of the knee and the biceps femoris muscle.
4. What is the natal cleft? What is its function?
5. What are the gluteal folds? What do they represent?

- *Click **LAYER 2** in the **LAYER CONTROLS** window, and you will see the following image:*

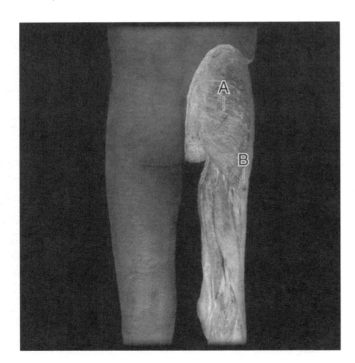

- *Mouse-over the pins on the screen to find the information necessary to identify the following structures:*

A. _____

B. _____

CHECK POINT:

Hip and Thigh, Posterior View, cont'd

6. Name a muscle of the posterior thigh not important in walking.
7. Name a muscle of the posterior thigh important for powerful extension of the femur as in running, climbing stairs, and rising from the seated position.
8. Name the structure that provides attachment for the tensor fascia lata and gluteus maximus muscles.

- *Click **LAYER 3** in the **LAYER CONTROLS** window, and you will see the following image:*

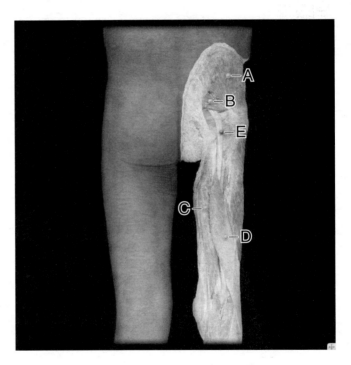

- *Mouse-over the green pins on the screen to find the information necessary to identify the following structures:*

A. _____

B. _____

C. _____

D. _____

Non-muscular System Structure (blue pin)

E. _____

CHECK POINT:

Hip and Thigh, Posterior View, cont'd

9. Name the two muscles that allow the non-weight-bearing limb to swing forward during walking.
10. Name the two heads of the biceps femoris.
11. Name the largest nerve in the body.

- *Click* **LAYER 4** *in the* **LAYER CONTROLS** *window, and you will see the following image:*

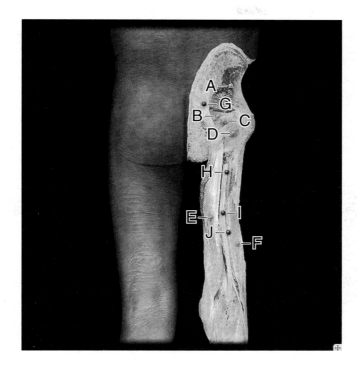

- *Mouse-over the green pins on the screen to find the information necessary to identify the following structures:*

A. _____

B. _____

C. _____

D. _____

E. _____

F. _____

Non-muscular System Structures (blue pins)

G. _____

H. _____

I. _____

J. _____

CHECK POINT:

Hip and Thigh, Posterior View, cont'd

12. Name the structure that is an important anchor of the sacrum to the hip bone.
13. Name the two components of the sciatic nerve.

- *Click* **LAYER 5** *in the* **LAYER CONTROLS** *window, and you will see the following image:*

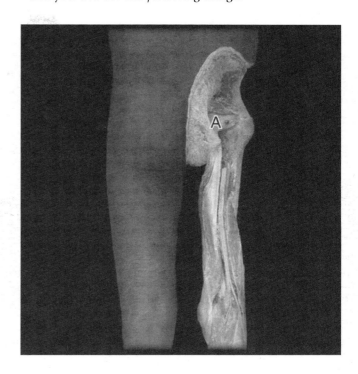

- *Mouse-over the blue pin on the screen to find the information necessary to identify the following non-muscular system structure:*

A. _____

CHECK POINT:

Hip and Thigh, Posterior View, cont'd

14. Name the thick fibrous band fused to the posterior surface of the hip joint capsule.
15. Name the ligament that resists hyperflexion of the hip.

Animations: Muscle Actions

 | **SELECT ANIMATION** **Adductor Magnus Muscle** | **PLAY**

- *After viewing this animation, select and view these additional animations:*

  – **Short head of biceps femoris muscle**
  – **Gluteus maximus muscle**

- **Gluteus medius muscle**
- **Hamstring muscles**
- **Quadriceps femoris muscle**
- **Sartorius muscle**

Self Test
Take this opportunity to quiz yourself by taking the **SELF TEST.** See page 15 for a reminder on how to access the self-test for this section.

I N R E V I E W

What Have I Learned?

The following questions cover the material that you have just learned, the muscles of the hip and thigh. Apply what you have learned to answer these questions on a separate piece of paper.

1. Name the four muscles of the quadriceps femoris.

2. Name the most powerful flexor of the hip joint.

3. Name the muscle of the thigh weak in humans and used in muscle transplants.

4. Name the muscle often involved in a "pulled groin."

5. Name the strongest ligament around the hip joint.

6. Name a muscle of the posterior thigh important for powerful extension of the femur as in running, climbing stairs, and rising from the seated position.

7. Name the two muscles that allow the non-weight-bearing limb to swing forward during walking.

8. Name the two heads of the biceps femoris.

9. Name the thick fibrous band fused to the posterior surface of the hip joint capsule.

EXERCISE 4.17:

Skeletal Muscles—Leg and Foot, Anterior View

SELECT TOPIC	SELECT VIEW	
Leg and Foot ▶	**Anterior**	GO

- *Click* **LAYER 1** *in the* **LAYER CONTROLS** *window, and you will see the following image:*

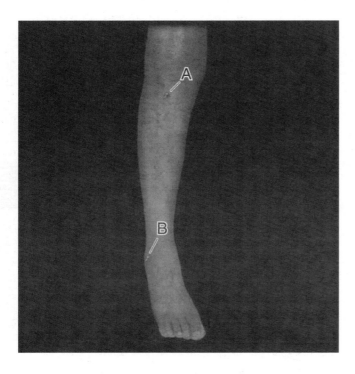

- *Mouse-over the blue pins on the screen to find the information necessary to identify the following non-muscular system structures:*

A. _____

B. _____

C H E C K P O I N T :

Leg and Foot, Anterior View

1. Name the bony elevation of the anterior proximal tibia.
2. Name the lateral subcutaneous projection that contributes to the ankle joint.

• *Click* **LAYER 2** *in the* **LAYER CONTROLS** *window, and you will see the following image:*

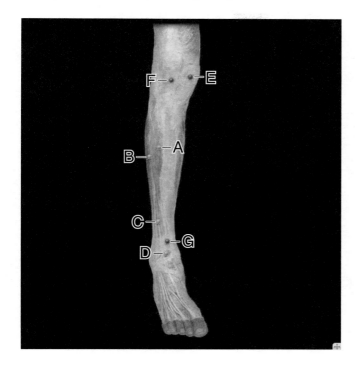

• *Mouse-over the green pins on the screen to find the information necessary to identify the following structures:*

A. _____

B. _____

C. _____

D. _____

Non-muscular System Structures (blue pins)

E. _____

F. _____

G. _____

CHECK POINT:

Leg and Foot, Anterior View, cont'd

3. What is the anatomical term for the first toe?
4. The tendons of which muscles are subcutaneous on the dorsum of the foot?
5. Name the structure that serves to bind in place the tendons from the anterior compartment of the leg as they cross the ankle joint.

• *Click* **LAYER 3** *in the* **LAYER CONTROLS** *window, and you will see the following image:*

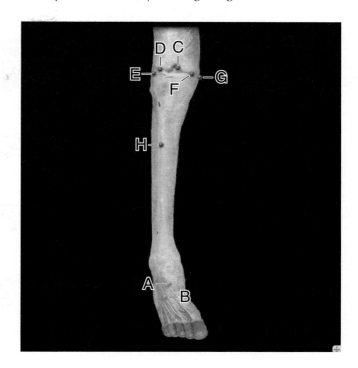

• *Mouse-over the green pins on the screen to find the information necessary to identify the following structures:*

A. _____

B. _____

Non-muscular System Structures (blue pins)

C. _____

D. _____

E. _____

F. _____

G. _____

H. _____

CHECK POINT:

Leg and Foot, Anterior View, cont'd

6. Name the thick sheet of connective tissue between the tibia and fibula.
7. What is the function of this sheet of connective tissue?
8. Name the structures referred to as the "unhappy triad."

EXERCISE 4.18:

Skeletal Muscles—Leg and Foot, Posterior View

SELECT TOPIC		SELECT VIEW	
Leg and Foot	▶	**Posterior**	**GO**

• *Click* **LAYER 1** *in the* **LAYER CONTROLS** *window, and you will see the following image:*

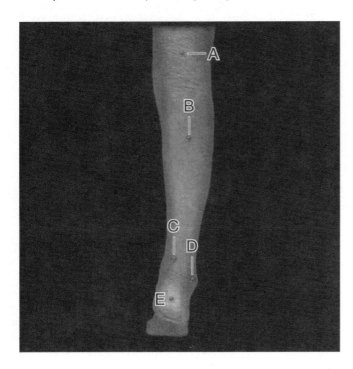

• *Mouse-over the blue pins on the screen to find the information necessary to identify the following non-muscular system structures:*

A. _____

B. _____

C. _____

D. _____

E. _____

CHECK POINT:

Leg and Foot, Posterior View

1. Name the strongest tendon in the body.
2. Give an example of this tendon's strength from the **STRUCTURE INFORMATION** window.
3. Name the tendon also known as the "Achilles" tendon.

• *Click* **LAYER 2** *in the* **LAYER CONTROLS** *window, and you will see the following image:*

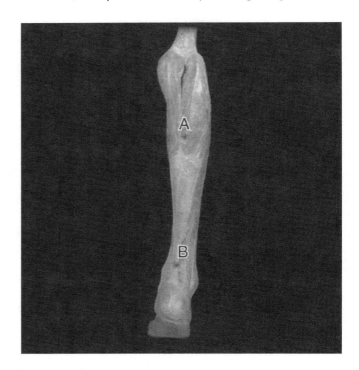

• *Mouse-over the pins on the screen to find the information necessary to identify the following structures:*

A. _____

B. _____

CHECK POINT:

Leg and Foot, Posterior View, cont'd

4. The tendons of which two muscles contribute to the calcaneal tendon?
5. Name the calf muscle that consists of a medial and a lateral belly.
6. Name the superficial calf muscle.

• *Click* **LAYER 3** *in the* **LAYER CONTROLS** *window, and you will see the following image:*

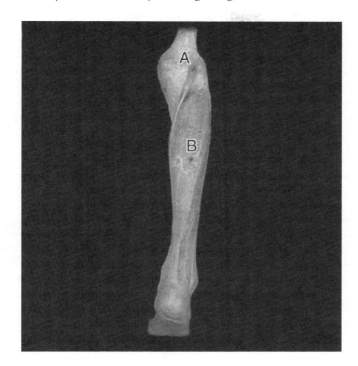

• *Mouse-over the green pins on the screen to find the information necessary to identify the following structures:*

A. _____

B. _____

CHECK POINT:

Leg and Foot, Posterior View, cont'd

7. Name the calf muscle deep to the gastrocnemius.
8. Name the long thin tendon that is a common source for tendon transplants.

• *Click* **LAYER 4** *in the* **LAYER CONTROLS** *window, and you will see the following image:*

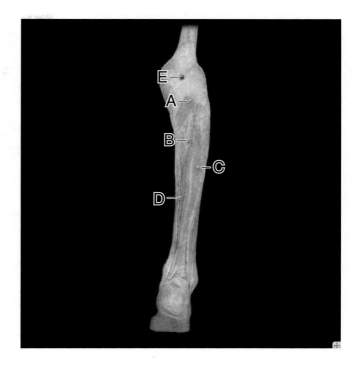

• *Mouse-over the green pins on the screen to find the information necessary to identify the following structures:*

A. _____

B. _____

C. _____

D. _____

Non-muscular System Structure (blue pin)

E. _____

CHECK POINT:

Leg and Foot, Posterior View, cont'd

9. Name the muscle that helps "unlock" the knee joint from full extension.
10. Name the two structures that maintain the position of the femur on the tibia in full knee flexion such as squatting.
11. Name the powerful muscle for "push-off" of the foot during walking or running.

- *Click **LAYER 5** in the **LAYER CONTROLS** window, and you will see the following image:*

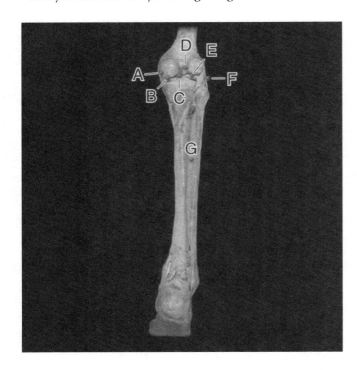

- *Mouse-over the blue pins on the screen to find the information necessary to identify the following non-muscular system structures:*

A. _____

B. _____

C. _____

D. _____

E. _____

F. _____

G. _____

CHECK POINT:

Leg and Foot, Posterior View, cont'd

12. Name the thinner and weaker of the cruciate ligaments.
13. Name the cartilaginous structure on the tibia that articulates with the medial condyle of the femur.
14. Name the structure that limits rotation between the femur and the tibia.

Animations: Muscle Actions

SELECT ANIMATION	
Fibularis Longus and Brevis Muscle	PLAY

- *After viewing this animation, select and view this additional animation:*
  - **Gastrocnemius muscle**

Self Test

Take this opportunity to quiz yourself by taking the **SELF TEST.** See page 15 for a reminder on how to access the self-test for this section.

I N   R E V I E W

What Have I Learned?

The following questions cover the material that you have just learned, the muscles of the leg and foot. Apply what you have learned to answer these questions on a separate piece of paper.

1. What is the anatomical term for the first toe?

2. Name the thick sheet of connective tissue between the tibia and fibula. What is its function?

3. Name the structure that serves to bind the tendons from the anterior compartment of the leg in place as they cross the ankle joint.

4. Name the strongest tendon in the body.

5. The tendons of which two muscles contribute to the calcaneal tendon?

6. Name the long thin tendon that is a common source for tendon transplants.

7. Name the powerful muscle for "push-off" of the foot during walking or running.

8. Name the cruciate ligaments. Which of the two is thinner and weaker?

• *Click* **LAYER 2** *in the* **LAYER CONTROLS** *window, and you will see the following image:*

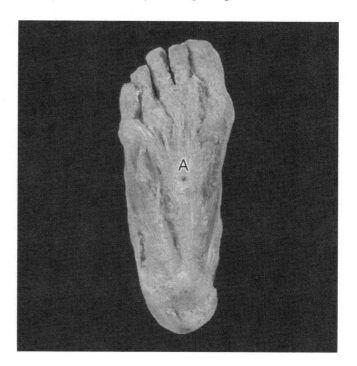

EXERCISE 4.19:

Skeletal Muscles—Foot, Plantar View

SELECT TOPIC	SELECT VIEW	
Foot	**Plantar**	**GO**

• *Click* **LAYER 1** *in the* **LAYER CONTROLS** *window, and you will see the following image:*

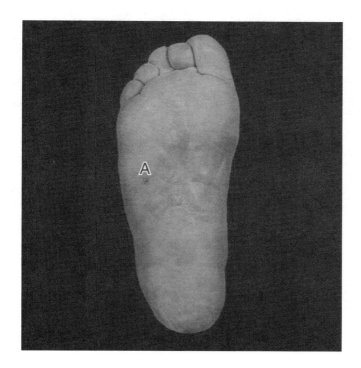

• *Mouse-over the pin on the screen to find the information necessary to identify the following structure:*

A. _____

CHECK POINT:

Foot, Plantar View, cont'd

2. Name the structure that protects the muscles, vessels, and nerves of plantar foot.

• *Mouse-over the pin on the screen to find the information necessary to identify the following non-muscular system structure:*

A. _____

CHECK POINT:

Foot, Plantar View

1. Name the structures that serve as contact points of the foot for weight bearing.

- *Click* **LAYER 3** *in the* **LAYER CONTROLS** *window, and you will see the following image:*

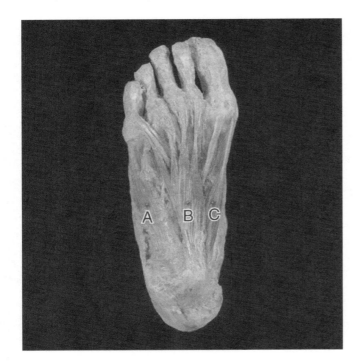

- *Mouse-over the pins on the screen to find the information necessary to identify the following structures:*

A. _____

B. _____

C. _____

CHECK POINT:

Foot, Plantar View, cont'd

3. What is the anatomical term for the first toe?
4. Name the muscle responsible for flexion of toes two–five.
5. Name the muscle that supports the medial longitudinal arch of the foot during weight bearing.

- *Click* **LAYER 4** *in the* **LAYER CONTROLS** *window, and you will see the following image:*

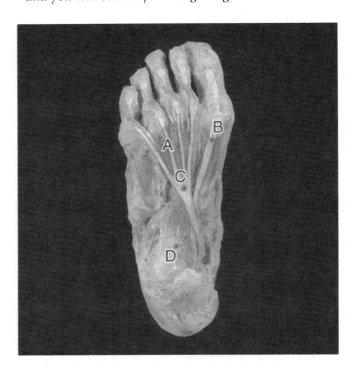

- *Mouse-over the pins on the screen to find the information necessary to identify the following structures:*

A. _____

B. _____

C. _____

D. _____

CHECK POINT:

Foot, Plantar View, cont'd

6. Name two muscles located on the posterior leg whose tendons run along the plantar foot.
7. Name a muscle of the foot that uses the tendons of another muscle to produce toe flexion.

- *Click **LAYER 5** in the **LAYER CONTROLS** window, and you will see the following image:*

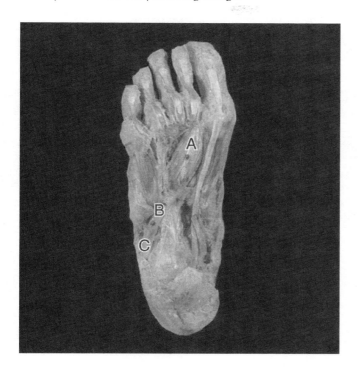

- *Mouse-over the pins on the screen to find the information necessary to identify the following structures:*

A. _____

B. _____

C. _____

CHECK POINT:

Foot, Plantar View, cont'd

8. Name a muscle that resists separation ("spreading") of the metatarsals during weight-bearing.
9. Name two muscles of the lateral leg whose tendons run along the plantar foot.
10. Name a muscle responsible for adduction of the first toe.

Self Test

Take this opportunity to quiz yourself by taking the **SELF TEST.** See page 15 for a reminder on how to access the self-test for this section.

I N   R E V I E W

What Have I Learned?

The following questions cover the material that you have just learned, the muscles of the plantar foot. Apply what you have learned to answer these questions on a separate piece of paper.

1. Name the structures that serve as contact points of the foot for weight-bearing.

2. Name the structure that protects the muscles, vessels, and nerves of plantar foot.

3. Name the muscle that supports the medial longitudinal arch of the foot during weight-bearing.

4. Name a muscle that resists separation ("spreading") of the metatarsals during weight-bearing.

Muscle Attachments (on Skeleton)

We will now concentrate on how and where the skeletal muscles attach to the skeleton. After all, without this attachment between the two, we would be unable to accomplish any movement. Don't be intimidated by this section of *Anatomy and Physiology | Revealed*®. It is very basic physics. Think of your computer mouse (assuming you are using one). You place your hand on the mouse, and by moving your arm, the mouse moves in the same direction. Likewise, when the insertion end of a muscle is attached to a bone, when the muscle moves (contracts), the bone it is attached to moves in the direction of contraction. Remember that the bone located at the **origin** of the muscle will not be moving, but will instead function as an anchor for the muscle during contraction. The bone at the **insertion** end of the muscle will be moving, as the contracting muscle exerts a force on it.

As we view the images in *Anatomy and Physiology | Revealed*®, remember that the origins of muscles are shaded in red and the insertions are shaded in blue. A box will be visible on the screen showing the overall region being viewed, with a key to remind you which color represents which end of the muscle.

EXERCISE 4.20:

Skull, Anterior View

SELECT TOPIC
**Muscle Attachments
(on skeleton)**

SELECT TOPIC
Skull

SELECT VIEW
Anterior

GO

- *Click* **LAYER 1** *in the* **LAYER CONTROLS** *window, and you will see the following image:*

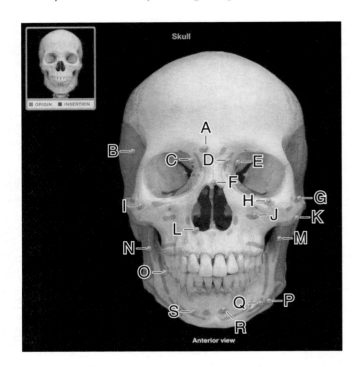

- *Mouse-over the pins on the screen to find the information necessary to identify the following attachments:*

A. _____

B. _____

C. _____

D. _____

E. _____

F. _____

G. _____

H. _____

I. _____

J. _____

K. _____

L. _____

M. _____

N. _____

O. _____

P. _____

Q. _____

R. _____

S. _____

EXERCISE 4.21:

Skull, Lateral View

SELECT TOPIC
Skull

SELECT VIEW
Lateral

GO

- *Click* **LAYER 1** *in the* **LAYER CONTROLS** *window, and you will see the following image:*

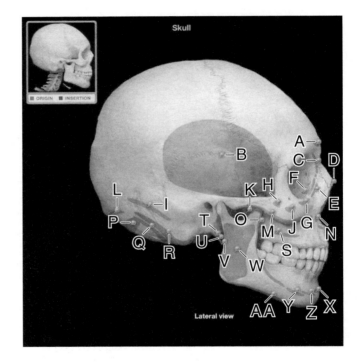

- *Mouse-over the pins on the screen to find the information necessary to identify the following attachments:*

A. _____

B. _____

C. _____

D. _____

E. _____

F. _____

G. _____

H. _____

I. _____

J. _____

K. _____

L. _____

M. _____

N. _____

O. _____

P. _____

Q. _____

R. _____

S. _____

T. _____

U. _____

V. _____

W. _____

X. _____

Y. _____

Z. _____

AA. _____

- *Mouse-over the pins on the screen to find the information necessary to identify the following attachments:*

A. _____

B. _____

C. _____

D. _____

E. _____

F. _____

G. _____

H. _____

I. _____

J. _____

K. _____

L. _____

EXERCISE 4.22:

Skull, Mid-sagittal View

| | SELECT TOPIC **Skull** | SELECT VIEW **Mid-sagittal** | GO |

- *Click **LAYER 1** in the **LAYER CONTROLS** window, and you will see the following image:*

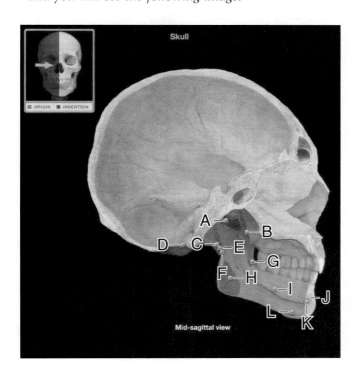

EXERCISE 4.23:

Skull, Posterior View

| | SELECT TOPIC **Skull** | SELECT VIEW **Posterior** | GO |

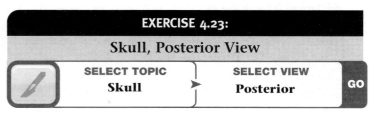

- *Click **LAYER 1** in the **LAYER CONTROLS** window, and you will see the following image:*

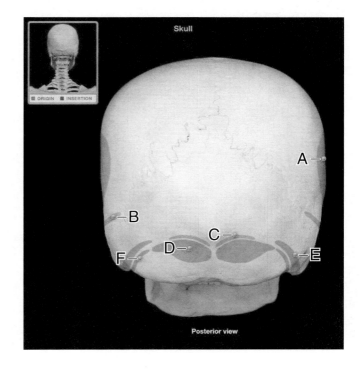

- *Mouse-over the pins on the screen to find the information necessary to identify the following attachments:*

A. _____

B. _____

C. _____

D. _____

E. _____

F. _____

I. _____

J. _____

K. _____

L. _____

M. _____

N. _____

O. _____

P. _____

Q. _____

EXERCISE 4.24:

Skull, Inferior View

SELECT TOPIC	SELECT VIEW	
Skull	Inferior	GO

- *Click **LAYER 1** in the **LAYER CONTROLS** window, and you will see the following image:*

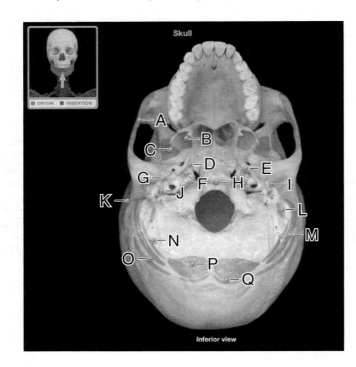

- *Mouse-over the pins on the screen to find the information necessary to identify the following attachments:*

A. _____

B. _____

C. _____

D. _____

E. _____

F. _____

G. _____

H. _____

EXERCISE 4.25:

Orbit, Anterolateral View

SELECT TOPIC	SELECT VIEW	
Orbit	Anterolateral	GO

- *Click **LAYER 1** in the **LAYER CONTROLS** window, and you will see the following image:*

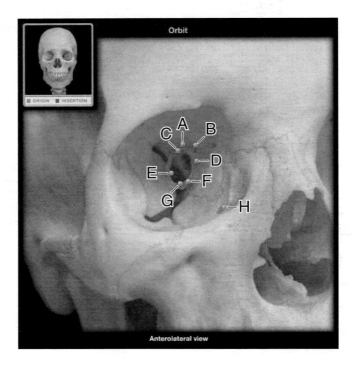

- *Mouse-over the pins on the screen to find the information necessary to identify the following attachments:*

A. _____

B. _____

C. _____

D. _____

E. _____

F. _____

G. _____

H. _____

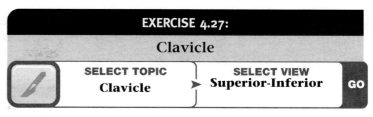

EXERCISE 4.26:	EXERCISE 4.27:
Hyoid Bone, Anterior and Lateral View	**Clavicle**

SELECT TOPIC Hyoid Bone ▸ **SELECT VIEW** Anterior-Lateral **GO**

SELECT TOPIC Clavicle ▸ **SELECT VIEW** Superior-Inferior **GO**

- *Click* **LAYER 1** *in the* **LAYER CONTROLS** *window, and you will see the following image:*

- *Click* **LAYER 1** *in the* **LAYER CONTROLS** *window, and you will see the following image:*

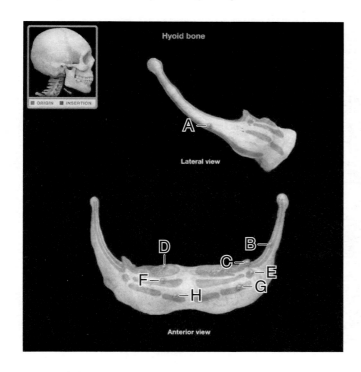

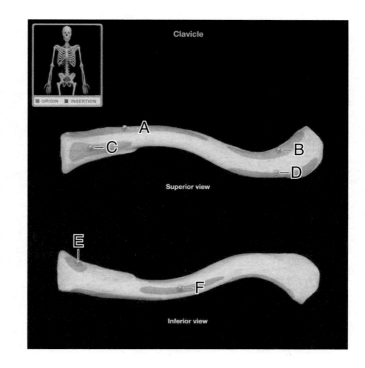

- *Mouse-over the pins on the screen to find the information necessary to identify the following attachments:*

A. _____

B. _____

C. _____

D. _____

E. _____

F. _____

G. _____

H. _____

- *Mouse-over the pins on the screen to find the information necessary to identify the following attachments:*

A. _____

B. _____

C. _____

D. _____

E. _____

F. _____

Self Test

Take this opportunity to quiz yourself by taking the **SELF TEST.** See page 15 for a reminder on how to access the self-test for this section.

EXERCISE 4.28:
Scapula

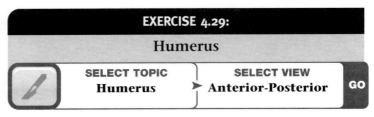

SELECT TOPIC	SELECT VIEW	
Scapula	▶ Anterior-Posterior-Lateral	GO

- *Click* **LAYER 1** *in the* **LAYER CONTROLS** *window, and you will see the following image:*

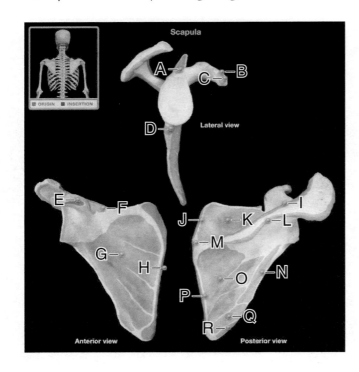

- *Mouse-over the pins on the screen to find the information necessary to identify the following attachments:*

A. _____

B. _____

C. _____

D. _____

E. _____

F. _____

G. _____

H. _____

I. _____

J. _____

K. _____

L. _____

M. _____

N. _____

O. _____

P. _____

Q. _____

R. _____

EXERCISE 4.29:
Humerus

SELECT TOPIC	SELECT VIEW	
Humerus	▶ Anterior-Posterior	GO

- *Click* **LAYER 1** *in the* **LAYER CONTROLS** *window, and you will see the following image:*

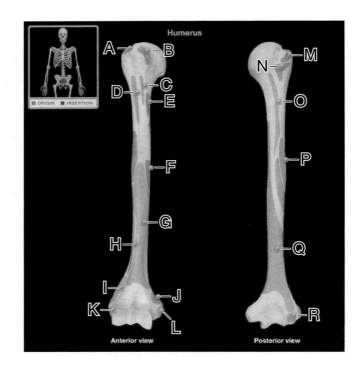

- *Mouse-over the pins on the screen to find the information necessary to identify the following attachments:*

A. _____

B. _____

C. _____

D. _____

E. _____

F. _____

G. _____

H. _____

I. _____

J. _____

K. _____

L. _____

M. _____

N. _____

O. _____

P. _____

Q. _____

R. _____

Self Test

Take this opportunity to quiz yourself by taking the **SELF TEST.** See page 15 for a reminder on how to access the self-test for this section.

EXERCISE 4.30:

Radius and Ulna

SELECT TOPIC	SELECT VIEW	
Radius and Ulna ➤	**Anterior-Posterior**	**GO**

- *Click* **LAYER 1** *in the* **LAYER CONTROLS** *window, and you will see the following image:*

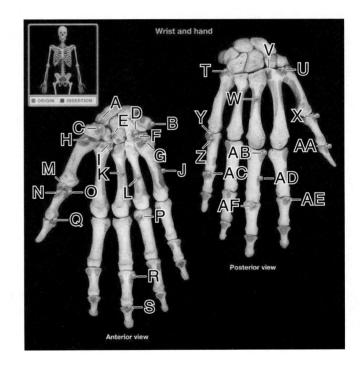

- *Mouse-over the pins on the screen to find the information necessary to identify the following attachments:*

A. _____

B. _____

C. _____

D. _____

E. _____

F. _____

G. _____

H. _____

I. _____

J. _____

K. _____

L. _____

M. _____

N. _____

O. _____

P. _____

Q. _____

R. _____

S. _____

T. _____

EXERCISE 4.31:

Wrist and Hand

SELECT TOPIC	SELECT VIEW	
Wrist and Hand ➤	**Anterior-Posterior**	**GO**

- *Click* **LAYER 1** *in the* **LAYER CONTROLS** *window, and you will see the following image:*

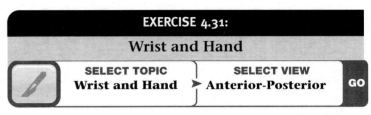

- *Mouse-over the pins on the screen to find the information necessary to identify the following attachments:*

A. _____

B. _____

C. _____

D. _____

E. _____

F. _____

G. _____

H. _____

I. _____

J. _____

K. _____

L. _____

M. _____

N. _____

O. _____

P. _____

Q. _____

R. _____

S. _____

T. _____

U. _____

V. _____

W. _____

X. _____

Y. _____

Z. _____

AA. _____

AB. _____

AC. _____

AD _____

AE. _____

AF. _____

Self Test

Take this opportunity to quiz yourself by taking the **SELF TEST.** See page 15 for a reminder on how to access the self-test for this section.

EXERCISE 4.32:

Thoracic Cage

SELECT TOPIC	SELECT VIEW	
Thoracic Cage	**Anterior**	GO

- *Click* **LAYER 1** *in the* **LAYER CONTROLS** *window, and you will see the following image:*

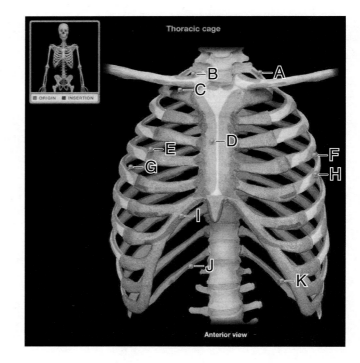

- *Mouse-over the pins on the screen to find the information necessary to identify the following attachments:*

A. _____

B. _____

C. _____

D. _____

E. _____

F. _____

G. _____

H. _____

I. _____

J. _____

K. _____

Self Test

Take this opportunity to quiz yourself by taking the **SELF TEST.** See page 15 for a reminder on how to access the self-test for this section.

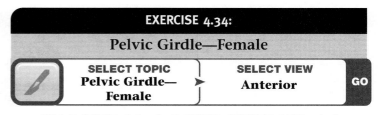

EXERCISE 4.33:
Vertebral Column

SELECT TOPIC	SELECT VIEW	
Vertebral Column	▸ **Anterior-Posterior-Lateral**	GO

- *Click **LAYER 1** in the **LAYER CONTROLS** window, and you will see the following image:*

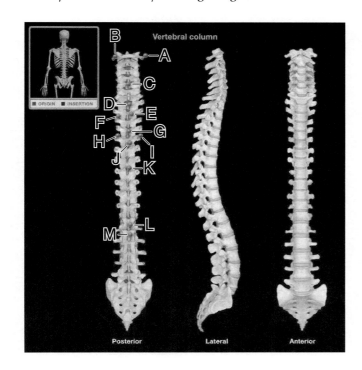

- *Mouse-over the pins on the screen to find the information necessary to identify the following attachments:*

A. _____

B. _____

C. _____

D. _____

E. _____

F. _____

G. _____

H. _____

I. _____

J. _____

K. _____

L. _____

M. _____

Self Test
Take this opportunity to quiz yourself by taking the **SELF TEST.** See page 15 for a reminder on how to access the self-test for this section.

EXERCISE 4.34:
Pelvic Girdle—Female

SELECT TOPIC	SELECT VIEW	
Pelvic Girdle—Female	▸ **Anterior**	GO

- *Click **LAYER 1** in the **LAYER CONTROLS** window, and you will see the following image:*

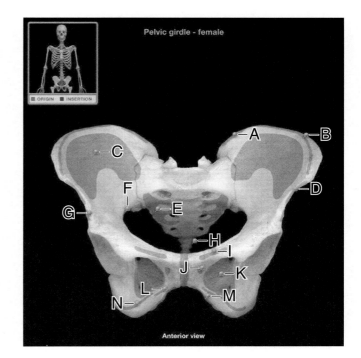

- *Mouse-over the pins on the screen to find the information necessary to identify the following attachments:*

A. _____

B. _____

C. _____

D. _____

E. _____

F. _____

G. _____

H. _____

I. _____

J. _____

K. _____

L. _____

M. _____

N. _____

EXERCISE 4.35:
Pelvic Girdle—Male

	SELECT TOPIC	SELECT VIEW	
	Pelvic Girdle—Male ▶	Anterior	GO

- *Click* **LAYER 1** *in the* **LAYER CONTROLS** *window, and you will see the following image:*

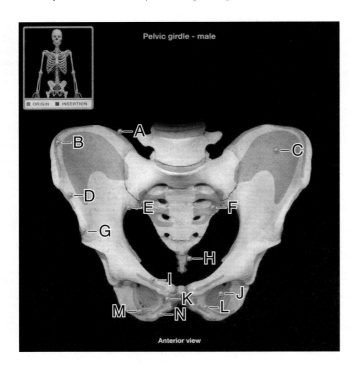

- *Mouse-over the pins on the screen to find the information necessary to identify the following attachments:*

A. _____

B. _____

C. _____

D. _____

E. _____

F. _____

G. _____

H. _____

I. _____

J. _____

K. _____

L. _____

M. _____

N. _____

EXERCISE 4.36:
Sacrum and Coccyx

	SELECT TOPIC	SELECT VIEW	
	Sacrum and Coccyx ▶	Anterior-Posterior	GO

- *Click* **LAYER 1** *in the* **LAYER CONTROLS** *window, and you will see the following image:*

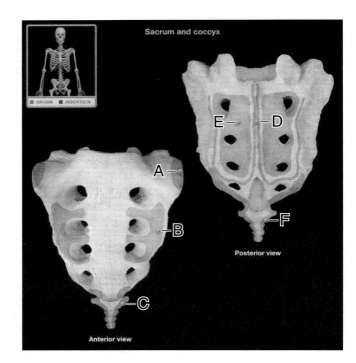

- *Mouse-over the pins on the screen to find the information necessary to identify the following attachments:*

A. _____

B. _____

C. _____

D. _____

E. _____

F. _____

Self Test

Take this opportunity to quiz yourself by taking the **SELF TEST.** See page 15 for a reminder on how to access the self-test for this section.

EXERCISE 4.37:

Hip Bone

SELECT TOPIC	SELECT VIEW	
Hip Bone ▸	**Medial-Lateral**	**GO**

- *Click* **LAYER 1** *in the* **LAYER CONTROLS** *window, and you will see the following image:*

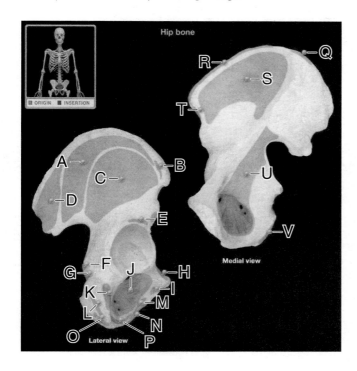

- *Mouse-over the pins on the screen to find the information necessary to identify the following attachments:*

A. _____

B. _____

C. _____

D. _____

E. _____

F. _____

G. _____

H. _____

I. _____

J. _____

K. _____

L. _____

M. _____

N. _____

O. _____

P. _____

Q. _____

R. _____

S. _____

T. _____

U. _____

V. _____

EXERCISE 4.38:

Femur

SELECT TOPIC	SELECT VIEW	
Femur ▸	**Anterior-Posterior**	**GO**

- *Click* **LAYER 1** *in the* **LAYER CONTROLS** *window, and you will see the following image:*

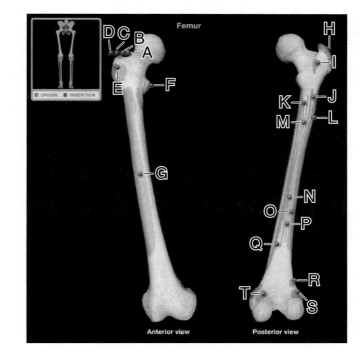

- *Mouse-over the pins on the screen to find the information necessary to identify the following attachments:*

A. _____

B. _____

C. _____

D. _____

E. _____

F. _____

G. _____

H. _____

I. _____

J. _____

K. _____

L. _____

M. _____

N. _____

O. _____

P. _____

Q. _____

R. _____

S. _____

T. _____

Self Test

Take this opportunity to quiz yourself by taking the **SELF TEST.** See page 15 for a reminder on how to access the self-test for this section.

G. _____

H. _____

I. _____

J. _____

K. _____

L. _____

M. _____

N. _____

O. _____

P. _____

Q. _____

R. _____

EXERCISE 4.39:

Tibia and Fibula

SELECT TOPIC	SELECT VIEW	
Tibia and Fibula ➤	**Anterior-Posterior**	**GO**

- *Click* **LAYER 1** *in the* **LAYER CONTROLS** *window, and you will see the following image:*

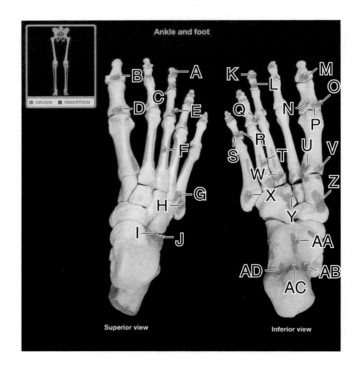

- *Mouse-over the pins on the screen to find the information necessary to identify the following attachments:*

A. _____

B. _____

C. _____

D. _____

E. _____

F. _____

EXERCISE 4.40:

Ankle and Foot

SELECT TOPIC	SELECT VIEW	
Ankle and Foot ➤	**Superior-Inferior**	**GO**

- *Click* **LAYER 1** *in the* **LAYER CONTROLS** *window, and you will see the following image:*

- *Mouse-over the pins on the screen to find the information necessary to identify the following attachments:*

A. _____

B. _____

C. _____

D. _____

E. _____

F. _____

G. _____

H. _____

I. _____

J. _____

K. _____

L. _____

M. _____

N. _____

O. _____

P. _____

Q. _____

R. _____

S. _____

T. _____

U. _____

V. _____

W. _____

X. _____

Y. _____

Z. _____

AA. _____

AB. _____

AC. _____

AD. _____

Self Test

Take this opportunity to quiz yourself by taking the **SELF TEST.** See page 15 for a reminder on how to access the self-test for this section.

EXERCISE 4.41:

Cardiac Muscle, Histology

SELECT TOPIC	SELECT VIEW	
Cardiac Muscle >		GO

- *Click the* **TURN TAGS ON** *button, and you will see the following image:*

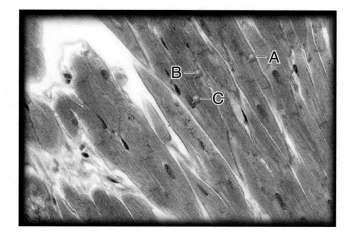

- *Mouse-over the pins on the screen to find the information necessary to identify the following structures:*

A. _____

B. _____

C. _____

CHECK POINT:

Cardiac Muscle, Histology

1. State the functions of the intercalated discs.
2. What are desmosomes?
3. How many nuclei are found in each cardiac muscle fiber?

EXERCISE 4.42:

Skeletal Muscle, Histology

SELECT TOPIC	SELECT VIEW	
Skeletal Muscle ▶		GO

- *Click the* **TURN TAGS ON** *button, and you will see the following image:*

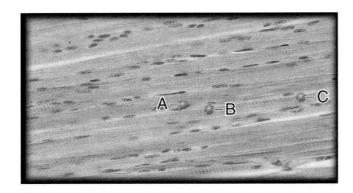

- *Mouse-over the pins on the screen to find the information necessary to identify the following structures:*

A. _____

B. _____

C. _____

CHECK POINT:

Skeletal Muscle, Histology

1. Describe the A bands of skeletal muscle fibers.
2. Describe the I bands of skeletal muscle fibers.
3. How many nuclei are found in each skeletal muscle fiber? Where are they located?

EXERCISE 4.43:

Smooth Muscle, Histology

SELECT TOPIC	SELECT VIEW	
Smooth Muscle ▶		GO

- *Click the* **TURN TAGS ON** *button and you will see the following image:*

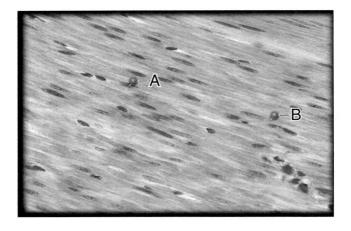

- *Mouse-over the pins on the screen to find the information necessary to identify the following structures:*

A. _____

B. _____

CHECK POINT:

Smooth Muscle, Histology

1. Describe the smooth muscle fibers.
2. What is another name for smooth muscle?
3. Where is this tissue found?

EXERCISE 4.44:

Neuromuscular Junction, Histology

| SELECT TOPIC
**Neuromuscular
Junction** | ► | SELECT VIEW | GO |

- *Click the* **TURN TAGS ON** *button and you will see the following image:*

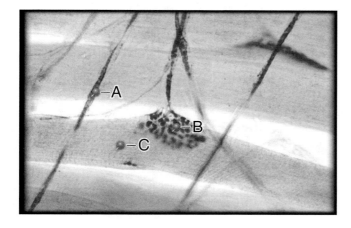

- *Mouse-over the pins on the screen to find the information necessary to identify the following structures:*

A. _____

B. _____

C. _____

CHECK POINT:

Neuromuscular Junction, Histology

1. Describe the motor end plate. Where is it located?
2. What neurotransmitter is found at the neuromuscular junction?
3. Describe the axon of the motor neuron. What is its function?

Self Test

Take this opportunity to quiz yourself by taking the **SELF TEST.** See page 15 for a reminder on how to access the self-test for this section.

I N   R E V I E W

What Have I Learned?

The following questions cover the material you have just learned, the histology of muscle tissue. Apply what you have learned to answer the following questions on a separate piece of paper.

1. State the functions of the intercalated discs.

2. What are desmosomes?

3. How many nuclei are found in each cardiac muscle fiber?

4. Describe the A bands of skeletal muscle fibers.

5. How did the A bands gain their designation?

6. Describe the I bands of skeletal muscle fibers.

7. How did the I bands gain their designation?

8. How many nuclei are found in each skeletal muscle fiber? Where are they located?

9. Describe the smooth muscle fibers.

10. What is another name for smooth muscle?

11. Where is this tissue found?

12. Describe the motor end plate. Where is it located?

13. What neurotransmitter is found at the neuromuscular junction?

14. What is the sole-plate?

15. Describe the axon of the motor neuron. What is its function?

EXERCISE 4.45:

Coloring Exercise

Look up the muscles of the upper extremity and then color them in with colored pens or pencils. Fill in the blank labels with the appropriate names.

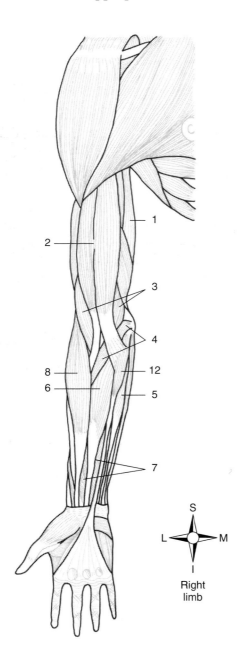

Right limb

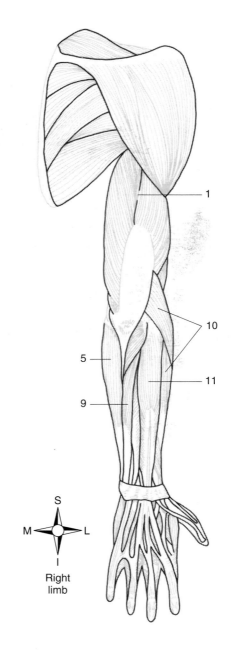

Right limb

❏ 1. _____

❏ 2. _____

❏ 3. _____

❏ 4. _____

❏ 5. _____

❏ 6. _____

❏ 7. _____

❏ 8. _____

❏ 9. _____

❏ 10. _____

❏ 11. _____

❏ 12. _____

CHAPTER 5

The Nervous System

Overview: Nervous System

The nervous system is the master controlling and communicating system of your body. It is divided into two major divisions. The **central nervous system** (CNS) consists of the brain and spinal cord, while the **peripheral nervous system** (PNS) is comprised of nerve pathways leading to and from the CNS.

The brain in the adult is one of our largest organs. The average weight for the adult human brain is 1.4 kg, or 3 pounds. It consists of roughly 100 billion neurons and 900 billion glial cells. This means that there are one-fourth as many neurons and 2½ times as many total cells in your brain as there are stars in our Milky Way galaxy! Think about that the next time you are out at night, looking at the stars.

We will look at the central nervous system first, and then continue with the peripheral nervous system and the special senses.

- *From the* **Home screen,** *click the drop-down box on the* **Select system** *menu.*
- *From the systems listed, click on the* **Nervous,** *and you will see the image to the right:*
- *This is the opening screen for the* **Nervous System.**

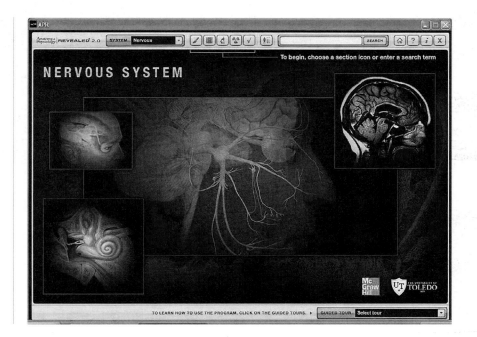

┌─ **HEADS UP!** ──────────────────
Be sure to click all **TAG** *buttons in the* **Dissection** *section of* **Anatomy & Physiology | Revealed**®*, even if it is not necessary for completion of the assignments. This will allow you to have a better feel for the location of all of the structures that you do need to know.*
└─────────────────────────────────

Animation: Divisions of the Brain

	SELECT ANIMATION **Divisions of Brain**	PLAY

- *After viewing the animation, answer these questions:*

1. Name the four regions of the brain.

2. Describe the cerebrum.

3. How many cerebral hemispheres exist?

4. Name the five lobes of the cerebral hemispheres.

5. The cerebrum is considered _____ responsible for _____.

6. The cerebellum is composed of . . .

7. The primary function of the cerebellum is . . .

8. The diencephalon is composed of three structures. What are they, and what are their functions?

9. Name and describe the three divisions of the brainstem.

10. Name the functions of each division of the brainstem.

11. Where do the cerebellar hemispheres attach to the brainstem?

12. The medulla oblongata extends from . . .

13. The medulla oblongata is continuous with . . .

The Brain

EXERCISE 5.1:
Nervous System—Brain, Coronal View

	SELECT TOPIC **Brain**	>	SELECT VIEW **Coronal**	GO

- *Click **LAYER 1** in the **LAYER CONTROLS** window, and you will see the following image:*

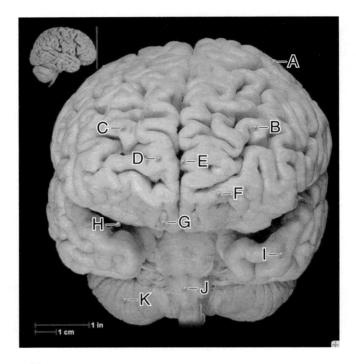

HEADS UP!

*There are two images visible in the **IMAGE AREA**, the large image with the identification pins and a smaller image in the top-left corner. This second, smaller image is to show you the plane of section, which allows you to keep the larger image in perspective.*

- *Mouse-over the pins on the screen to find the information necessary to identify the following structures:*

A. _____

B. _____

C. _____

D. _____

E. _____

F. _____

G. _____

H. _____

I. _____

J. _____

K. _____

K. _____

L. _____

CHECK POINT:

Brain, Coronal View

1. Name the deep grooves that separate the temporal lobes from the frontal and parietal lobes.
2. Name the deep groove that separates the right and left cerebral hemispheres.
3. Name the site of synapse for the olfactory neurons after they pass through the cribriform plate.

CHECK POINT:

Brain, Coronal View, cont'd

4. What is the term for unmyelinated nervous tissue?
5. What is the term for the collection of myelinated axons in the brain?
6. What is the large myelinated fiber tract that connects the right and left cerebral hemispheres?

• _Click_ **LAYER 2** _in the_ **LAYER CONTROLS** _window, and you will see the following image:_

• _Click_ **LAYER 3** _in the_ **LAYER CONTROLS** _window, and you will see the following image:_

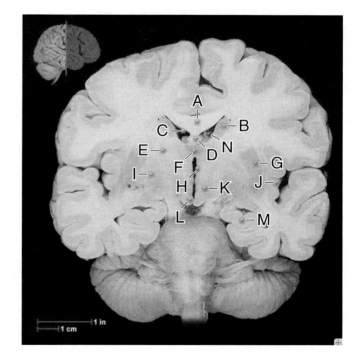

• _Mouse-over the pins on the screen to find the information necessary to identify the following structures:_

• _Mouse-over the green pins on the screen to find the information necessary to identify the following structures:_

A. _____

A. _____

B. _____

B. _____

C. _____

C. _____

D. _____

D. _____

E. _____

E. _____

F. _____

F. _____

G. _____

G. _____

H. _____

H. _____

I. _____

I. _____

J. _____

J. _____

K. _____

L. _____

M. _____

Non-nervous System Structure (blue pin)

N. _____

D. _____

E. _____

F. _____

G. _____

CHECK POINT:

Brain, Coronal View, cont'd

7. Name the paired, rounded projections involved in regulation of autonomic functions, emotional behavior, and memory.
8. Name the structure located in the cerebral ventricles that is the site of the production of cerebrospinal fluid (CSF).
9. Name the structure that is primarily for the relay of sensory information to the cerebral cortex.

CHECK POINT:

Brain, Coronal View, cont'd

10. Name the structure that coordinates orienting movements of the eyes and head.
11. Name the narrow midline channel between the third and fourth ventricles.
12. Name the structure involved with suppression and modulation of pain.

- *Click **LAYER 5** in the **LAYER CONTROLS** window, and you will see the following image:*

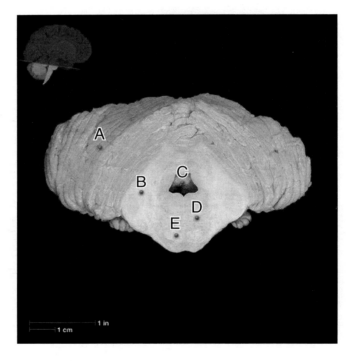

- *Click **LAYER 4** in the **LAYER CONTROLS** window, and you will see the following image:*

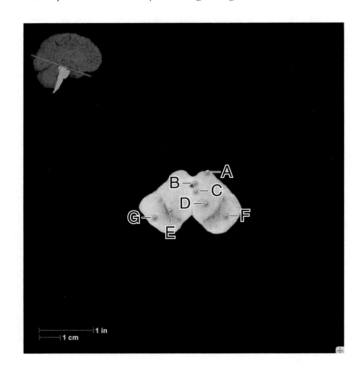

- *Mouse-over the pins on the screen to find the information necessary to identify the following structures:*

A. _____

B. _____

C. _____

D. _____

E. _____

- *Mouse-over the pins on the screen to find the information necessary to identify the following structures:*

A. _____

B. _____

C. _____

CHECK POINT:

Brain, Coronal View, cont'd

13. Name the structure that controls voluntary movement.
14. Name the major afferent pathway for information from the motor cortex to the cerebellum.
15. Name the cerebrospinal fluid-filled triangular cavity that is continuous with the cerebral aqueduct and the central canal of the spinal cord.

- *Click **LAYER 6** in the **LAYER CONTROLS** window, and you will see the following image:*

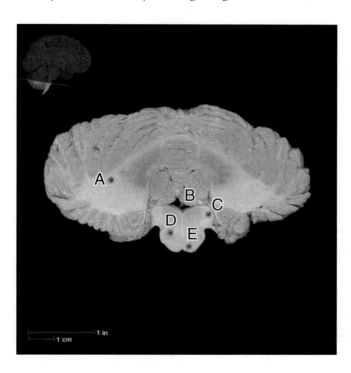

- *Mouse-over the pins on the screen to find the information necessary to identify the following structures:*

A. _____

B. _____

C. _____

D. _____

E. _____

CHECK POINT:

Brain, Coronal View, cont'd

16. Name the structure that processes and sends information to the cerebellum from many CNS nuclei and skeletal muscle proprioceptors.
17. Name the structure of the medulla oblongata that controls voluntary movement.
18. Name the structure that carries information about muscle performance from the spinal cord to the cerebellum.

Animation: Meninges

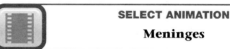

- *After viewing the animation, answer these questions:*

1. List the three meninges in order from superficial to deep.

2. Name the two layers of the most superficial of the meninges.

3. Name the structures formed where these two layers split.

4. Name the space located between the middle and deepest meninges. What fills this space?

Self Test
Take this opportunity to quiz yourself by taking the **SELF TEST**. See page 15 for a reminder on how to access the self-test for this section.

EXERCISE 5.2:

Nervous System—Brain, Lateral View

SELECT TOPIC	SELECT VIEW	
Brain ▶	**Lateral**	**GO**

• *Click* **LAYER 1** *in the* **LAYER CONTROLS** *window, and you will see the following image:*

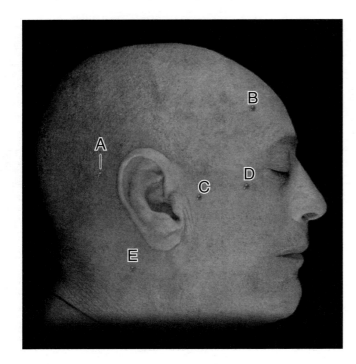

• *Mouse-over the blue, non-nervous system pins on the screen to find the information necessary to identify the following structures:*

A. _____

B. _____

C. _____

D. _____

E. _____

CHECK POINT:

Brain, Lateral View

1. What is a dermatome?
2. The skin of the superior face is innervated by which nerve?
3. The external ear is innervated by which nerve?

• *Click* **LAYER 3** *in the* **LAYER CONTROLS** *window, and you will see the following image:*

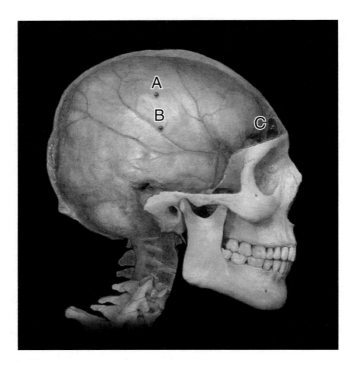

• *Mouse-over the blue, non-nervous system pins on the screen to find the information necessary to identify the following structures:*

A. _____

B. _____

C. _____

CHECK POINT:

Brain, Lateral View, cont'd

4. Name the paired mucous membrane-lined cavities within the frontal bone.
5. Name the most external of the meninges.
6. Name an artery that courses between the dura mater and the cranium.

• *Click* **LAYER 4** *in the* **LAYER CONTROLS** *window, and you will see the following image:*

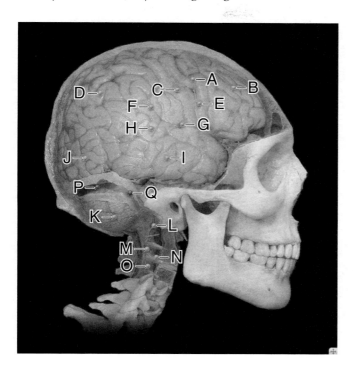

• *Mouse-over the green pins on the screen to find the information necessary to identify the following structures:*

A. _____

B. _____

C. _____

D. _____

E. _____

F. _____

G. _____

H. _____

I. _____

J. _____

K. _____

L. _____

M. _____

N. _____

O. _____

Non-nervous System Structures (blue pins)

P. _____

Q. _____

CHECK POINT:

Brain, Lateral View, cont'd

7. Name the distinct fold at the posterior border of the frontal lobe that controls voluntary movement.
8. Name the distinct fold at the anterior border of the parietal lobe that receives somatosensory information from the body.
9. Name the groove that forms the boundary between the frontal and parietal lobes.

• *Click* **LAYER 5** *in the* **LAYER CONTROLS** *window, and you will see the following image:*

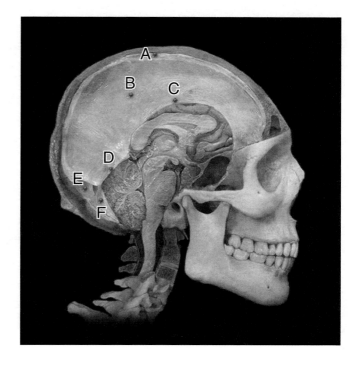

• *Mouse-over the blue pins on the screen to find the information necessary to identify the following non-nervous system structures:*

A. _____

B. _____

C. _____

D. _____

E. _____

F. _____

C H E C K   P O I N T :

Brain, Lateral View, cont'd

10. Name the large, crescent-shaped fold of the dura mater that separates the two cerebral hemispheres.
11. Name the structure that contains arachnoid granulations. What is the function of these granulations?
12. The confluence of the sinuses is the meeting point for four different sinuses. What are they?

• *Click* **LAYER 6** *in the* **LAYER CONTROLS** *window, and you will see the following image:*

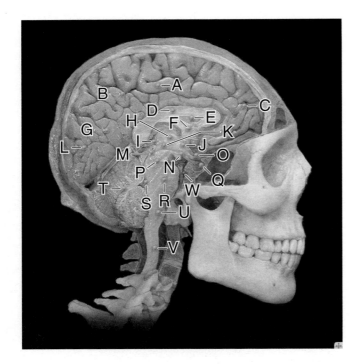

• *Mouse-over the green pins on the screen to find the information necessary to identify the following structures:*

A. _____

B. _____

C. _____

D. _____

E. _____

F. _____

G. _____

H. _____

I. _____

J. _____

K. _____

L. _____

M. _____

N. _____

O. _____

P. _____

Q. _____

R. _____

S. _____

T. _____

U. _____

V. _____

Non-nervous System Structure (blue pin)

W. _____

C H E C K   P O I N T :

Brain, Lateral View, cont'd

13. Name the groove that separates the parietal and occipital lobes of the brain.
14. Name the pea-sized endocrine gland attached to the roof of the third ventricle. What hormone does it secrete?
15. Name the narrow cerebrospinal fluid-filled channel between the third and fourth ventricles.

Self Test

Take this opportunity to quiz yourself by taking the **SELF TEST.** See page 15 for a reminder on how to access the self-test for this section.

EXERCISE 5.3:

Nervous System—Brain, Superior View

SELECT TOPIC	SELECT VIEW	
Brain	**Superior**	GO

- *Click* **LAYER 1** *in the* **LAYER CONTROLS** *window, and you will see the following image:*

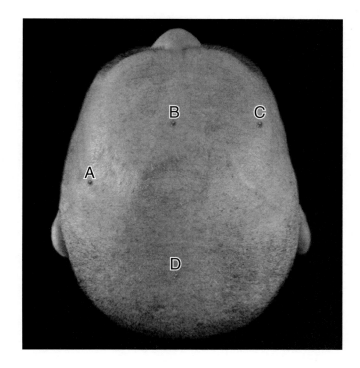

- *Mouse-over the blue pins on the screen to find the information necessary to identify the following non-nervous system structures:*

A. _____

B. _____

C. _____

D. _____

CHECK POINT:

Brain, Superior View

1. Name the spinal nerve that innervates the posterior scalp.
2. Name the nerve that innervates the skin over the temple and the anterior portion of the external ear.
3. Name the nerve that innervates the skin over the anterior scalp and forehead.

- *Click* **LAYER 2** *in the* **LAYER CONTROLS** *window, and you will see the following image:*

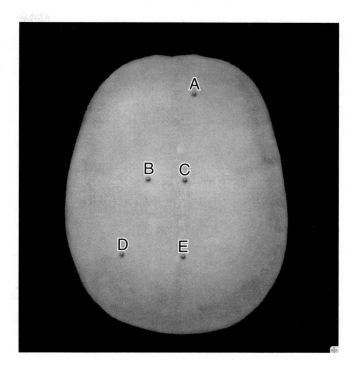

- *Mouse-over the blue pins on the screen to find the information necessary to identify the following non-nervous system structures:*

A. _____

B. _____

C. _____

D. _____

E. _____

- *Click* **LAYER 3** *in the* **LAYER CONTROLS** *window, and you will see the following image:*

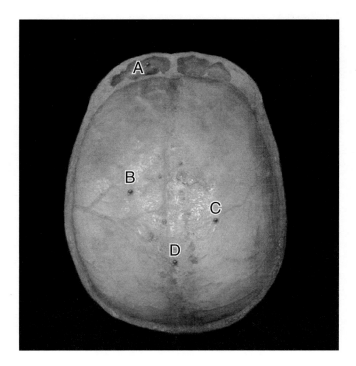

• *Mouse-over the blue pins on the screen to find the information necessary to identify the following non-nervous system structures:*

A. _____

B. _____

C. _____

D. _____

CHECK POINT:

Brain, Superior View, cont'd

4. Name the structure that allows the return of cerebrospinal fluid to the venous circulation.
5. Name an artery that courses between the dura mater and the cranium.

• *Click **LAYER 4** in the **LAYER CONTROLS** window, and you will see the following image:*

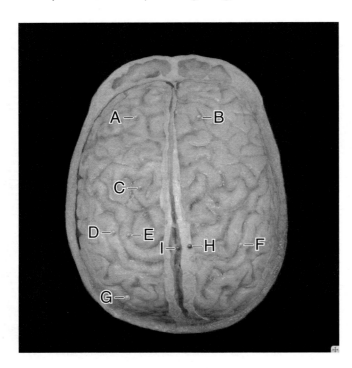

• *Mouse-over the green pins on the screen to find the information necessary to identify the following structures:*

A. _____

B. _____

C. _____

D. _____

E. _____

F. _____

G. _____

Non-nervous System Structures (blue pins)

H. _____

I. _____

CHECK POINT:

Brain, Superior View, cont'd

6. Name an unpaired dural venous sinus that terminates at the confluence of sinuses.
7. Name the structure that is also called the primary motor cortex.
8. Name the structure that is also called the primary somatosensory cortex.

• *Click **LAYER 5** in the **LAYER CONTROLS** window, and you will see the following image:*

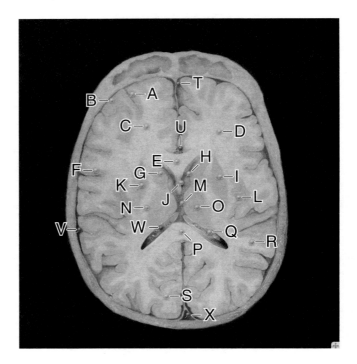

• *Mouse-over the green pins on the screen to find the information necessary to identify the following structures:*

A. _____

B. _____

C. _____

D. _____

E. _____

F. _____

G. _____

H. _____

I. _____

J. _____

K. _____

L. _____

M. _____

N. _____

O. _____

P. _____

Q. _____

R. _____

S. _____

Non-nervous System Structures (blue pins)

T. _____

U. _____

V. _____

W. _____

X. _____

CHECK POINT:

Brain, Superior View, cont'd

9. The cerebral ventricles are lined with tufts of capillaries covered by specialized ependymal cells. What are these cells called?
10. Both the superior sagittal sinus and the inferior sagittal sinus are located in the margins of the _____.
11. Name the slitlike cavity that separates the right and left halves of the diencephalon.

• *Click* **LAYER 6** *in the* **LAYER CONTROLS** *window, and you will see the following image:*

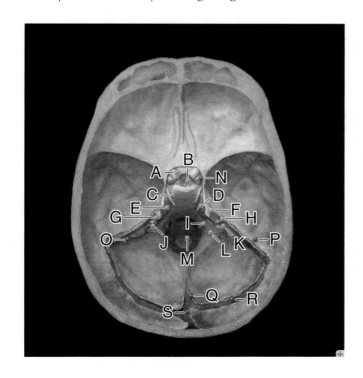

• *Mouse-over the pins on the screen to find the information necessary to identify the following structures:*

A. _____

B. _____

C. _____

D. _____

E. _____

F. _____

G. _____

H. _____

I. _____

J. _____

K. _____

L. _____

M. _____

Non-nervous System Structures (blue pins)

N. _____

O. _____

P. _____

Q. _____

R. _____

S. _____

CHECK POINT:

Brain, Superior View, cont'd

12. Name the point of attachment of the pituitary gland to the hypothalamus.
13. Name the S-shaped groove on the inner aspect of the temporal bone.
14. Name the cranial nerve that controls the superior oblique muscle.

Self Test

Take this opportunity to quiz yourself by taking the **SELF TEST.** See page 15 for a reminder on how to access the self-test for this section.

Animation: CSF Flow

SELECT ANIMATION	
CSF Flow	PLAY

- *After viewing the animation, answer these questions:*

1. In what brain structures would you expect to find (CSF)?

2. Where is this CSF produced?

3. What structure produces the CSF?

4. Beginning in the lateral ventricles, trace the flow of CSF.

5. What are arachnoid granulations?

Animation: Dural Sinus Blood Flow

SELECT ANIMATION	
Dural Sinus Blood Flow	PLAY

- *After viewing the animation, answer these questions:*

1. What are the dural venous sinuses?

2. Where are they located?

3. Name the two dural sinuses located along the midline.

4. Name the three sinuses that unite at the confluence of sinuses.

5. What vessels do the sigmoid sinuses become?

EXERCISE 5.4:

Nervous System—Brain, Inferior View

SELECT TOPIC	SELECT VIEW	
Brain >	**Inferior**	GO

- *Click **LAYER 1** in the **LAYER CONTROLS** window, and you will see the following image:*

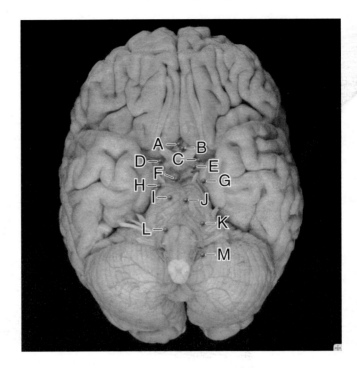

- *Mouse-over the blue pins on the screen to find the information necessary to identify the following non-nervous system structures:*

A. _____

B. _____

C. _____

D. _____

E. _____

F. _____

G. _____

H. _____

I. _____

J. _____

K. _____

L. _____

M. _____

CHECK POINT:

Brain, Inferior View

1. Name the circular anastomosis on the ventral surface of the brain also referred to as the "Circle of Willis."
2. Name the artery that passes through the transverse foramina of the cervical vertebrae.
3. Name the unpaired midline artery that ascends on the anterior surface of the pons.

• *Click* **LAYER 2** *in the* **LAYER CONTROLS** *window, and you will see the following image:*

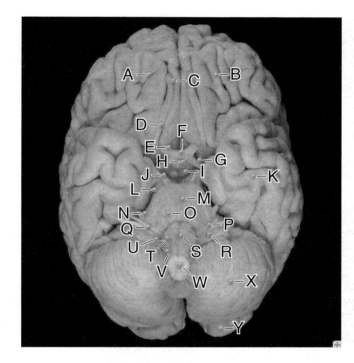

• *Mouse-over the pins on the screen to find the information necessary to identify the following structures:*

A. _____

B. _____

C. _____

D. _____

E. _____

F. _____

G. _____

H. _____

I. _____

J. _____

K. _____

L. _____

M. _____

N. _____

O. _____

P. _____

Q. _____

R. _____

S. _____

T. _____

U. _____

V. _____

W. _____

X. _____

Y. _____

CHECK POINT:

Brain, Inferior View, cont'd

4. Name the crossing white-matter tract between the optic nerve and the optic tracts.
5. Name the brain structure whose name means "bridge."
6. Name the most caudal portion of the brain. What are its functions?

Self Test

Take this opportunity to quiz yourself by taking the **SELF TEST.** See page 15 for a reminder on how to access the self-test for this section.

EXERCISE 5.5:

Nervous System—Brain, Inferior View (close-up)

SELECT TOPIC	SELECT VIEW	
Brain	▶ **Inferior (close-up)**	**GO**

• *Click* **LAYER 1** *in the* **LAYER CONTROLS** *window, and you will see the following image:*

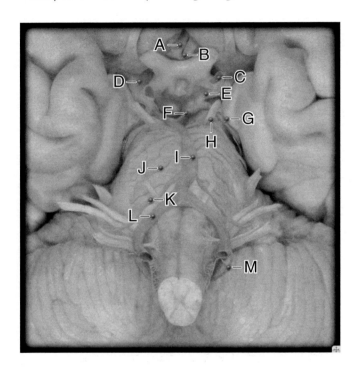

• *Mouse-over the blue pins on the screen to find the information necessary to identify the following non-nervous system structures:*

A. _____

B. _____

C. _____

D. _____

E. _____

F. _____

G. _____

H. _____

I. _____

J. _____

K. _____

L. _____

M. _____

CHECK POINT:

Brain, Inferior View (close-up)

1. Name the artery whose significant branches include the ophthalmic, anterior cerebral, and middle cerebral arteries.
2. Name the artery whose major branches include the pontine and superior cerebellar arteries.
3. Name the artery whose numerous branches course laterally across the surface of the pons.

• *Click* **LAYER 2** *in the* **LAYER CONTROLS** *window, and you will see the following image:*

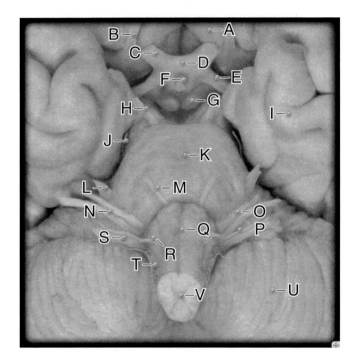

• *Mouse-over the pins on the screen to find the information necessary to identify the following structures:*

A. _____

B. _____

C. _____

D. _____

E. _____

F. _____

G. _____

H. _____

I. _____

J. _____

K. _____

L. _____

M. _____

N. _____

O. _____

P. _____

Q. _____

R. _____

S. _____

T. _____

U. _____

V. _____

CHECK POINT:

Brain, Inferior View (close-up), cont'd

4. Name the paired, small, rounded projections of the hypothalamus involved in the regulation of autonomic functions, emotional behavior, and memory.
5. Name the "funnel-shaped" extension of the floor of the third ventricle that continues as the pituitary stalk.
6. Name the structure that gives rise to eight pairs of spinal nerves and contains the cervical enlargement.

Self Test

Take this opportunity to quiz yourself by taking the **SELF TEST.** See page 15 for a reminder on how to access the self-test for this section.

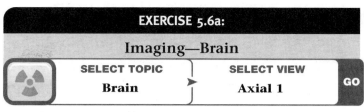

* Click the **TURN TAGS ON** *button, and you will see the following image:*

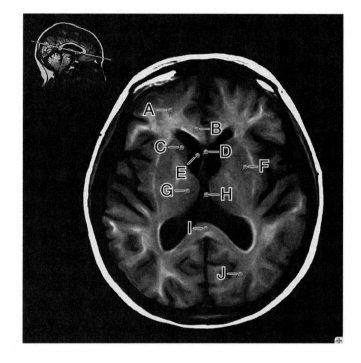

* *Mouse-over the pins on the screen to find the information necessary to identify the following structures:*

A. _____

B. _____

C. _____

D. _____

E. _____

F. _____

G. _____

H. _____

I. _____

J. _____

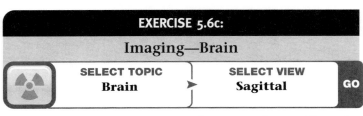

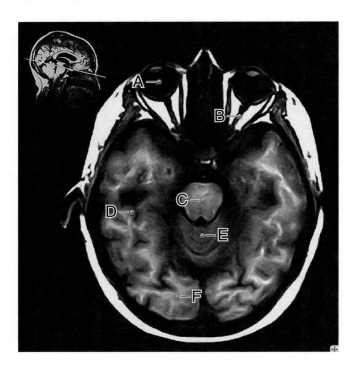

- *Click the* **TURN TAGS ON** *button, and you will see the following image:*

- *Mouse-over the pins on the screen to find the information necessary to identify the following structures:*

A. _____

B. _____

C. _____

D. _____

E. _____

F. _____

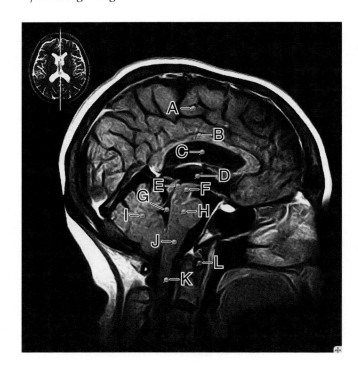

- *Click the* **TURN TAGS ON** *button, and you will see the following image:*

- *Mouse-over the green pins on the screen to find the information necessary to identify the following structures:*

A. _____

B. _____

C. _____

D. _____

E. _____

F. _____

G. _____

H. _____

I. _____

J. _____

K. _____

Non-nervous System Structure (blue pin)

L. _____

IN REVIEW

What Have I Learned?

The following questions cover the material that you have just learned, the brain. Apply what you have learned to answer these questions on a separate piece of paper.

1. Name the division of the brain that includes the midbrain, pons, and medulla oblongata.

2. Name the division of the brain that coordinates complex movements and smooths muscle contractions.

3. The primary hearing and smell areas are located in which lobe of the brain?

4. Memory is located in which lobe of the brain?

5. Speech perception and recognition areas are located in which hemisphere of which lobe of the brain?

6. What is the term that refers to the superficial gray matter of the cerebrum?

7. The reception of general sensory information from the body occurs in which lobe of the brain?

8. Tactile object recognition occurs in which lobe of the brain?

9. Language and verbatim repetition of terms occurs in which lobe of which hemisphere of the brain?

10. Name the structure that forms the floor of the longitudinal fissure.

11. Name the paired cavities containing cerebrospinal fluid in the brain.

12. Where is the falx cerebri located?

13. Name the crossing white-matter tracts between the optic nerves and the optic tracts.

14. Name two structures responsible for planning and execution of movement, muscle tone, and posture.

15. Name the slitlike, fluid-filled cavity that separates the right and left halves of the diencephalon.

16. Name the site of emotional behavior.

17. Name the structure that regulates body temperature, eating, and drinking.

18. Name the structure that controls the autonomic nervous system.

19. Name the lobe that is the primary visual area.

20. Name the lobe that controls voluntary motor activity.

21. Name the lobe that is the site of higher mental processing.

22. Name the structure continuous with the brain at the foramen magnum.

Cranial Nerves

EXERCISE 5.7:

Nervous System—Cranial Nerves, Inferior Brain (CN I-XII)

SELECT TOPIC
Cranial Nerves ▸

SELECT VIEW
Inferior Brain (CN I-XII)   GO

- *Click* **LAYER 1** *in the* **LAYER CONTROLS** *window, and you will see the image to the right:*

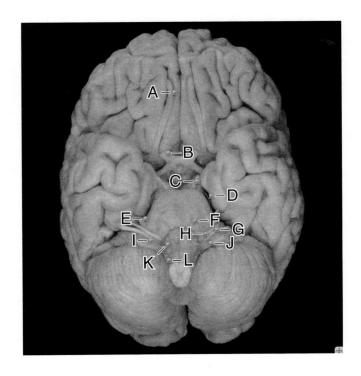

- *Mouse-over the pins on the screen to find the information necessary to identify the following structures:*

A. _____

B. _____

C. _____

D. _____

E. _____

F. _____

G. _____

H. _____

I. _____

J. _____

K. _____

L. _____

EXERCISE 5.8:

Nervous System—Cranial Nerves, CN I Olfactory

SELECT TOPIC	SELECT VIEW	
Cranial Nerves	**CN I Olfactory**	**GO**

- *Click **LAYER 3** in the **LAYER CONTROLS** window, and you will see the following image:*

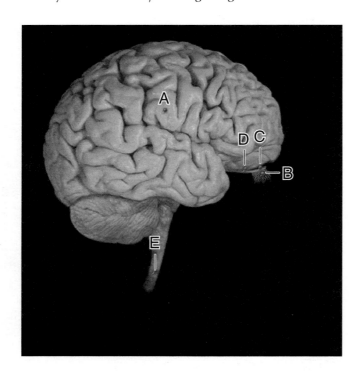

- *Mouse-over the pins on the screen to find the information necessary to identify the following structures:*

A. _____

B. _____

C. _____

D. _____

E. _____

- *Click **LAYER 4** in the **LAYER CONTROLS** window, and you will see the following image:*

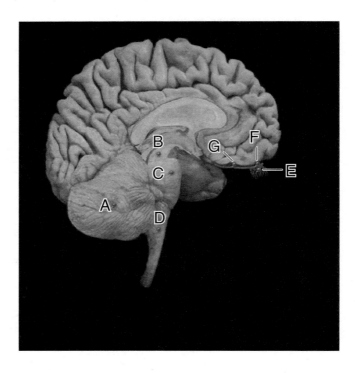

- *Mouse-over the pins on the screen to find the information necessary to identify the following structures:*

A. _____

B. _____

C. _____

D. _____

E. _____

F. _____

G. _____

EXERCISE 5.9:

Nervous System—Cranial Nerves, CN II Optic

SELECT TOPIC	SELECT VIEW	
Cranial Nerves >	**CN II Optic**	**GO**

• *Click **LAYER 3** in the **LAYER CONTROLS** window, and you will see the following image:*

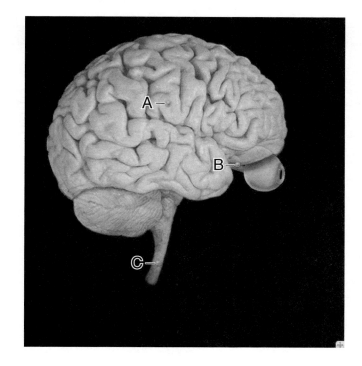

• *Mouse-over the pins on the screen to find the information necessary to identify the following structures:*

A. _____

B. _____

C. _____

• *Click **LAYER 4** in the **LAYER CONTROLS** window, and you will see the following image:*

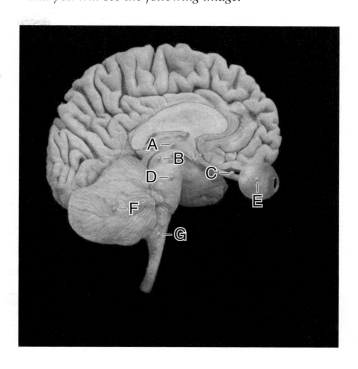

• *Mouse-over the pins on the screen to find the information necessary to identify the following structures:*

A. _____

B. _____

C. _____

D. _____

E. _____

F. _____

G. _____

Animation: Vision

SELECT ANIMATION	
Vision	**PLAY**

• *After viewing the animation, answer these questions:*

1. What are the three structures involved in vision?

2. The optic nerve runs between what two structures?

3. Describe the two parts of the retina.

4. What neurotransmitter do the photoreceptors release in the dark?

5. Describe the light pathway from the eye to the brain.

6. Name the structure where the optic nerves converge.

7. Images are perceived in which lobe of the brain?

C. _____

D. _____

E. _____

F. _____

G. _____

H. _____

I. _____

EXERCISE 5.11:

Nervous System—Cranial Nerves, CN IV Trochlear

SELECT TOPIC	SELECT VIEW	
Cranial Nerves	CN IV Trochlear	GO

- *Click **LAYER 4** in the **LAYER CONTROLS** window, and you will see the following image:*

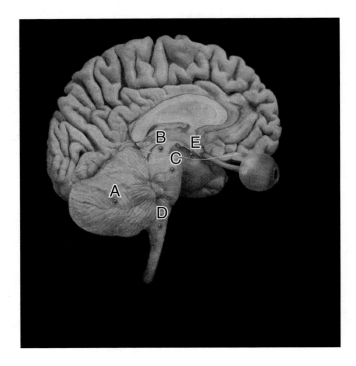

- *Mouse-over the pins on the screen to find the information necessary to identify the following structures:*

A. _____

B. _____

C. _____

D. _____

E. _____

EXERCISE 5.10:

Nervous System—Cranial Nerves, CN III Oculomotor

SELECT TOPIC	SELECT VIEW	
Cranial Nerves	CN III Oculomotor	GO

- *Click **LAYER 4** in the **LAYER CONTROLS** window, and you will see the following image:*

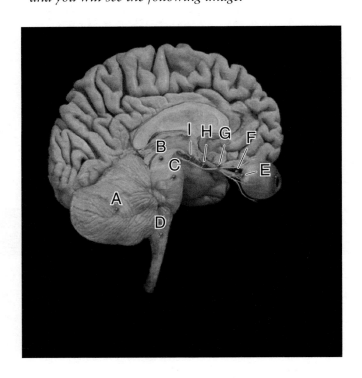

- *Mouse-over the pins on the screen to find the information necessary to identify the following structures:*

A. _____

B. _____

EXERCISE 5.12:

Nervous System—Cranial Nerves, CN V Trigeminal

SELECT TOPIC	SELECT VIEW	
Cranial Nerves	**CN V Trigeminal**	GO

- *Click* **LAYER 4** *in the* **LAYER CONTROLS** *window, and you will see the following image:*

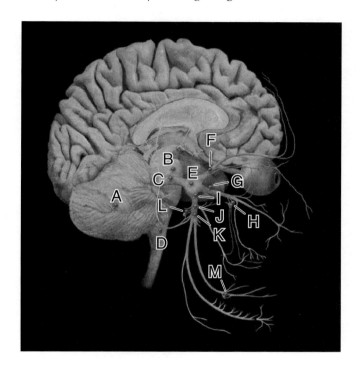

- *Mouse-over the pins on the screen to find the information necessary to identify the following structures:*

A. _____

B. _____

C. _____

D. _____

E. _____

F. _____

G. _____

H. _____

I. _____

J. _____

K. _____

L. _____

M. _____

EXERCISE 5.13:

Nervous System—Cranial Nerves, CN VI Abducens

SELECT TOPIC	SELECT VIEW	
Cranial Nerves	**CN VI Abducens**	GO

- *Click* **LAYER 4** *in the* **LAYER CONTROLS** *window, and you will see the following image:*

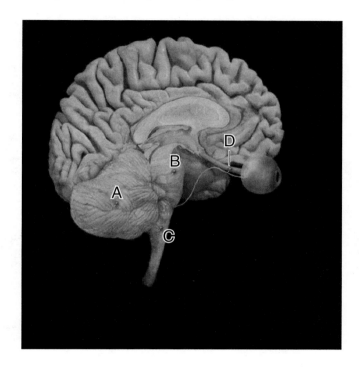

- *Mouse-over the pins on the screen to find the information necessary to identify the following structures:*

A. _____

B. _____

C. _____

D. _____

EXERCISE 5.14:

Nervous System—Cranial Nerves, CN VII Facial

SELECT TOPIC	SELECT VIEW	
Cranial Nerves	CN VII Facial	GO

- *Click* **LAYER 4** *in the* **LAYER CONTROLS** *window, and you will see the following image:*

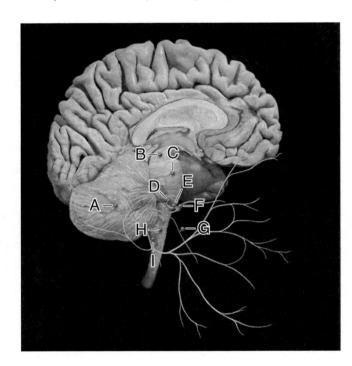

- *Mouse-over the pins on the screen to find the information necessary to identify the following structures:*

A. _____

B. _____

C. _____

D. _____

E. _____

F. _____

G. _____

H. _____

I. _____

EXERCISE 5.15:

Nervous System—Cranial Nerves, CN VIII Vestibulocochlear

SELECT TOPIC	SELECT VIEW	
Cranial Nerves	CN VIII Vestibulocochlear	GO

- *Click* **LAYER 4** *in the* **LAYER CONTROLS** *window, and you will see the following image:*

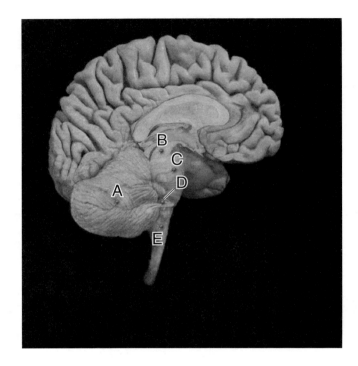

- *Mouse-over the pins on the screen to find the information necessary to identify the following structures:*

A. _____

B. _____

C. _____

D. _____

E. _____

Animation: Hearing

| | SELECT ANIMATION **Hearing** | PLAY |

- *After viewing the animation, answer these questions:*

1. What structure do sound waves strike and cause to vibrate?

2. The vibrations are transferred to the three bones of the middle ear. From lateral to medial, what are those bones?

3. What structure transfers this vibration to the oval window?

4. The vibrations are then transferred to a fluid-filled chamber of the inner ear. What is that fluid, and what is the name of that chamber?

5. What is the difference in location for the detection of high-pitched and low-pitched sounds?

- *Mouse-over the pins on the screen to find the information necessary to identify the following structures:*

A. _____

B. _____

C. _____

D. _____

E. _____

F. _____

G. _____

H. _____

I. _____

J. _____

EXERCISE 5.16:

Nervous System—Cranial Nerves, CN IX Glossopharyngeal

| | SELECT TOPIC **Cranial Nerves** ▶ | SELECT VIEW **CN IX Glossopharyngeal** | GO |

- *Click **LAYER 4** in the **LAYER CONTROLS** window, and you will see the following image:*

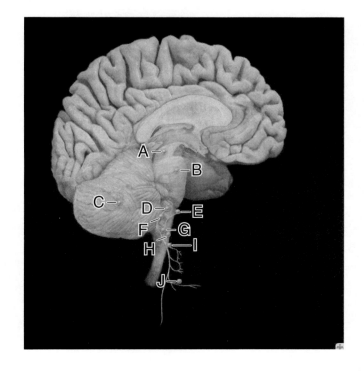

EXERCISE 5.17:

Nervous System—Cranial Nerves, CN X Vagus

| | SELECT TOPIC **Cranial Nerves** ▶ | SELECT VIEW **CN X Vagus** | GO |

- *Click **LAYER 4** in the **LAYER CONTROLS** window, and you will see the following image:*

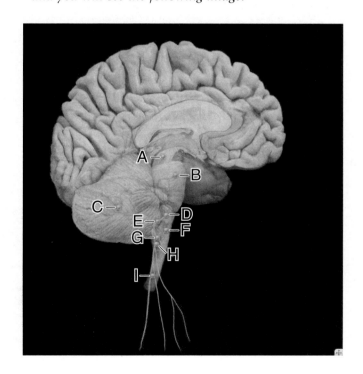

- *Mouse-over the pins on the screen to find the information necessary to identify the following structures:*

A. _____

B. _____

C. _____

D. _____

E. _____

F. _____

G. _____

H. _____

I. _____

EXERCISE 5.18:

Cranial Nerves, CN XI Accessory

SELECT TOPIC	SELECT VIEW	
Cranial Nerves	**CN XI Accessory**	GO

- *Click **LAYER 4** in the **LAYER CONTROLS** window, and you will see the following image:*

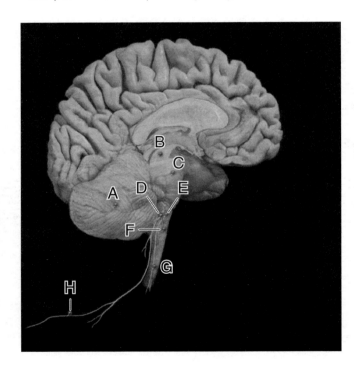

- *Mouse-over the pins on the screen to find the information necessary to identify the following structures:*

A. _____

B. _____

C. _____

D. _____

E. _____

F. _____

G. _____

H. _____

EXERCISE 5.19:

Nervous System—Cranial Nerves, CN XII Hypoglossal

SELECT TOPIC	SELECT VIEW	
Cranial Nerves	**CN XII Hypoglossal**	GO

- *Click **LAYER 4** in the **LAYER CONTROLS** window, and you will see the following image:*

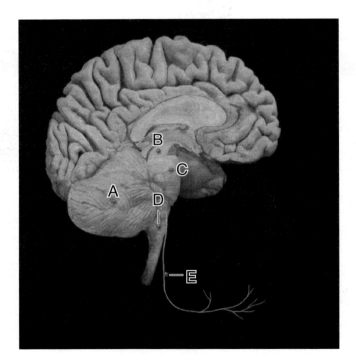

- *Mouse-over the pins on the screen to find the information necessary to identify the following structures:*

A. _____

B. _____

C. _____

D. _____

E. _____

IN REVIEW

What Have I Learned?

The following questions cover the material that you have just learned, the cranial nerves. Apply what you have learned to answer these questions on a separate piece of paper.

1. Which cranial nerve is composed of the ophthalmic, the maxillary, and the mandibular nerves?

2. Name the cranial nerve responsible for the intrinsic and extrinsic muscles of the tongue.

3. Which cranial nerve has sensory fibers that monitor blood pressure at the carotid sinus?

4. Which cranial nerve is responsible for the constriction of the pupillary sphincter and accommodation of the lens for near vision?

5. Which cranial nerve is responsible for taste from the anterior two-thirds of the tongue and the muscles of facial expression?

6. Which cranial nerve has olfactory neurons on the mucosa of the anterosuperior nasal cavity?

7. Which cranial nerve controls the lateral rectus muscle of the eye?

8. Which cranial nerve is involved with hearing and balance?

9. Which cranial nerve is responsible for vision?

10. Which cranial nerve controls the extraocular muscle of the superior oblique?

11. Which cranial nerve controls all but one of the muscles of the palate, pharynx, and the intrinsic muscles of the larynx?

12. Which cranial nerve is the only one that extends beyond the head and neck?

Animation: Action Potential Generation

SELECT ANIMATION
Action Potential Generation   **PLAY**

- *After viewing the animation, answer these questions:*

1. When the cell membrane is at its resting membrane potential, in what position are the voltage-gated sodium and potassium channels?

2. _____ is initiated by a stimulus that makes the membrane _____.

3. What affect does this have on the voltage-gated sodium channels? What occurs at threshold?

4. What causes depolarization? What happens to the voltage-gated potassium channels?

5. Therefore, depolarization occurs because . . .

6. What causes the diffusion of sodium ions to decrease? What happens to the potassium ions?

7. What causes the membrane potential to become slightly more negative than the resting value?

8. What causes reestablishment of resting membrane potential?

Animation: Action Potential Propagation

SELECT ANIMATION **Action Potential Propagation**	**PLAY**

- *After viewing the animation, answer these questions:*

1. An action potential is propagated . . .

2. During an action potential, the inside of the cell membrane . . .

3. What happens to the membrane immediately adjacent to the action potential?

4. What occurs when depolarization caused by the local currents reaches threshold?

5. Why does action potential propagation occur in only one direction?

Animation: Chemical Synapse

SELECT ANIMATION **Chemical Synapse**	**PLAY**

- *After viewing the animation, answer these questions:*

1. What is caused by action potentials arriving at the presynaptic terminal?

2. What occurs when calcium ions diffuse into the cell?

3. What do the acetylcholine molecules do? What does this cause?

4. If the membrane potential reaches threshold level, . . .

Spinal Cord

EXERCISE 5.20:

Nervous System—Spinal Cord, Overview

SELECT TOPIC	SELECT VIEW	
Spinal Cord ▶	**Overview**	**GO**

- *Click **LAYER 1** in the **LAYER CONTROLS** window, and you will see the following image:*

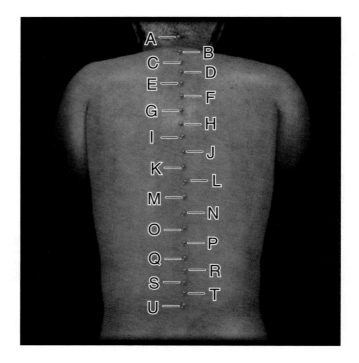

- *Mouse-over the pins on the screen to find the information necessary to identify the following structures:*

A. _____

B. _____

C. _____

D. _____

E. _____

F. _____

G. _____

H. _____

I. _____

J. _____

K. _____

L. _____

M. _____

N. _____

O. _____

P. _____

Q. _____

R. _____

S. _____

T. _____

U. _____

CHECK POINT:

Spinal Cord, Overview

1. What is a dermatome?
2. Which dermatome includes the skin of the foot, including the middle three toes?
3. Which dermatome includes the skin over the knee and on the medial foot, including the great toe?

• *Click* **LAYER 3** *in the* **LAYER CONTROLS** *window, and you will see the following image:*

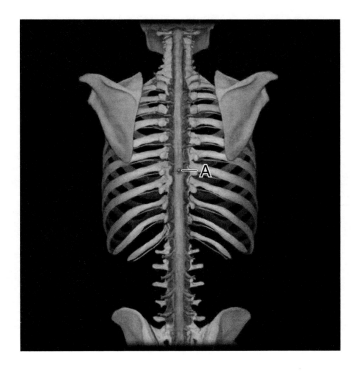

• *Mouse-over the blue pin on the screen to find the information necessary to identify the following non-nervous system structure:*

A. _____

CHECK POINT:

Spinal Cord, Overview, cont'd

4. Name the outermost tough connective tissue that surrounds the brain and spinal cord.

• *Click* **LAYER 4** *in the* **LAYER CONTROLS** *window, and you will see the following image:*

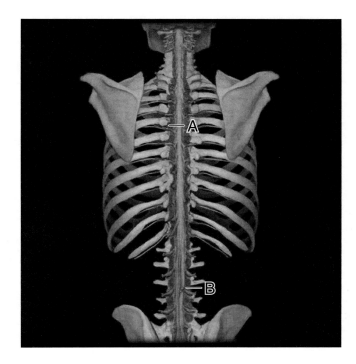

• *Mouse-over the pins on the screen to find the information necessary to identify the following structures:*

A. _____

B. _____

CHECK POINT:

Spinal Cord, Overview, cont'd

5. Name the large bundle of dorsal and ventral roots for spinal nerves below L2.
6. Name the structure that contains the sensory ganglion for each dorsal root.
7. Name the structure whose Latin name means "horse tail."

• Click **LAYER 5** in the **LAYER CONTROLS** window, and you will see the following image:

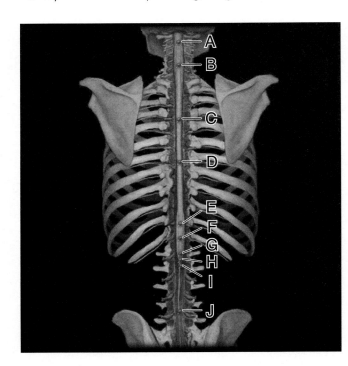

• Mouse-over the green pins on the screen to find the information necessary to identify the following structures:

A. _____

B. _____

C. _____

D. _____

E. _____

F. _____

G. _____

H. _____

I. _____

Non-nervous System Structure (blue pin)

J. _____

CHECK POINT:

Spinal Cord, Overview, cont'd

8. What is the cervical enlargement?
9. Name the structure that contains neurons for lower-limb innervation.
10. Name the tapered inferior end of the spinal cord.

EXERCISE 5.21:

Nervous System—Spinal Cord, Typical Spinal Nerve

SELECT TOPIC	SELECT VIEW	
Spinal Cord	▶ Typical Spinal Nerve	GO

• Click **LAYER 1** in the **LAYER CONTROLS** window, and you will see the following image:

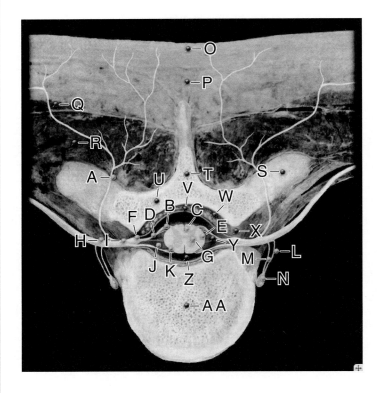

• Mouse-over the green pins on the screen to find the information necessary to identify the following structures:

A. _____

B. _____

C. _____

D. _____

E. _____

F. _____

G. _____

H. _____

I. _____

J. _____

K. _____

L. _____

M. _____

N. _____

Non-nervous System Structures (blue pins)

O. _____

P. _____

Q. _____

R. _____

S. _____

T. _____

U. _____

V. _____

W. _____

X. _____

Y. _____

Z. _____

AA. _____

CHECK POINT:

Typical Spinal Nerve

1. In what respect is the spinal cord structurally opposite to the brain?
2. What structure makes up the core of the spinal cord?
3. Name the structure that consists of ascending and descending bundles of myelinated axons.

Animation: Typical Spinal Nerve

 SELECT ANIMATION
Typical Spinal Nerve   PLAY

- *After viewing the animation, answer these questions:*

1. How many spinal nerves exist for each vertebral level?

2. What structures connect each spinal nerve to the spinal cord?

3. What structure is made up of bundles of nerve fibers carrying sensory information from the skin to the spinal cord?

4. Where are the cell bodies (somas) of these sensory nerve fibers located?

5. Name the structure consisting of bundles of motor (efferent) fibers carrying impulses away from the spinal cord to the skeletal muscles.

6. Where are the cell bodies (somas) associated with these nerve fibers located?

7. What structures unite to form the spinal nerve?

8. The spinal nerves exit the vertebral column through what structure?

9. What is a mixed nerve?

10. Name the two branches that form from each spinal nerve.

11. What structures innervate the muscles and skin of the back?

12. What structures innervate the muscles and skin of the lateral and ventral trunk and the limbs?

EXERCISE 5.22:

Nervous System—Spinal Cord, Cervical Region

SELECT TOPIC	SELECT VIEW	
Spinal Cord	▶ **Cervical Region**	**GO**

• *Click* **LAYER 1** *in the* **LAYER CONTROLS** *window, and you will see the following image:*

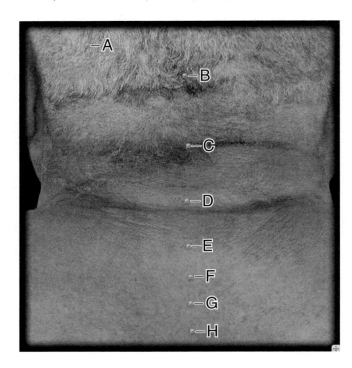

• *Mouse-over the pins on the screen to find the information necessary to identify the following structures:*

A. _____

B. _____

C. _____

D. _____

E. _____

F. _____

G. _____

H. _____

• *Click* **LAYER 3** *in the* **LAYER CONTROLS** *window, and you will see the following image:*

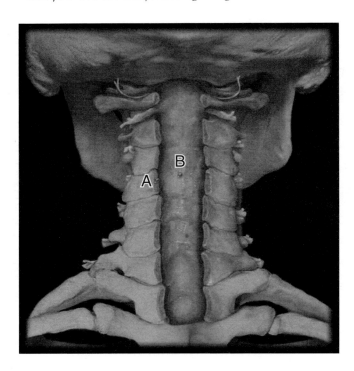

• *Mouse-over the blue pins on the screen to find the information necessary to identify the following non-nervous system structures:*

A. _____

B. _____

• *Click* **LAYER 4** *in the* **LAYER CONTROLS** *window, and you will see the following image:*

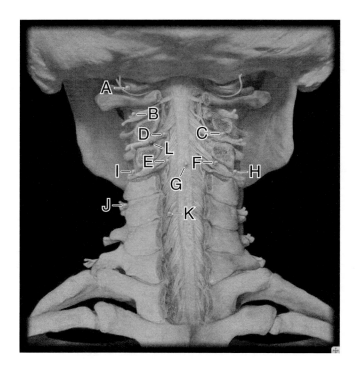

- *Mouse-over the green pins on the screen to find the information necessary to identify the following structures:*

A. _____

B. _____

C. _____

D. _____

E. _____

F. _____

G. _____

H. _____

I. _____

J. _____

K. _____

Non-nervous System Structure (blue pin)

L. _____

C H E C K   P O I N T :

Spinal Cord, Cervical Region

1. Name the series of small nerves branching from the dorsal length of the spinal cord.
2. Name the afferent (sensory) limb of each spinal nerve.
3. Name the two components of each spinal nerve.

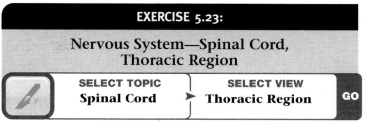

- *Click **LAYER 1** in the **LAYER CONTROLS** window, and you will see the following image:*

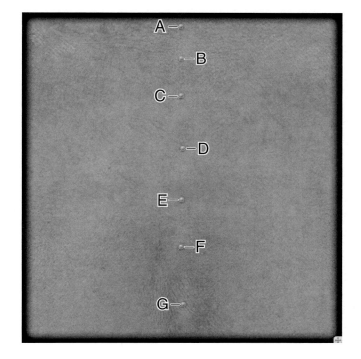

- *Mouse-over the pins on the screen to find the information necessary to identify the following structures:*

A. _____

B. _____

C. _____

D. _____

E. _____

F. _____

G. _____

- *Click* **LAYER 3** *in the* **LAYER CONTROLS** *window, and you will see the following image:*

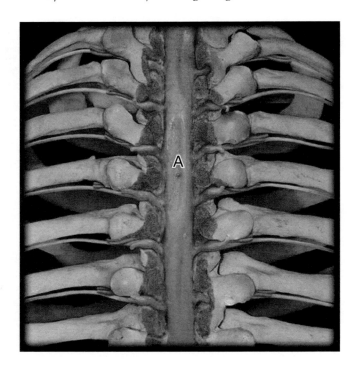

- *Mouse-over the blue pin on the screen to find the information necessary to identify the following non-nervous system structure:*

A. _____

- *Click* **LAYER 4** *in the* **LAYER CONTROLS** *window, and you will see the following image:*

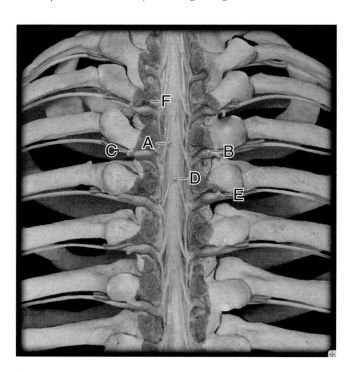

- *Mouse-over the green pins on the screen to find the information necessary to identify the following structures:*

A. _____

B. _____

C. _____

D. _____

E. _____

Non-nervous System Structure (blue pin)

F. _____

EXERCISE 5.24:

Nervous System—Spinal Cord, Lumbar Region

SELECT TOPIC	SELECT VIEW	
Spinal Cord ▶	**Lumbar Region**	GO

- *Click* **LAYER 1** *in the* **LAYER CONTROLS** *window, and you will see the following image:*

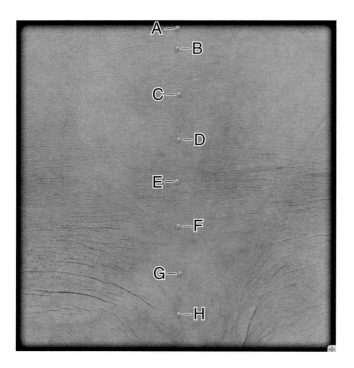

- *Mouse-over the pins on the screen to find the information necessary to identify the following structures:*

A. _____

B. _____

C. _____

D. _____

E. _____

F. _____

G. _____

H. _____

- *Click* **LAYER 3** *in the* **LAYER CONTROLS** *window, and you will see the following image:*

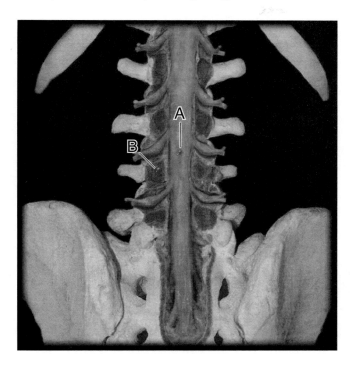

- *Mouse-over the blue pins on the screen to find the information necessary to identify the following non-nervous system structures:*

A. _____

B. _____

- *Click* **LAYER 4** *in the* **LAYER CONTROLS** *window, and you will see the following image:*

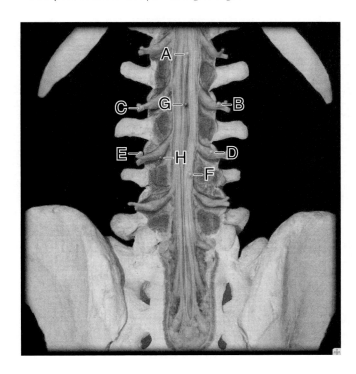

- *Mouse-over the green pins on the screen to find the information necessary to identify the following structures:*

A. _____

B. _____

C. _____

D. _____

E. _____

F. _____

Non-nervous System Structures (blue pins)

G. _____

H. _____

CHECK POINT:

Spinal Cord, Lumbar Region

1. Name the structure of the pia mater caudal to the spinal cord.

Self Test

Take this opportunity to quiz yourself by taking the **Self Test.** See page 15 for a reminder on how to access the self-test for this section.

IN REVIEW

What Have I Learned?

The following questions cover the material that you have just learned, the spinal cord. Apply what you have learned to answer these questions on a separate piece of paper.

1. Name the structure that consists of a filament of pia mater that forms at the conus medullaris, passes through the sacral hiatus, and attaches to the coccyx.

2. In adults, where does the spinal cord end?

3. Name the structures that contribute to the lumbosacral enlargement.

4. Name the bony structure through which each spinal nerve passes.

5. Name the structure that is the sensory ganglion of each dorsal root.

6. Which terminal branch of the spinal nerves contains motor innervation to muscles of the suboccipital triangle?

7. What do the dorsal and ventral roots at the same spinal cord level unite to form?

Peripheral Nerves

> EXERCISE 5.25:
>
> ## Nervous System—Peripheral Nerves, Cervical Plexus
>
> | SELECT TOPIC | SELECT VIEW | |
> | Peripheral Nerves ▶ | Cervical Plexus | GO |

- *Click **LAYER 1** in the **LAYER CONTROLS** window, and you will see the following image:*

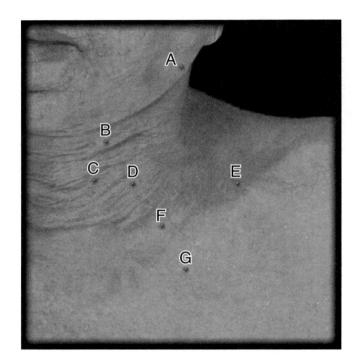

- *Mouse-over the pins on the screen to find the information necessary to identify the following structures:*

A. _____

B. _____

C. _____

D. _____

E. _____

F. _____

G. _____

- *Click **LAYER 2** in the **LAYER CONTROLS** window, and you will see the following image:*

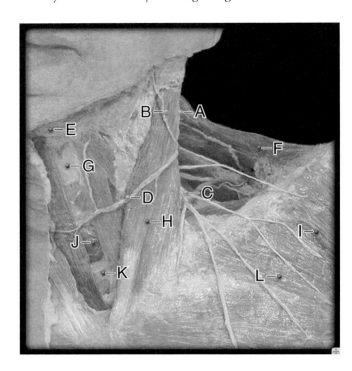

- *Mouse-over the green pins on the screen to find the information necessary to identify the following structures:*

A. _____

B. _____

C. _____

D. _____

Non-nervous System Structures (blue pins)

E. _____

F. _____

G. _____

H. _____

I. _____

J. _____

K. _____

L. _____

CHECK POINT:

Peripheral Nerves, Cervical Plexus

1. Name the four sensory branches of the cervical plexus.
2. Which of these four innervates the skin over the anterior and lateral neck?
3. Which of these branches innervates the lateral scalp and posterior auricle of the ear?

- *Click* **LAYER 3** *in the* **LAYER CONTROLS** *window, and you will see the following image:*

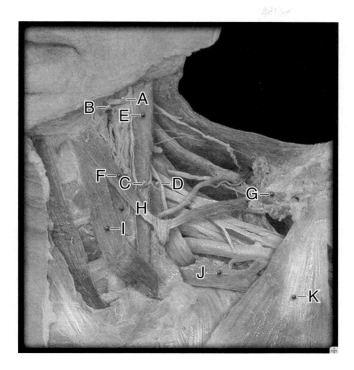

- *Mouse-over the green pins on the screen to find the information necessary to identify the following structures:*

A. _____

B. _____

C. _____

D. _____

Non-nervous System Structures (blue pins)

E. _____

F. _____

G. _____

H. _____

I. _____

J. _____

K. _____

CHECK POINT:

Peripheral Nerves, Cervical Plexus, cont'd

4. Which nerve innervates the genioglossus, hyoglossus, styloglossus, and intrinsic muscles of the tongue?

5. Which nerve fibers "hitchhike" on this nerve?

6. Which nerve's Latin name means "neck loop"?

- *Click* **LAYER 4** *in the* **LAYER CONTROLS** *window, and you will see the following image:*

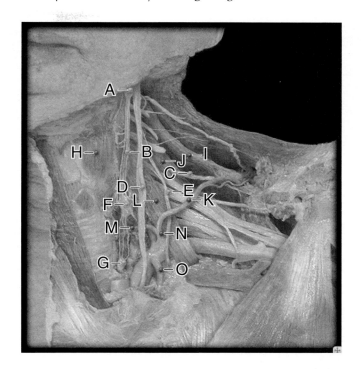

- *Mouse-over the pins on the screen to find the information necessary to identify the following structures:*

A. _____

B. _____

C. _____

D. _____

E. _____

F. _____

G. _____

Non-nervous System Structures (blue pins)

H. _____

I. _____

J. _____

K. _____

L. _____

M. _____

N. _____

O. _____

CHECK POINT:

Peripheral Nerves, Cervical Plexus, cont'd

7. Name the three sympathetic ganglia in the neck area.
8. Which of these distributes all postganglionic sympathetic nerve fibers to the head?

- *Click LAYER 5 in the LAYER CONTROLS window, and you will see the following image:*

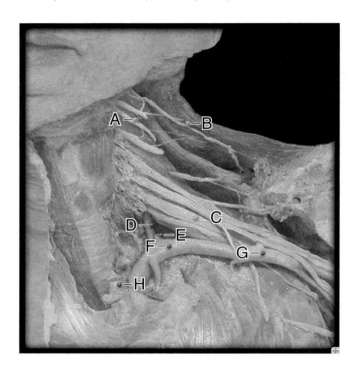

- *Mouse-over the green pins on the screen to find the information necessary to identify the following structures:*

A. _____

B. _____

C. _____

Non-nervous System Structures (blue pins)

D. _____

E. _____

F. _____

G. _____

H. _____

CHECK POINT:

Peripheral Nerves, Cervical Plexus, cont'd

9. Name the structure that consists of nerves distributed to the upper limb.
10. What are the ventral rami of spinal nerves C1–4 referred to as?

Self Test

Take this opportunity to quiz yourself by taking the **Self Test.** See page 15 for a reminder on how to access the self-test for this section.

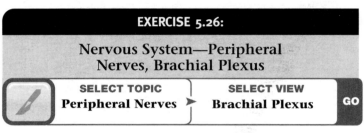

EXERCISE 5.26:

Nervous System—Peripheral Nerves, Brachial Plexus

| SELECT TOPIC | SELECT VIEW | |
| Peripheral Nerves ▶ | Brachial Plexus | GO |

- *Click LAYER 1 in the LAYER CONTROLS window, and you will see the following image:*

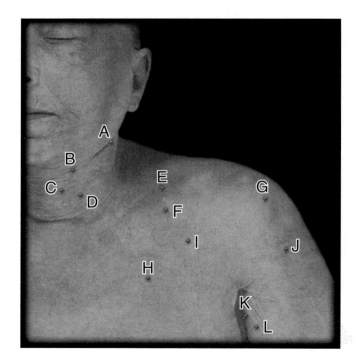

- *Mouse-over the pins on the screen to find the information necessary to identify the following structures:*

A. _____

B. _____

C. _____

D. _____

E. _____

F. _____

G. _____

H. _____

I. _____

J. _____

K. _____

L. _____

CHECK POINT:

Peripheral Nerves, Brachial Plexus

1. Name the nerve that supplies cutaneous innervation to the skin of and around the auricle of the ear.

• *Click* **LAYER 3** *in the* **LAYER CONTROLS** *window,
and you will see the following image:*

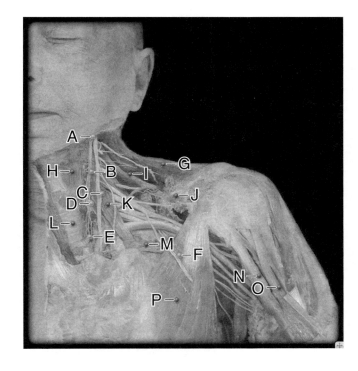

• *Mouse-over the green pins on the screen to find the information necessary to identify the following structures:*

A. _____

B. _____

C. _____

D. _____

E. _____

F. _____

Non-nervous System Structures (blue pins)

G. _____

H. _____

I. _____

J. _____

K. _____

L. _____

M. _____

N. _____

O. _____

P. _____

• *Click* **LAYER 4** *in the* **LAYER CONTROLS** *window,
and you will see the following image:*

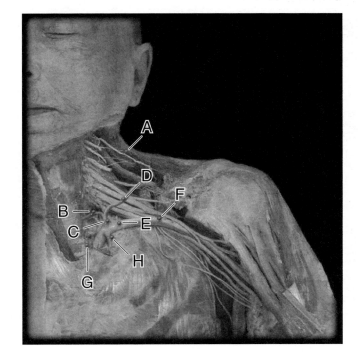

- *Mouse-over the green pin on the screen to find the information necessary to identify the following structure:*

A. _____

Non-nervous System Structures (blue pins)

B. _____

C. _____

D. _____

E. _____

F. _____

G. _____

H. _____

- *Click **LAYER 5** in the **LAYER CONTROLS** window, and you will see the following image:*

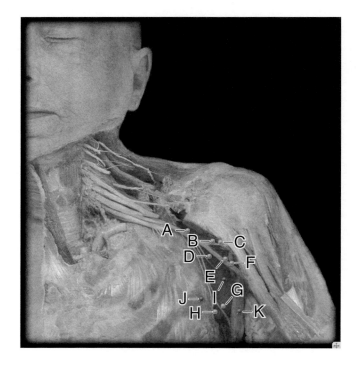

- *Mouse-over the green pins on the screen to find the information necessary to identify the following structures:*

A. _____

B. _____

C. _____

D. _____

E. _____

F. _____

G. _____

H. _____

I. _____

J. _____

K. _____

L. _____

M. _____

N. _____

O. _____

P. _____

Q. _____

Non-nervous System Structure (blue pin)

R. _____

C H E C K   P O I N T :

Pheripheral Nerves, Brachial Plexus, cont'd

2. Which nerve is stimulated when you strike your "funny bone"?
3. Name a nerve that innervates the pectoralis major and minor muscles.
4. Which brachial plexus roots contribute to the long thoracic nerve?

- *Click **LAYER 6** in the **LAYER CONTROLS** window, and you will see the following image:*

- *Mouse-over the green pins on the screen to find the information necessary to identify the following structures:*

A. _____

B. _____

C. _____

D. _____

E. _____

F. _____

G. _____

H. _____

Non-nervous System Structures (blue pins)

I. _____

J. _____

K. _____

Self Test

Take this opportunity to quiz yourself by taking the **Self Test.** See page 15 for a reminder on how to access the self-test for this section.

EXERCISE 5.27:

Nervous System—Peripheral Nerves, Upper Limb, Anterior View

SELECT TOPIC	SELECT VIEW	
Peripheral Nerves ▶	Upper Limb: Anterior	GO

- *Click* **LAYER 1** *in the* **LAYER CONTROLS** *window, and you will see the following image:*

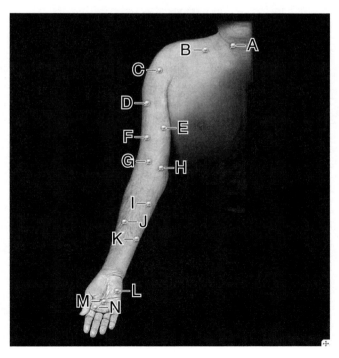

- *Mouse-over the pins on the screen to find the information necessary to identify the following structures:*

A. _____

B. _____

C. _____

D. _____

E. _____

F. _____

G. _____

H. _____

I. _____

J. _____

K. _____

L. _____

M. _____

N. _____

CHECK POINT:

Pheripheral Nerves, Upper Limb, Anterior View

1. Which nerve supplies the cutaneous innervation to the skin over the medial arm and forearm?
2. Which nerve supplies the cutaneous innervation to the skin over finger V?
3. Name two nerves responsible for the cutaneous innervation of the shoulder.

- *Click* **LAYER 3** *in the* **LAYER CONTROLS** *window, and you will see the following image:*

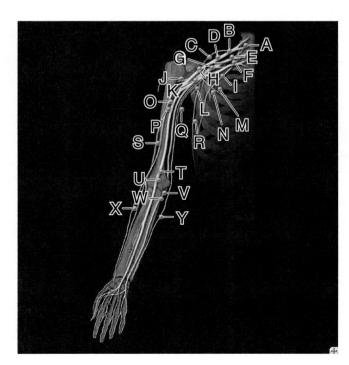

• *Mouse-over the pins on the screen to find the information necessary to identify the following structures:*

A. _____

B. _____

C. _____

D. _____

E. _____

F. _____

G. _____

H. _____

I. _____

J. _____

K. _____

L. _____

M. _____

N. _____

O. _____

P. _____

Q. _____

R. _____

S. _____

T. _____

U. _____

V. _____

W. _____

X. _____

Y. _____

C H E C K   P O I N T:

Pheripheral Nerves, Upper Limb, Anterior View, cont'd

4. Name the nerve that innervates the joints of the hand.
5. Which division of the brachial plexus gives rise to nerves that distribute to the anterior aspect of the forearm?
6. Name the nerve responsible for the innervation of the glenohumeral joint.

Self Test
Take this opportunity to quiz yourself by taking the **SELF TEST.** See page 15 for a reminder on how to access the self-test for this section.

EXERCISE 5.28:

Nervous System—Peripheral Nerves, Upper Limb, Posterior View

| SELECT TOPIC **Peripheral Nerves** ▶ | SELECT VIEW **Upper Limb: Posterior** | GO |

• *Click **LAYER 1** in the **LAYER CONTROLS** window, and you will see the following image:*

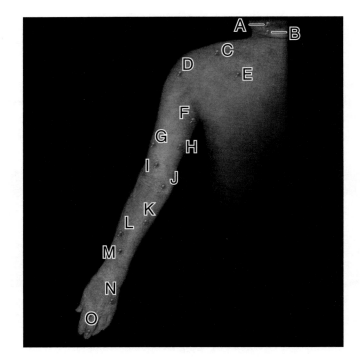

• *Mouse-over the pins on the screen to find the information necessary to identify the following structures:*

A. _____

B. _____

C. _____

D. _____

E. _____

F. _____

G. _____

H. _____

I. _____

J. _____

K. _____

L. _____

M. _____

N. _____

O. _____

- *Click* **LAYER 3** *in the* **LAYER CONTROLS** *window, and you will see the following image:*

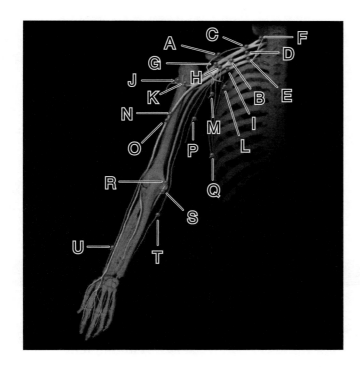

- *Mouse-over the pins on the screen to find the information necessary to identify the following structures:*

A. _____

B. _____

C. _____

D. _____

E. _____

F. _____

G. _____

H. _____

I. _____

J. _____

K. _____

L. _____

M. _____

N. _____

O. _____

P. _____

Q. _____

R. _____

S. _____

T. _____

U. _____

CHECK POINT:

Peripheral Nerves, Upper Limb, Posterior View

1. Name a nerve that innervates the serratus anterior muscle.
2. Name a nerve that innervates the supraspinatus and infraspinatus muscles.
3. Name a nerve that innervates the subscapularis and teres major muscles.

Self Test

Take this opportunity to quiz yourself by taking the **SELF TEST.** See page 15 for a reminder on how to access the self-test for this section.

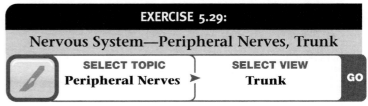

EXERCISE 5.29:

Nervous System—Peripheral Nerves, Trunk

SELECT TOPIC	SELECT VIEW	
Peripheral Nerves ❯	**Trunk**	GO

- *Click* **LAYER 1** *in the* **LAYER CONTROLS** *window, and you will see the following image:*

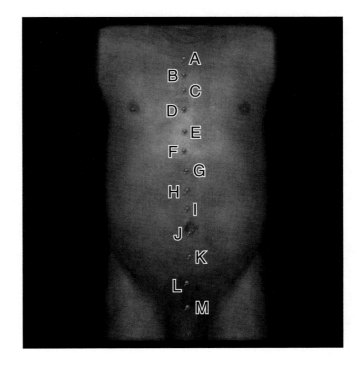

- *Mouse-over the pins on the screen to find the information necessary to identify the following structures:*

A. _____

B. _____

C. _____

D. _____

E. _____

F. _____

G. _____

H. _____

I. _____

J. _____

K. _____

L. _____

M. _____

• Click **LAYER 3** in the **LAYER CONTROLS** window, and you will see the following image:

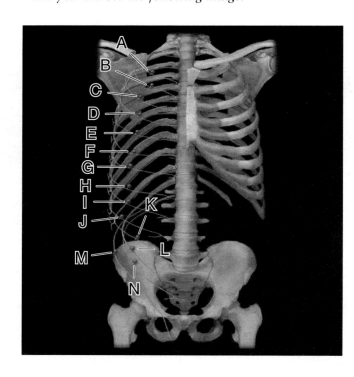

• Mouse-over the pins on the screen to find the information necessary to identify the following structures:

A. _____

B. _____

C. _____

D. _____

E. _____

F. _____

G. _____

H. _____

I. _____

J. _____

K. _____

L. _____

M. _____

N. _____

CHECK POINT:

Peripheral Nerves, Trunk

1. Name five nerves that innervate the skin of the abdomen.
2. Which group of nerves are located within the intercostal space between the innermost and the internal intercostal muscles?
3. Name five of these nerves.

Self Test

Take this opportunity to quiz yourself by taking the **SELF TEST.** See page 15 for a reminder on how to access the self-test for this section.

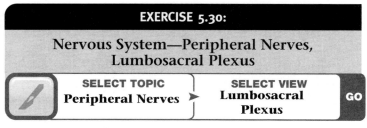

• Click **LAYER 1** in the **LAYER CONTROLS** window, and you will see the following image:

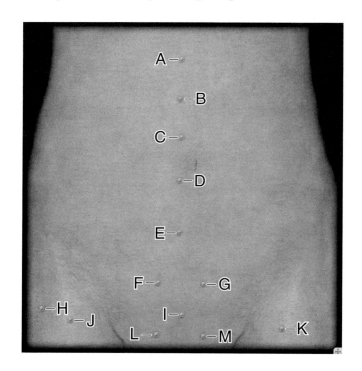

• Mouse-over the pins on the screen to find the information necessary to identify the following structures:

A. _____

B. _____

C. _____

D. _____

E. _____

F. _____

G. _____

H. _____

I. _____

J. _____

K. _____

L. _____

M. _____

- *Click **LAYER 2** in the **LAYER CONTROLS** window, and you will see the following image:*

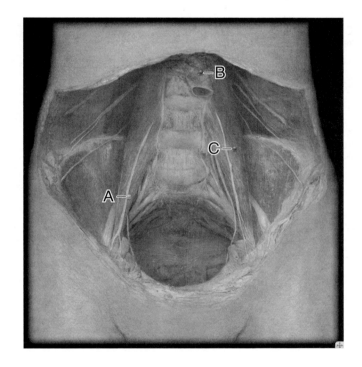

- *Mouse-over the green pin on the screen to find the information necessary to identify the following structure:*

A. _____

Non-nervous System Structures (blue pins)

B. _____

C. _____

- *Click **LAYER 3** in the **LAYER CONTROLS** window, and you will see the following image:*

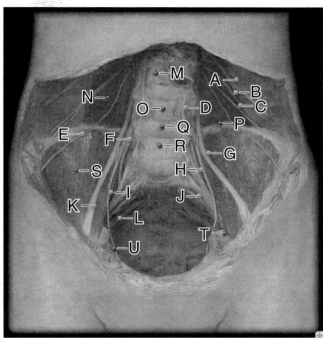

- *Mouse-over the green pins on the screen to find the information necessary to identify the following structures:*

A. _____

B. _____

C. _____

D. _____

E. _____

F. _____

G. _____

H. _____

I. _____

J. _____

K. _____

L. _____

Non-nervous System Structures (blue pins)

M. _____

N. _____

O. _____

P. _____

Q. _____

R. _____

S. _____

T. _____

U. _____

CHECK POINT:

Peripheral Nerves, Lumbosacral Plexus

1. Nerves derived from what structure distribute to the lower anterior abdominal wall, spermatic cord, thigh, medial leg and foot, and the sacral plexus?
2. Name a nerve that supplies sensory innervation to the lateral thigh.
3. Name the seven terminal branches of the lumbar plexus.

Self Test

Take this opportunity to quiz yourself by taking the **SELF TEST.** See page 15 for a reminder on how to access the self-test for this section.

EXERCISE 5.31:

Nervous System—Peripheral Nerves, Lower Limb, Anterior View

SELECT TOPIC	SELECT VIEW	
Peripheral Nerves ➤	**Lower Limb: Anterior**	**GO**

- Click **LAYER 1** in the **LAYER CONTROLS** window, and you will see the following image:

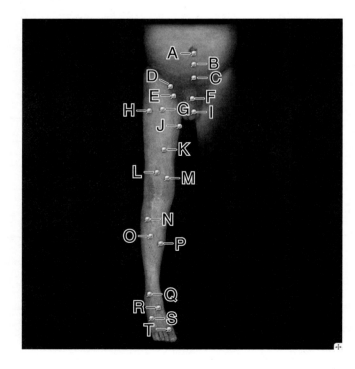

- *Mouse-over the pins on the screen to find the information necessary to identify the following structures:*

A. _____

B. _____

C. _____

D. _____

E. _____

F. _____

G. _____

H. _____

I. _____

J. _____

K. _____

L. _____

M. _____

N. _____

O. _____

P. _____

Q. _____

R. _____

S. _____

T. _____

CHECK POINT:

Peripheral Nerves, Lower Limb, Anterior View

1. Name the nerve providing cutaneous innervation to the lateral thigh.
2. Name the nerve providing cutaneous innervation to the anterior thigh, medial leg, and the medial margin of the foot.
3. Name the nerve providing cutaneous innervation to the medial leg and medial margin of the foot.

- *Click* **LAYER 2** *in the* **LAYER CONTROLS** *window, and you will see the following image:*

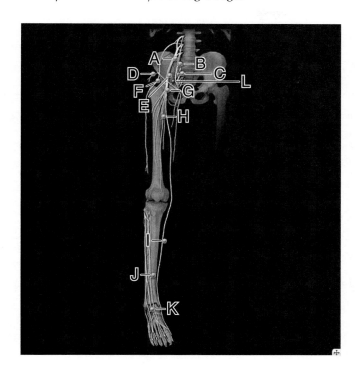

- *Mouse-over the green pins on the screen to find the information necessary to identify the following structures:*

A. _____

B. _____

C. _____

D. _____

E. _____

F. _____

G. _____

H. _____

I. _____

J. _____

K. _____

Non-nervous System Structure (blue pin)

L. _____

CHECK POINT:

Peripheral Nerves, Lower Limb, Anterior View, cont'd

4. Name the branch of the femoral nerve that provides sensory innervation to the medial leg and the medial margin of the foot.
5. Name the nerve providing motor innervation to the gluteus medius, gluteus minimus, and the tensor fascia latae muscles.
6. Name the nerve providing motor and sensory innervation to the muscles of the lateral leg.

Self Test

Take this opportunity to quiz yourself by taking the **SELF TEST.** See page 15 for a reminder on how to access the self-test for this section.

EXERCISE 5.32:
Nervous System—Peripheral Nerves, Lower Limb, Posterior View

SELECT TOPIC	SELECT VIEW	
Peripheral Nerves ▶	**Lower Limb: Posterior**	GO

- *Click* **LAYER 1** *in the* **LAYER CONTROLS** *window, and you will see the following image:*

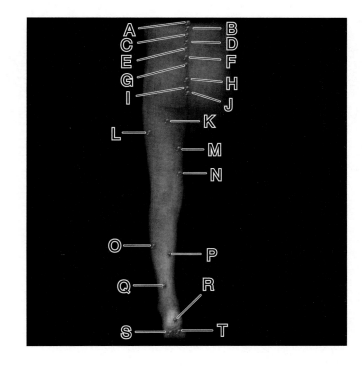

- *Mouse-over the pins on the screen to find the information necessary to identify the following structures:*

A. _____

B. _____

C. _____

D. _____

E. _____

F. _____

G. _____

H. _____

I. _____

J. _____

K. _____

L. _____

M. _____

N. _____

O. _____

P. _____

Q. _____

R. _____

S. _____

T. _____

CHECK POINT:

Peripheral Nerves, Lower Limb, Posterior View

1. Name the nerve providing cutaneous innervation to the posterior distal and lateral proximal leg and the lateral margin of the foot.
2. Name the two nerves providing cutaneous innervation to the sole of the foot.
3. Name the nerve providing innervation to the skin of the medial thigh.

- *Click* **LAYER 2** *in the* **LAYER CONTROLS** *window, and you will see the following image:*

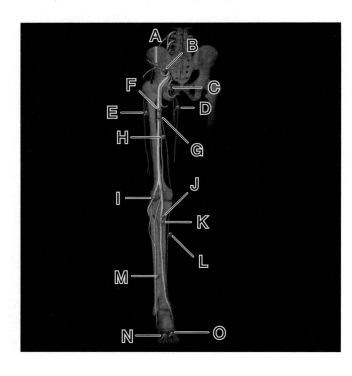

- *Mouse-over the pins on the screen to find the information necessary to identify the following structures:*

A. _____

B. _____

C. _____

D. _____

E. _____

F. _____

G. _____

H. _____

I. _____

J. _____

K. _____

L. _____

M. _____

N. _____

O. _____

CHECK POINT:

Peripheral Nerves, Lower Limb, Posterior View, cont'd

4. Name the nerve supplying motor innervation to the anterior and lateral leg muscles and the muscles of the dorsum of the foot.
5. Name the nerve supplying motor innervation to the adductor muscles of the medial thigh.
6. Name the nerve supplying sensory innervation to the medial leg and the medial margin of the foot.

Self Test

Take this opportunity to quiz yourself by taking the **SELF TEST.** See page 15 for a reminder on how to access the self-test for this section.

Autonomic Nervous System

EXERCISE 5.33:
Nervous System—Sympathetic (ANS), Overview

	SELECT TOPIC	SELECT VIEW	
🔪	**Sympathetic (ANS)** ▶	**Overview**	GO

- *Click* **LAYER 1** *in the* **LAYER CONTROLS** *window, and you will see the following image:*

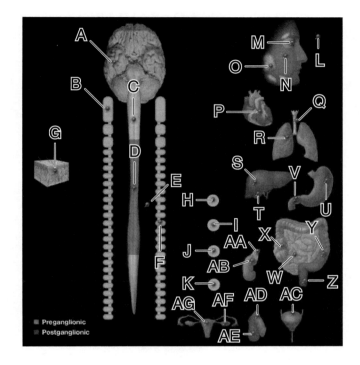

- *Mouse-over the pins on the screen to find the information necessary to identify the following structures:*

A. _____

B. _____

C. _____

D. _____

E. _____

F. _____

G. _____

H. _____

I. _____

J. _____

K. _____

L. _____

M. _____

N. _____

O. _____

P. _____

Q. _____

R. _____

S. _____

T. _____

U. _____

V. _____

W. _____

X. _____

Y. _____

Z. _____

AA. _____

AB. _____

AC. _____

AD. _____

AE. _____

AF. _____

AG. _____

CHECK POINT:

Sympathetic (ANS), Overview

1. Name the structure also known as the sympathetic chain.
2. Name the sympathetic ganglion that distributes postganglionic neuronal processes to the stomach, duodenum, and spleen.
3. Describe the sympathetic postganglionic neuron and pathway of the adrenal medulla.

EXERCISE 5.34:

Nervous System—Sympathetic (ANS), Thoracic Region

SELECT TOPIC	SELECT VIEW	
Sympathetic (ANS)	**Thoracic Region**	GO

• *Click* **LAYER 4** *in the* **LAYER CONTROLS** *window, and you will see the following image:*

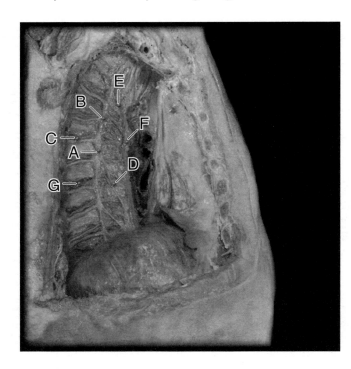

• *Mouse-over the green pins on the screen to find the information necessary to identify the following structures:*

A. _____

B. _____

C. _____

D. _____

Non-nervous System Structures (blue pins)

E. _____

F. _____

G. _____

CHECK POINT:

Sympathetic (ANS), Thoracic Region

1. Name the location of the postganglionic sympathetic neuronal cell bodies (somas).
2. Which nerves are responsible for motor innervation of all thoracic muscles?
3. Name the three nerves that contain preganglionic sympathetic nerve fibers that enter the abdomen to synapse in prevertebral ganglia.

Self Test

Take this opportunity to quiz yourself by taking the **SELF TEST.** See page 15 for a reminder on how to access the self-test for this section.

EXERCISE 5.35:

Nervous System—Parasympathetic (ANS), Overview

SELECT TOPIC	SELECT VIEW	
Parasympathetic (ANS)	**Overview**	GO

• *Click* **LAYER 1** *in the* **LAYER CONTROLS** *window, and you will see the following image:*

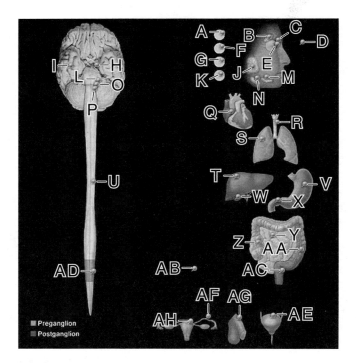

• *Mouse-over the pins on the screen to find the information necessary to identify the following structures:*

A. _____

B. _____

C. _____

D. _____

E. _____

F. _____

G. _____

H. _____

I. _____

J. _____

K. _____

L. _____

M. _____

N. _____

O. _____

P. _____

Q. _____

R. _____

S. _____

T. _____

U. _____

V. _____

W. _____

X. _____

Y. _____

Z. _____

AA. _____

AB. _____

AC. _____

AD. _____

AE. _____

AF. _____

AG. _____

AH. _____

CHECK POINT:

Parasympathetic (ANS), Overview

1. Where do the parasympathetic fibers terminate in the heart?
2. Name the nerves that distribute to the pelvic viscera via blood vessels.
3. The postganglionic fibers from which ganglion are distributed along branches of the auriculotemporal nerve to the parotid gland?

EXERCISE 5.36:

Nervous System—Parasympathetic (ANS), Inferior Brain

SELECT TOPIC	SELECT VIEW	
Parasympathetic (ANS)	Inferior Brain	GO

- *Click* **LAYER 1** *in the* **LAYER CONTROLS** *window, and you will see the following image:*

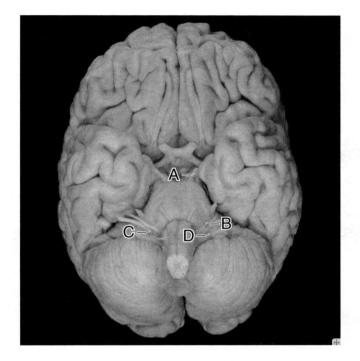

- *Mouse-over the pins on the screen to find the information necessary to identify the following structures:*

A. _____

B. _____

C. _____

D. _____

CHECK POINT:

Parasympathetic (ANS), Inferior Brain

1. Which cranial nerve has its postganglionic parasympathetic cell bodies (somas) located in the ciliary ganglia?

IN REVIEW

What Have I Learned?

The following questions cover the material that you have just learned, the peripheral nerves and autonomic nervous system. Apply what you have learned to answer these questions on a separate piece of paper.

1. What structure is also known as the sympathetic chain ganglia?

2. What nerve carries parasympathetic impulses to the smooth muscle of the thoracic and abdominal viscera?

3. Which nerve supplies all motor innervation to the diaphragm?

4. Name the sympathetic ganglion that distributes postganglionic neuronal processes to the kidneys and gonads.

5. Which division of the ANS has its terminal ganglia near or in the wall of the innervated organ?

6. Which division of the ANS has its terminal ganglia near the spinal cord?

7. Name a network of nerves with a branch that passes through the spermatic cord.

The Senses

EXERCISE 5.37:

Nervous System—Taste, Inferior Brain

SELECT TOPIC
Taste

SELECT VIEW
Inferior Brain

GO

- *Click **LAYER 1** in the **LAYER CONTROLS** window, and you will see the following image:*

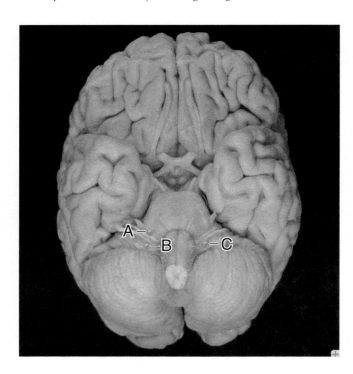

- *Mouse-over the pins on the screen to find the information necessary to identify the following structures:*

A. _____

B. _____

C. _____

EXERCISE 5.38:

Nervous System—Taste, Tongue, Superior View

SELECT TOPIC
Taste

SELECT VIEW
Tongue: Superior

GO

- *Click **LAYER 2** in the **LAYER CONTROLS** window, and you will see the following image:*

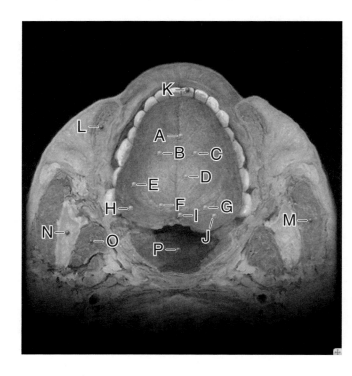

- *Mouse-over the green pins on the screen to find the information necessary to identify the following structures:*

A. _____

B. _____

C. _____

D. _____

E. _____

F. _____

G. _____

H. _____

I. _____

J. _____

Non-nervous System Structures (blue pins)

K. _____

L. _____

M. _____

N. _____

O. _____

P. _____

CHECK POINT:

Taste

1. Name the shallow median longitudinal groove on the anterior part of the tongue.
2. Name the nerve that innervates the posterior tongue, pharynx, and middle ear.
3. Name the tonsil associated with the tongue.

EXERCISE 5.39:

Nervous System—Taste, Histology— Vallate Papilla

| SELECT TOPIC Taste—Vallate Papilla | ▶ | SELECT VIEW | GO |

- *Click the **TURN TAGS ON** button, and you will see the following image:*

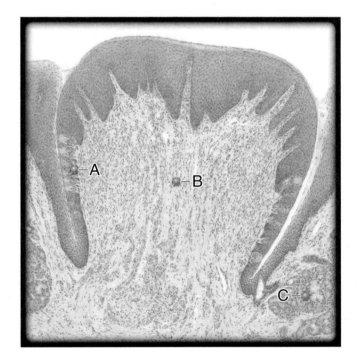

- *Mouse-over the pins on the screen to find the information necessary to identify the following structures:*

A. _____

B. _____

C. _____

CHECK POINT:

Taste, Histology—Vallate Papilla

1. Name the nerve that receives special sensory information from the taste buds.
2. In what structure are the taste buds located?
3. Name the five primary taste sensations.

EXERCISE 5.40:

Nervous System—Taste, Histology—Taste Bud

SELECT TOPIC	SELECT VIEW	
Taste—Taste Bud	>	GO

- *Click the **TURN TAGS ON** button, and you will see the following image:*

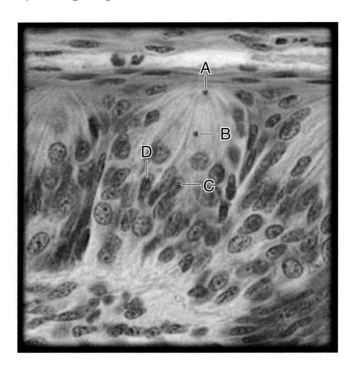

- *Mouse-over the pins on the screen to find the information necessary to identify the following structures:*

A. _____

B. _____

C. _____

D. _____

CHECK POINT:

Taste, Histology—Taste Bud

1. Name the structure that serves as a receptor surface for taste molecules.
2. What type of cells are taste cells?
3. What is the site of taste reception?

EXERCISE 5.41:

Nervous System—Smell, Inferior Brain

SELECT TOPIC	SELECT VIEW	
Smell	> Inferior Brain	GO

- *Click **LAYER 1** in the **LAYER CONTROLS** window, and you will see the following image:*

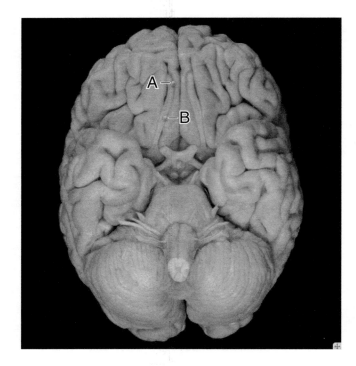

- *Mouse-over the pins on the screen to find the information necessary to identify the following structures:*

A. _____

B. _____

CHECK POINT:

Smell, Inferior Brain

1. Name the type of neurons found in the mucous membranes of the nasal cavities.
2. What bony structure do these neurons pass through to synapse with the olfactory nerve?
3. With what specific portion of the olfactory neuron do these neurons synapse?

EXERCISE 5.42:

Nervous System—Smell, Nasal Cavity, Lateral View

SELECT TOPIC	SELECT VIEW	
Smell	▶ **Nasal Cavity: Lateral**	**GO**

- *Click **LAYER 3** in the **LAYER CONTROLS** window, and you will see the following image:*

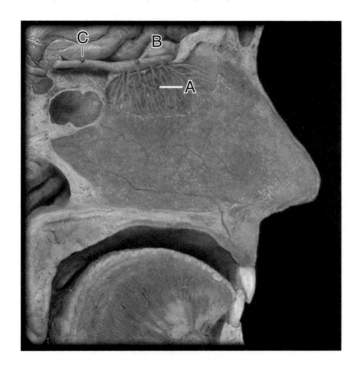

- *Mouse-over the pins on the screen to find the information necessary to identify the following structures:*

A. _____

B. _____

C. _____

CHECK POINT:

Smell, Nasal Cavity, Lateral View

1. Where in the brain does the olfactory nerve connect?
2. What structure detects odorants in the nasal cavities?
3. Name the expanded anterior end of the olfactory tract.

EXERCISE 5.43:

Nervous System—Smell, Histology—Olfactory Mucosa

SELECT TOPIC	SELECT VIEW	
Smell—Olfactory Mucosa	▶	**GO**

- *Click the **TURN TAGS ON** button, and you will see the following image:*

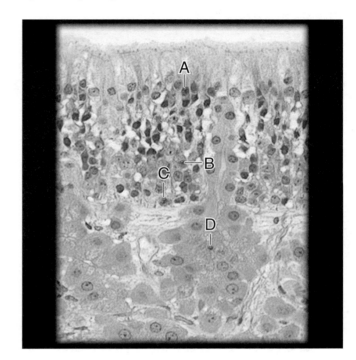

- *Mouse-over the pins on the screen to find the information necessary to identify the following structures:*

A. _____

B. _____

C. _____

D. _____

CHECK POINT:

Smell, Histology—Olfactory Mucosa

1. Name the structures that produce mucus in the nasal cavities.
2. What is the function of the mucus?
3. Name the cells that produce new olfactory receptors.

EXERCISE 5.44:

Nervous System—Hearing/Balance, Inferior Brain

SELECT TOPIC	SELECT VIEW	
Hearing/Balance ▶	Inferior Brain	GO

- *Click the flashing* **GO** *button.*

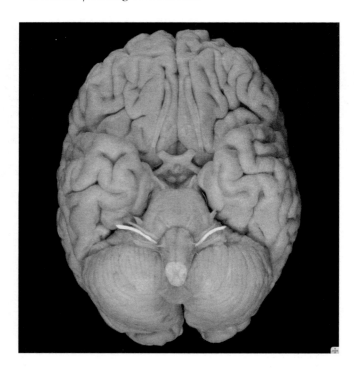

- *There are no pins with this image, so identify the nerve highlighted in yellow:*

_____

CHECK POINT:

Hearing/Balance, Inferior Brain

1. What two special senses are provided by the vestibulocochlear nerve?
2. What bony structure allows internal passage of the vestibulocochlear nerve?
3. What is the CNS connection for the vestibulocochlear nerve?

EXERCISE 5.45:

Nervous System—Hearing/Balance, Ear, Anterior View

SELECT TOPIC	SELECT VIEW	
Hearing/Balance ▶	Ear: Anterior	GO

- *Click* **LAYER 1** *in the* **LAYER CONTROLS** *window, and you will see the following image:*

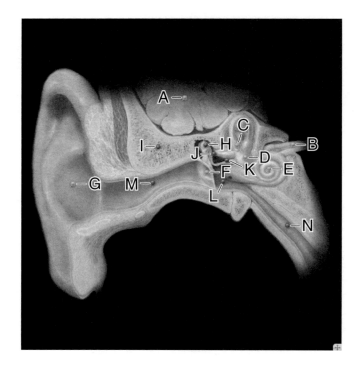

- *Mouse-over the green pins on the screen to find the information necessary to identify the following structures:*

A. _____

B. _____

C. _____

D. _____

E. _____

F. _____

G. _____

Non-nervous System Structures (blue pins)

H. _____

I. _____

J. _____

K. _____

L. _____

M. _____

N. _____

CHECK POINT:

Hearing/Balance, Ear, Anterior View

1. Name the organ of hearing.
2. Name the structure of the ear that senses linear movement.
3. Name the organ of equilibrium.

EXERCISE 5.46:

Nervous System—Hearing/Balance, Histology—Cochlea

 SELECT TOPIC Hearing/Balance—Cochlea ▸ **SELECT VIEW** High Magnification **GO**

- Click the **TURN TAGS ON** button, and you will see the following image:

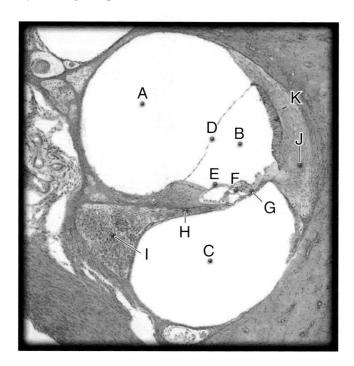

- Mouse-over the pins on the screen to find the information necessary to identify the following structures:

A. _____

B. _____

C. _____

D. _____

E. _____

F. _____

G. _____

H. _____

I. _____

J. _____

K. _____

CHECK POINT:

Hearing/Balance, Histology—Cochlea

1. Name the three fluid-filled chambers of the cochlea, from superior to inferior.
2. What specific structure of the cochlea is the site of conversion of sound vibrations into electrochemical signals?
3. Name the structure that supports this structure.

EXERCISE 5.47:

Nervous System—Hearing/Balance, Histology—Spiral Organ

 SELECT TOPIC Hearing/Balance—Spiral Organ ▸ **SELECT VIEW** Low Magnification **GO**

- Click the **TURN TAGS ON** button, and you will see the following image:

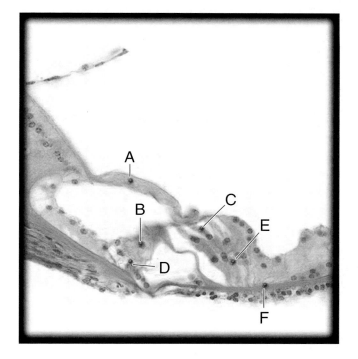

- Mouse-over the pins on the screen to find the information necessary to identify the following structures:

A. _____

B. _____

C. _____

D. _____

E. _____

F. _____

Animation: Hearing

SELECT ANIMATION	
Hearing	PLAY

- *After viewing the animation, answer these questions:*

1. What structure do sound waves strike and cause to vibrate?

2. The vibrations are transferred to the three bones of the middle ear. From lateral to medial, what are those bones?

3. What structure transfers this vibration to the oval window?

4. The vibrations are then transferred to a fluid-filled chamber of the inner ear. What is that fluid, and what is the name of that chamber?

5. What is the difference in location for the detection of high-pitched and low-pitched sounds?

CHECK POINT:

Hearing/Balance, Histology—Spiral Organ

1. Name the structure with the function of transmitting vibrations to embedded stereocilia of sensory cells.
2. What specific structures function as the receptors for hearing?
3. Describe the function of stereocilia.

EXERCISE 5.48:

Nervous System—Vision, Inferior Brain

SELECT TOPIC	SELECT VIEW	
Vision	**Inferior Brain**	GO

- *Click **LAYER 1** in the **LAYER CONTROLS** window, and you will see the following image:*

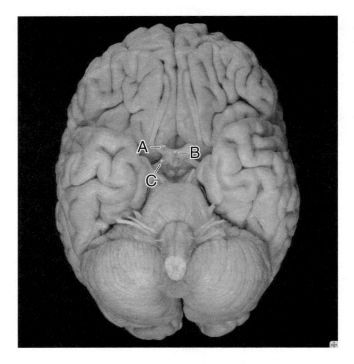

- *Mouse-over the pins on the screen to find the information necessary to identify the following structures:*

A. _____

B. _____

C. _____

CHECK POINT:

Vision, Inferior Brain

1. Name the cranial nerve of vision.
2. What is this nerve's origin?
3. Where does this nerve terminate?

EXERCISE 5.49:

Nervous System—Vision, Orbit, Lateral View

	SELECT TOPIC	SELECT VIEW	
🔪	**Vision**	▶ **Orbit: Lateral**	**GO**

- *Click **LAYER 3** in the **LAYER CONTROLS** window, and you will see the following image:*

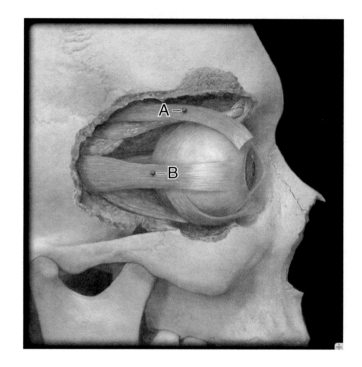

- *Mouse-over the blue pins on the screen to find the information necessary to identify the following non-nervous system structures:*

A. _____

B. _____

- *Click **LAYER 4** in the **LAYER CONTROLS** window, and you will see the following image:*

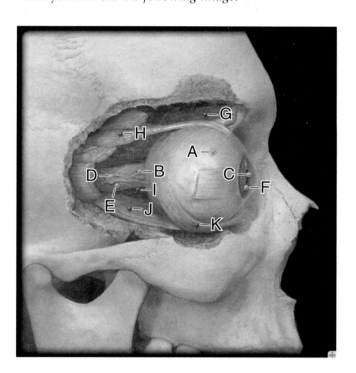

- *Mouse-over the green pins on the screen to find the information necessary to identify the following structures:*

A. _____

B. _____

C. _____

D. _____

E. _____

F. _____

Non-nervous System Structures (blue pins)

G. _____

H. _____

I. _____

J. _____

K. _____

CHECK POINT:

Vision, Orbit, Lateral View

1. Name and describe the colored part of the eye.
2. What is its function?
3. Describe the short ciliary nerves.

- *Click* **LAYER 5** *in the* **LAYER CONTROLS** *window, and you will see the following image:*

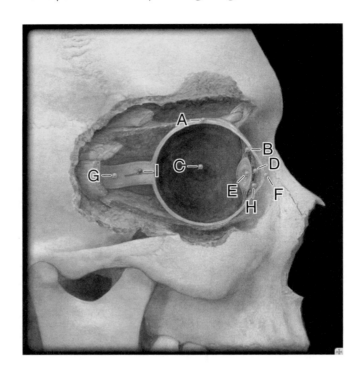

- *Mouse-over the green pins on the screen to find the information necessary to identify the following structures:*

A. _____

B. _____

C. _____

D. _____

E. _____

F. _____

G. _____

H. _____

Non-nervous System Structure (blue pin)

I. _____

C H E C K   P O I N T :

Vision, Orbit, Lateral View, cont'd

4. Name the nerve responsible for the special sensation of vision.
5. Name the outer layer of the posterior five-sixths of the eye.
6. Name the outer layer of the anterior one-sixth of the eye.

EXERCISE 5.50:

Nervous System—Vision, Eye, Lateral View

SELECT TOPIC	SELECT VIEW	
Vision ▶	**Eye: Lateral**	**GO**

- *Click* **LAYER 1** *in the* **LAYER CONTROLS** *window, and you will see the following image:*

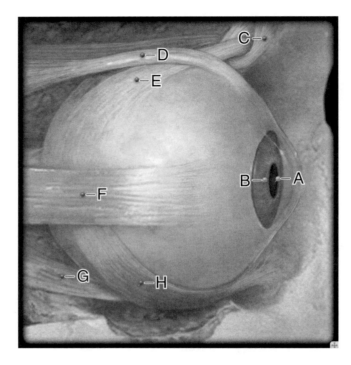

- *Mouse-over the green pins on the screen to find the information necessary to identify the following structures:*

A. _____

B. _____

Non-nervous System Structures (blue pins)

C. _____

D. _____

E. _____

F. _____

G. _____

H. _____

CHECK POINT:

Vision, Eye, Lateral View

1. Name the muscle responsible for elevation of the upper eyelid.
2. What cranial nerve supplies the innervation for this muscle?
3. Name the muscle responsible for abduction of the eyeball.
4. Name the muscle responsible for depression and medial rotation of the eyeball.

- *Click **LAYER 2** in the **LAYER CONTROLS** window, and you will see the following image:*

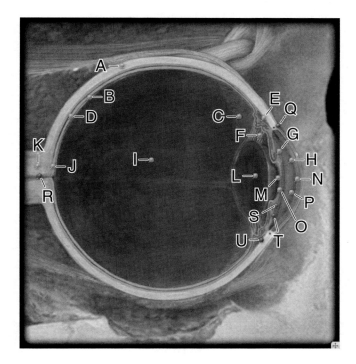

- *Mouse-over the pins on the screen to find the information necessary to identify the following structures:*

A. _____

B. _____

C. _____

D. _____

E. _____

F. _____

G. _____

H. _____

I. _____

J. _____

K. _____

L. _____

M. _____

N. _____

O. _____

P. _____

Non-nervous System Structures (blue pins)

Q. _____

R. _____

S. _____

T. _____

U. _____

CHECK POINT:

Vision, Eye, Lateral View, cont'd

5. Name the structure referred to as the "white" of the eye.
6. Name the artery that, when blocked, may cause blindness.
7. Name the internal structure of the eye that focuses incoming light.

IN REVIEW

What Have I Learned?

The following questions cover the material that you have just learned, vision, the orbit, and the eye. Apply what you have learned to answer the following questions on a separate piece of paper.

1. Name and describe the colored part of the eye.

2. What is its function?

3. Name the opening in the middle of this structure.

4. Describe the short ciliary nerves.

Continued

Continued

5. Describe the ciliary ganglion.

6. Name the muscle that allows looking upward and outward.

7. Name the muscle that allows looking downward and medially.

8. Name the nerve responsible for the special sensation of vision.

9. Name the outer layer of the posterior five-sixths of the eye.

10. Describe the tissue of this structure.

11. Name the outer layer of the anterior one-sixth of the eye.

12. Describe the tissue of this structure.

EXERCISE 5.51:

Nervous System—Vision, Histology—Retina

SELECT TOPIC   SELECT VIEW
Vision—Retina   GO

- *Click the* **TURN TAGS ON** *button, and you will see the following image:*

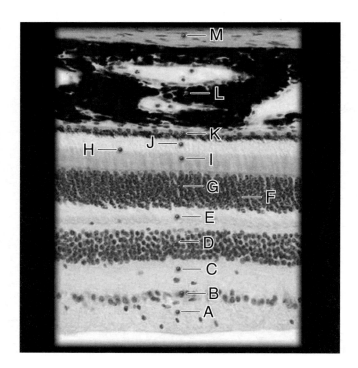

- *Mouse-over the pins on the screen to find the information necessary to identify the following structures:*

A. _____

B. _____

C. _____

D. _____

E. _____

F. _____

G. _____

H. _____

I. _____

J. _____

K. _____

L. _____

M. _____

Animation: Vision

SELECT ANIMATION
Vision   PLAY

- *After viewing the animation, answer these questions:*

1. What are the three structures involved in vision?

2. The optic nerve runs between what two structures?

3. Describe the two parts of the retina.

4. What neurotransmitter is released in the dark by the photoreceptors?

5. Describe the light pathway from the eye to the brain.

6. Name the structure where the optic nerves converge.

7. Images are perceived in which lobe of the brain?

CHECK POINT:

Vision, Histology—Retina

1. Which retinal cells detect color vision?
2. Name the highly vascular layer of the retina containing melanin.
3. What is the function of the ganglionic layer of the retina?

IN REVIEW

What Have I Learned?

The following questions cover the material that you have just learned, the senses. Apply what you have learned to answer these questions on a separate piece of paper.

1. Name the cranial nerves responsible for taste. Name the structures innervated by each of these nerves.

2. Name the ovoid collections of lymphoid tissue covered with mucous membrane on the posterior aspect of the tongue.

3. Name the three subdivisions of the pharynx.

4. Name the three types of cells found in taste buds.

5. What is another name for olfactory neurons?

6. Name the cranial nerve responsible for hearing and balance.

7. Name the organ of hearing.

8. Name the two organs of balance.

9. Name the bony canal of the external ear.

10. What is the anatomical name for the eardrum?

11. Name the three smallest bones in the body from lateral to medial.

12. What is the function of these three bones?

13. What is the name for the passage between the tympanic cavity and nasopharynx?

14. What is the function of this passage?

15. The auditory ossicles articulate with what structure of the inner ear?

16. What are the functions of the three fluid-filled chambers of the cochlea?

17. Where does the lens of the eye focus the incoming light?

18. What term refers to changes of the lens' shape when focusing?

19. Name the structure that prevents light scatter in the eye.

EXERCISE 5.52:

Multipolar Neuron - Golgi Stain, Histology

SELECT TOPIC	SELECT VIEW	
Multipolar Neuron ▶		**GO**

- *Click the* **TURN TAGS ON** *button, and you will see the following image:*

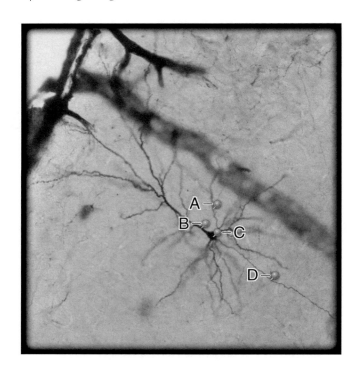

- *Mouse-over the pins on the screen to find the information necessary to identify the following structures:*

A. _____

B. _____

C. _____

D. _____

CHECK POINT:

Multipolar Neuron-Golgi Stain, Histology

1. Name the two types of cell processes.
2. Which of the cell processes convey efferent nerve impulses?
3. Which of the cell processes convey afferent nerve impulses?

EXERCISE 5.53:

Axon Hillock, Histology

SELECT TOPIC	SELECT VIEW	
Axon Hillock ▶		**GO**

- *Click the* **TURN TAGS ON** *button, and you will see the following image:*

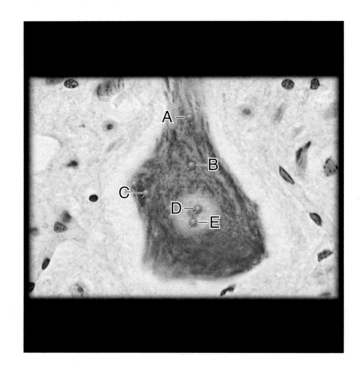

- *Mouse-over the pins on the screen to find the information necessary to identify the following structures:*

A. _____

B. _____

C. _____

D. _____

E. _____

CHECK POINT:

Axon Hillock, Histology

1. Describe the axon hillock.
2. Name the largest membrane-bound organelle of the neuron.
3. What is located within this organelle?

EXERCISE 5.54:
Unmyelinated Axon, Histology

| SELECT TOPIC **Unmyelinated Axon** | SELECT VIEW | GO |

- *Click the* **TURN TAGS ON** *button, and you will see the following image:*

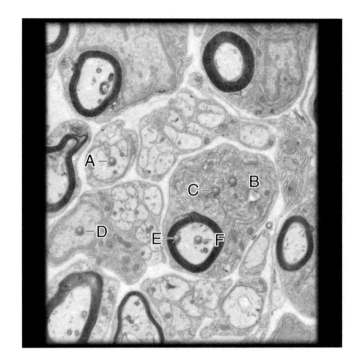

- *Mouse-over the pins on the screen to find the information necessary to identify the following structures:*

A. _____

B. _____

C. _____

D. _____

E. _____

F. _____

C H E C K   P O I N T :

Unmyelinated Axon, Histology

1. Name the structure that surrounds the myelinated neuron.
2. Name the myelinating cells for the CNS.
3. Name the myelinating cells for the PNS.

EXERCISE 5.55:
Synapse, Histology

| SELECT TOPIC **Synapses** | SELECT VIEW | GO |

- *Click the* **TURN TAGS ON** *button, and you will see the following image:*

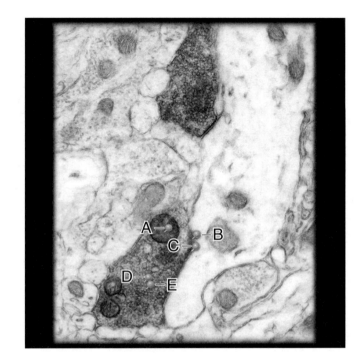

- *Mouse-over the pins on the screen to find the information necessary to identify the following structures:*

A. _____

B. _____

C. _____

D. _____

E. _____

C H E C K   P O I N T :

Synapse, Histology

1. Describe the synapse.
2. Where is it located?
3. What is its function?

EXERCISE 5.56:

Neuromuscular Junction, Histology

SELECT TOPIC	SELECT VIEW	
Neuromuscular Junction		GO

- *Click the **TURN TAGS ON** button, and you will see the following image:*

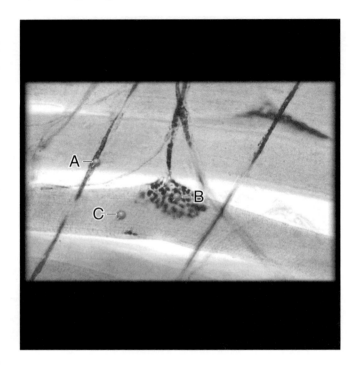

- *Mouse-over the pins on the screen to find the information necessary to identify the following structures:*

A. _____

B. _____

C. _____

CHECK POINT:

Neuromuscular Junction, Histology

1. Define neuromuscular junction.
2. Describe the neuromuscular junction.
3. What neurotransmitter is released at the neuromuscular junction?

EXERCISE 5.57:

Schwann Cell, Histology

SELECT TOPIC	SELECT VIEW	
Schwann Cell		GO

- *Click the **TURN TAGS ON** button, and you will see the following image:*

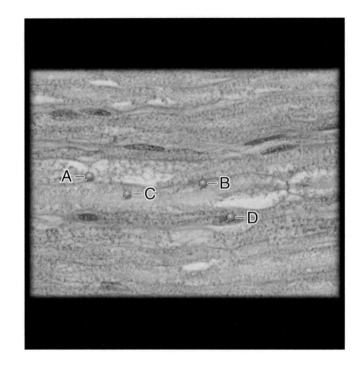

- *Mouse-over the pins on the screen to find the information necessary to identify the following structures:*

A. _____

B. _____

C. _____

D. _____

CHECK POINT:

Schwann Cell, Histology

1. What is the name for the cleft between the internodes of the myelin sheath?
2. What is saltatory conduction?
3. What is lacking between myelin layers?

HEADS UP!

Take this time to review the following animations we covered earlier in this chapter:

- *Action potential generation*
- *Action potential propagation*
- *Chemical synapse*

Self Test

Take this opportunity to quiz yourself by taking the **SELF TEST.** See page 15 for a reminder on how to access the self-test for this section.

IN REVIEW

What Have I Learned?

The following questions cover the material that you have just learned, histology. Apply what you have learned to answer these questions on a separate piece of paper.

1. Name the two types of cell processes.

2. Which of the cell processes convey efferent nerve impulses?

3. Which of the cell processes convey afferent nerve impulses?

4. What are Nissl bodies?

5. What structures of the soma support a neuron's structure and function?

6. Which cell processes are not myelinated?

7. Which specific cell processes are usually myelinated?

8. Which cell processes arise from the axon hillock?

9. Which cell processes are usually without branches near the cell of origin?

10. Which cell processes receive information from other neurons?

11. Which cell processes convey information to other neurons or effectors?

12. Name a type of neuron composed of a soma, multiple dendrites, and a single axon.

13. Describe the axon hillock.

14. Name the largest membrane-bound organelle of the neuron.

15. What is located within this organelle?

16. Name the distinct round, dark-staining organelle in the nucleus of the neuron. What is its function?

17. Name the structure that surrounds the myelinated neuron.

18. Name the myelinating cells for the CNS.

19. Name the myelinating cells for the PNS.

20. At what speeds do nerve impulses travel along unmyelinated neurons?

21. At what speeds do nerve impulses travel along comparable myelinated neurons?

22. Where do unmyelinated axons rest?

23. Describe the myelin sheath.

24. What is cytosol?

25. Describe the synapse.

26. Where is it located?

27. What is its function?

Continued

Continued

28. Describe the synaptic cleft.

29. Where is it located?

30. What is its function?

31. Describe synaptic vesicles.

32. Where are they located?

33. What is their function?

34. Define a neurotransmitter. Give two examples of neurotransmitters.

35. What is the function of the mitochondria?

36. Describe the presynaptic terminals.

37. Where are they located?

38. What is their function?

39. Define neuromuscular junction.

40. Describe the neuromuscular junction.

41. What neurotransmitter is released at the neuromuscular junction?

42. Describe the axon of a motor neuron.

43. What is its function?

44. What is the name for the cleft between the internodes of the myelin sheath?

45. What is salutatory conduction?

46. What is lacking between myelin layers?

EXERCISE 5.58:

Coloring Exercise

Look up the features of the ventral surface of the brain and the cranial nerves. Then, using colored pens or pencils, color in the figure.

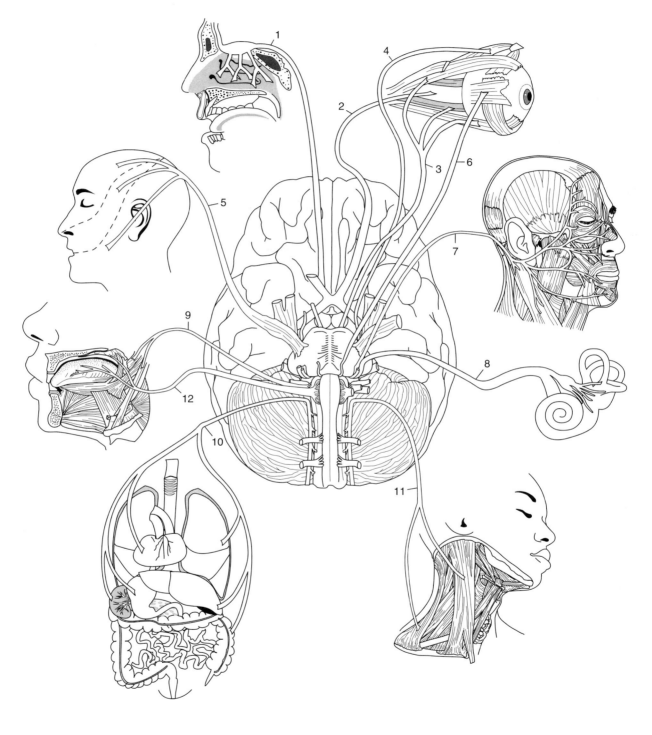

❏ 1. _____

❏ 2. _____

❏ 3. _____

❏ 4. _____

❏ 5. _____

❏ 6. _____

❏ 7. _____

❏ 8. _____

❏ 9. _____

❏ 10. _____

❏ 11. _____

❏ 12. _____

EXERCISE 5.59:

Coloring Exercise

Look up the parts of the eye and accessory structures. Then color them with colored pencils.

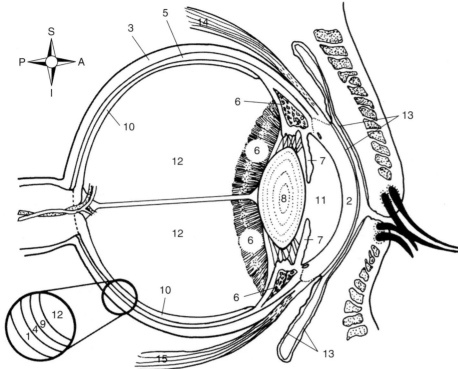

❑ 1. _____

❑ 2. _____

❑ 3. _____

❑ 4. _____

❑ 5. _____

❑ 6. _____

❑ 7. _____

❑ 8. _____

❑ 9. _____

❑ 10. _____

❑ 11. _____

❑ 12. _____

❑ 13. _____

❑ 14. _____

❑ 15. _____

❑ 16. _____

❑ 17. _____

❑ 18. _____

❑ 19. _____

CHAPTER 6

The Cardiovascular System

Overview: Cardiovascular System

Your heart and circulatory system accomplish amazing feats! Think about these accomplishments:

- Your heart, roughly the size of your two hands nested together, beats over 100,000 times per day. This equals approximately 36 million times per year.

- Your body contains an average volume of 5 liters of blood, which circulates through your heart, and thus your body, once every minute.

- It has been estimated that your heart pumps 1,900 gallons (7,200 liters) of blood through approximately 60,000–100,000 miles (96,560–63,730 kilometers) of blood vessels every day. That's 2½ to 3 times around the earth at the equator—every day.

- If you live 70 years, your heart will beat some 2.5 billion times. In this 70-year lifetime, it will pump roughly 1 million barrels of blood, enough to fill more than three super tankers.

Pretty impressive accomplishments, wouldn't you say? Let's begin our exploration of the wonders of this amazing cardiovascular system.

We'll begin with an overview of the entire system, and then we'll look at the structures in detail region by region, beginning with the heart. Let's get started.

- *From the* **Home screen,** *click the drop-down box on the* **Select system** *menu.*

- *From the systems listed, click on* **Cardiovascular,** *and you will see the image to the right:*

- *This is the opening screen for the* **Cardiovascular System.**

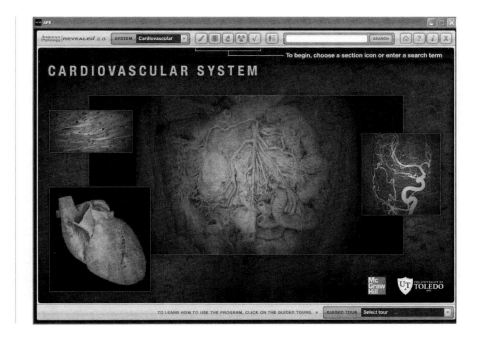

Animation: Cardiovascular System

SELECT ANIMATION
Cardiovascular System Overview   PLAY

- *After viewing the animation, answer these questions:*

1. The cardiovascular consists of what three structures?

2. The heart distributes and receives blood through which structures?

3. Define an artery.

4. Gases and nutrients are exchanged with the tissues through which blood vessels?

5. Define a vein.

Self Test

Take this opportunity to quiz yourself by taking the **SELF TEST.** See page 15 for a reminder on how to access the self-test for this section.

Animation: Blood Flow Through Heart

SELECT ANIMATION
Blood Flow Through Heart
PLAY

- *After viewing the animation, answer these questions:*

1. Name the four chambers of the heart. What are the differences between them?

2. Name the two major vessels that deliver oxygen-poor blood to the heart.

3. What areas of the body do these two vessels drain?

4. What is the route of venous blood from the heart?

5. Trace the route of blood as it flows through the heart. Be sure to list *all* structures involved.

6. Name the only arteries to carry oxygen-poor blood.

7. Name the only veins to carry oxygen-rich blood.

Self Test

Take this opportunity to quiz yourself by taking the **SELF TEST.** See page 15 for a reminder on how to access the self-test for this section.

> **H E A D S   U P !**
>
> *The hearts in the following exercise have been removed from the cadaver, and thus will not have all of the surrounding structures visible for reference. The subsequent section of the blood vessels of the thorax will cover the anatomy surrounding the heart in detail.*

The Heart

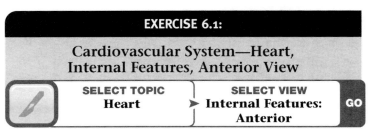

EXERCISE 6.1:

Cardiovascular System—Heart, Internal Features, Anterior View

SELECT TOPIC
Heart

SELECT VIEW
► **Internal Features: Anterior**

GO

- *Click* **LAYER 1** *in the* **LAYER CONTROLS** *window, and you will see the following image:*

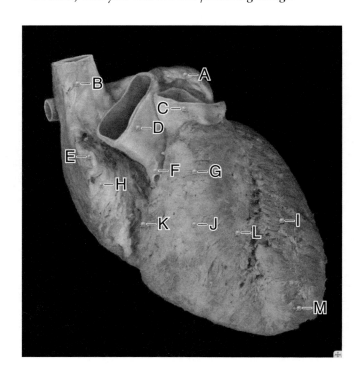

HEADS UP! ─────────

When you click on the pins for blood vessels of the cardiovascular system, they will be highlighted red for arteries and blue for veins. The heart muscle will also be highlighted in red when the corresponding pins are clicked. Nerves will be highlighted in yellow.

• *Mouse-over the pins on the screen to find the information necessary to identify the following structures:*

A. _____

B. _____

C. _____

D. _____

E. _____

F. _____

G. _____

H. _____

I. _____

J. _____

K. _____

L. _____

M. _____

CHECK POINT: ─────────

Heart, Internal Features, Anterior View

1. Name the major blood vessel between the heart and the aortic arch.
2. What are its two branches?
3. Name the branches of the right coronary artery.

• *Click **LAYER 2** in the **LAYER CONTROLS** window, and you will see the following image:*

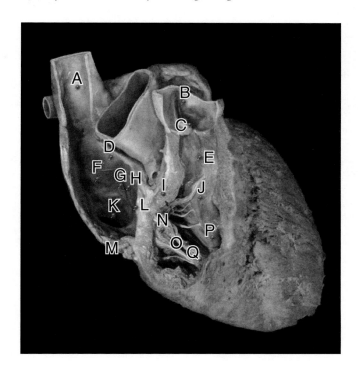

• *Mouse-over the pins on the screen to find the information necessary to identify the following structures:*

A. _____

B. _____

C. _____

D. _____

E. _____

F. _____

G. _____

H. _____

I. _____

J. _____

K. _____

L. _____

M. _____

N. _____

O. _____

P. _____

Q. _____

CHECK POINT:

Heart, Internal Features, Anterior View, cont'd

4. What are the two terms for the valve between the right atrium and right ventricle?
5. Name the structure that prevents reflux of blood into the right ventricle.
6. Name the two branches of the pulmonary trunk.

- *Click **LAYER 3** in the **LAYER CONTROLS** window, and you will see the following image:*

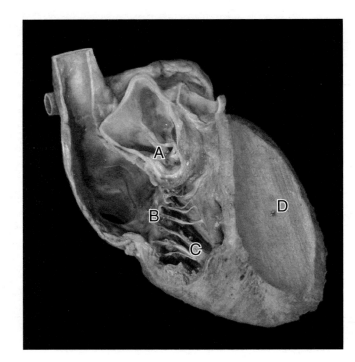

- *Mouse-over the pins on the screen to find the information necessary to identify the following structures:*

A. _____

B. _____

C. _____

D. _____

CHECK POINT:

Heart, Internal Features, Anterior View, cont'd

7. Name the structure that prevents reflux of blood into the left ventricle.
8. Name the muscle layer of the heart wall.
9. Where is that muscle layer thinner and thicker?

- *Click **LAYER 4** in the **LAYER CONTROLS** window, and you will see the following image:*

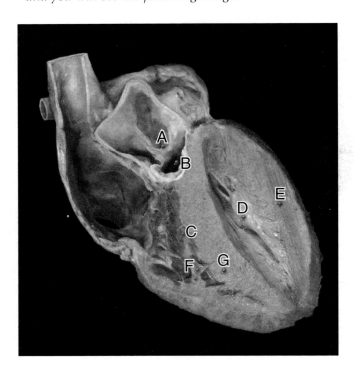

- *Mouse-over the pins on the screen to find the information necessary to identify the following structures:*

A. _____

B. _____

C. _____

D. _____

E. _____

F. _____

G. _____

CHECK POINT:

Heart, Internal Features, Anterior View, cont'd

10. What heart chamber is responsible for pumping oxygen-rich blood to the body (except the lungs)?
11. What heart chamber is responsible for pumping oxygen-poor blood to the lungs?
12. Name the irregular, muscular elevations on the internal surface of both ventricles.

- *Click* **LAYER 5** *in the* **LAYER CONTROLS** *window, and you will see the following image:*

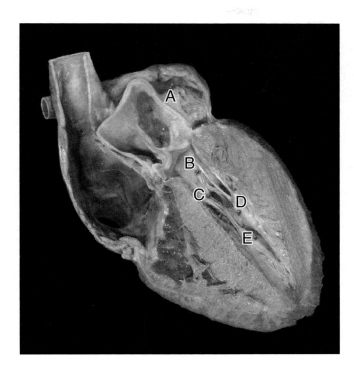

- *Mouse-over the pins on the screen to find the information necessary to identify the following structures:*

A. _____

B. _____

C. _____

D. _____

E. _____

CHECK POINT:

Heart, Internal Features, Anterior View, cont'd

13. Name the fibrous strands that attach the free edges of the atrioventricular valve cusps to the papillary muscles.
14. Name the heart chamber that receives oxygen-rich blood from the lungs.
15. Name the blood vessels that bring this blood from the lungs to the heart. How many are there?

- *Click* **LAYER 6** *in the* **LAYER CONTROLS** *window, and you will see the following image:*

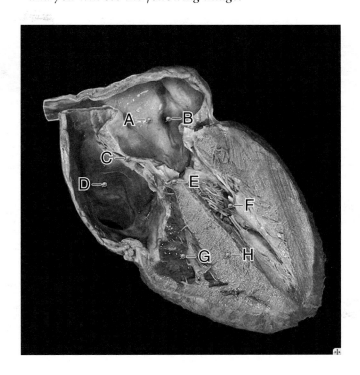

- *Mouse-over the pins on the screen to find the information necessary to identify the following structures:*

A. _____

B. _____

C. _____

D. _____

E. _____

F. _____

G. _____

H. _____

CHECK POINT:

Heart, Internal Features, Anterior View, cont'd

16. Name the structure that separates the right and left atria.
17. Name the structure that separates the right and left ventricles.
18. What is the name for the superior membranous part of the structure that separates the right and left ventricles?

EXERCISE 6.2:

Cardiovascular System—Heart, Vasculature, Anterior View

SELECT TOPIC		SELECT VIEW	
Heart	▶	**Vasculature: Anterior**	GO

• *Click **LAYER 1** in the **LAYER CONTROLS** window, and you will see the following image:*

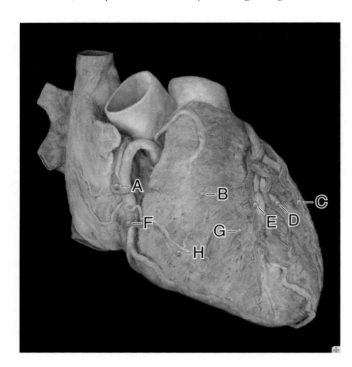

• *Mouse-over the pins on the screen to find the information necessary to identify the following structures:*

A. _____

B. _____

C. _____

D. _____

E. _____

F. _____

G. _____

H. _____

CHECK POINT:

Heart, Vasculature, Anterior View

1. Name the vein that ascends the anterior interventricular sulcus.
2. Name the vein that ascends across the anterior surface of the left ventricle.
3. Name the numerous veins that course across the anterior surface of the right ventricle.

• *Click **LAYER 2** in the **LAYER CONTROLS** window, and you will see the following image:*

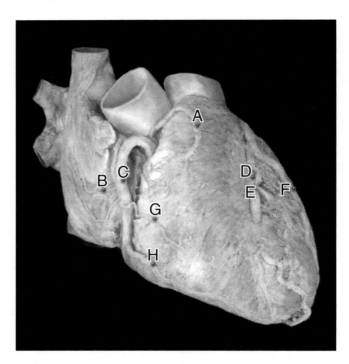

• *Mouse-over the pins on the screen to find the information necessary to identify the following structures:*

A. _____

B. _____

C. _____

D. _____

E. _____

F. _____

G. _____

H. _____

CHECK POINT:

Heart, Vasculature, Anterior View, cont'd

4. Name the artery that lies in the right coronary sulcus.
5. What are the branches of this artery?
6. Name the artery that descends along the margin of the left ventricle.

• *Click* **LAYER 3** *in the* **LAYER CONTROLS** *window, and you will see the following image:*

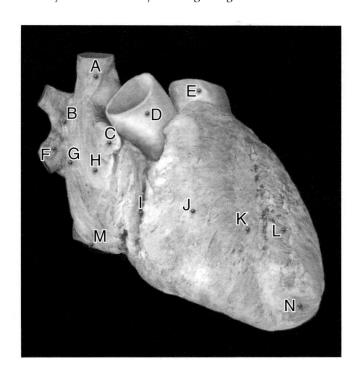

• *Mouse-over the pins on the screen to find the information necessary to identify the following structures:*

A. _____

B. _____

C. _____

D. _____

E. _____

F. _____

G. _____

H. _____

I. _____

J. _____

K. _____

L. _____

M. _____

N. _____

CHECK POINT:

Heart, Vasculature, Anterior View, cont'd

7. Name the major vein that drains everything inferior to the diaphragm.
8. Name the tributaries of the superior vena cava.
9. Name the inferolateral point of the heart.

EXERCISE 6.3:

Cardiovascular System—Heart, Vasculature, Posterior View

SELECT TOPIC	SELECT VIEW	
Heart	**Vasculature: Posterior**	GO

• *Click* **LAYER 1** *in the* **LAYER CONTROLS** *window, and you will see the following image:*

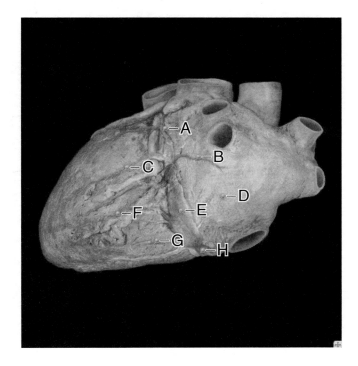

• *Mouse-over the pins on the screen to find the information necessary to identify the following structures:*

A. _____

B. _____

C. _____

D. _____

E. _____

F. _____

G. _____

H. _____

CHECK POINT:

Heart, Vasculature, Posterior View

1. Name the structure that receives venous blood from the heart.
2. What are the three major tributaries of this structure?
3. List the structures drained by the great cardiac vein.

• *Click* **LAYER 2** *in the* **LAYER CONTROLS** *window, and you will see the following image:*

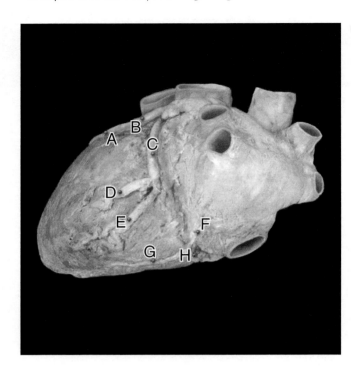

• *Mouse-over the pins on the screen to find the information necessary to identify the following structures:*

A. _____

B. _____

C. _____

D. _____

E. _____

F. _____

G. _____

H. _____

CHECK POINT:

Heart, Vasculature, Posterior View, cont'd

4. What is the term that refers to the end-to-end union of blood vessels?
5. Name the arteries located in the posterior interventricular sulcus.
6. Name the artery that descends in the anterior interventricular sulcus.

• *Click* **LAYER 3** *in the* **LAYER CONTROLS** *window, and you will see the following image:*

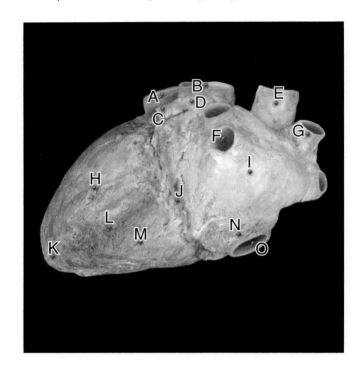

• *Mouse-over the pins on the screen to find the information necessary to identify the following structures:*

A. _____

B. _____

C. _____

D. _____

E. _____

F. _____

G. _____

H. _____

I. _____

J. _____

K. _____

L. _____

M. _____

N. _____

O. _____

Animation: Conducting System of Heart

SELECT ANIMATION
Conducting System of Heart
PLAY

- *After viewing the animation, answer these questions:*

1. Where do action potentials associated with heartbeat regulation originate?

2. From their origin, the action potentials travel across the wall of the atrium to the . . .

3. Name the structure in the interventricular septum where the action potentials pass upon leaving the atria.

4. The structure in question 3 then divides into two _____.

5. After passing the apex of the ventricles, the action potentials pass along what structure to the ventricle walls?

6. What causes the ventricular muscle cells to contract in unison, providing a strong contraction?

Self Test
Take this opportunity to quiz yourself by taking the **SELF TEST.** See page 15 for a reminder on how to access the self-test for this section.

Animation: Cardiac Cycle

SELECT ANIMATION
Cardiac Cycle
PLAY

- *After viewing the animation, answer these questions:*

1. A single cardiac cycle is made up of . . .

2. What is the term that refers to the relaxation of a heart chamber?

3. What physically occurs in a heart chamber during relaxation?

4. The term that refers to the contraction of a heart chamber is . . .

5. Atrial depolarization is represented by which wave on an electrocardiogram?

6. What is initiated by atrial depolarization?

7. This phenomenon is represented by what portion of the ECG?

8. Which section of the ECG represents ventricular depolarization?

9. This same section of the ECG in question 8 masks what portion of the cardiac cycle?

10. What is S1? How is it often described?

11. What is represented by the S-T segment of the ECG?

12. The T-wave on the ECG represents . . .

13. What happens in the ventricles at this time?

14. What is S2? How is it often described?

15. What causes the beginning of the next cardiac cycle?

Self Test

Take this opportunity to quiz yourself by taking the **SELF TEST.** See page 15 for a reminder on how to access the self-test for this section.

IN   REVIEW

What Have I Learned?

The following questions cover the material that you have just learned, the heart. Apply what you have learned to answer these questions on a separate piece of paper.

1. Name the body areas drained by the superior vena cava.

2. Name the small pouchlike extensions of the atria.

3. Name the vessel that conveys oxygen-poor blood from the right ventricle.

4. Name the three blood vessels that drain into the right atrium.

5. What external structure of the heart marks the position of the junction between the atria and the ventricles?

6. What external structure of the heart marks the position of the interventricular septum?

7. What is the term that refers to the blunt tip of the left ventricle?

8. Name the conical elevations of myocardium in the ventricular walls. What is their function?

9. Name the remnant of the fetal blood shunt from the right atrium to the left atrium.

10. Name the muscle layer of the heart wall.

The Thorax

EXERCISE 6.4:

Cardiovascular System—Thorax, Arteries, Anterior View

SELECT TOPIC	SELECT VIEW	
Thorax	▶ **Arteries: Anterior**	**GO**

- *Click* **LAYER 1** *in the* **LAYER CONTROLS** *window, and you will see the following image:*

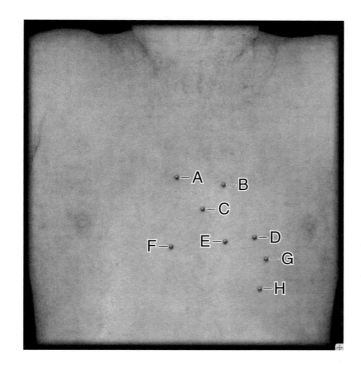

- *Mouse-over the blue pins on the screen to find the information necessary to identify the following non-cardiovascular system structures:*

A. _____

B. _____

C. _____

D. _____

E. _____

F. _____

G. _____

H. _____

CHECK POINT:

Thorax, Arteries, Anterior View

1. Where do heart valve sounds resonate?
2. What structure forms the right border of the heart?
3. What structure forms the apex of the heart?

HEADS UP!
Click on all of the blue pins in each exercise for **Anatomy & Physiology | Revealed**®. *The "What Have I Learned?" questions may cover the information for the blue pins, as well as the green ones.*

- *Click* **LAYER 2** *in the* **LAYER CONTROLS** *window, and you will see the following image:*

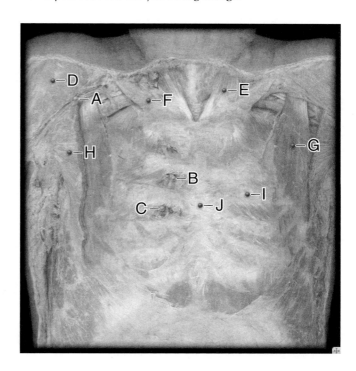

- *Mouse-over the green pins on the screen to find the information necessary to identify the following structures:*

A. _____

B. _____

C. _____

Non-cardiovascular System Structures (blue pins)

D. _____

E. _____

F. _____

G. _____

H. _____

I. _____

J. _____

Thorax, Arteries, Anterior View, cont'd

4. What artery is also called the internal mammary artery?
5. What vein is also called the internal mammary vein?

- *Click* **LAYER 3** *in the* **LAYER CONTROLS** *window, and you will see the following image:*

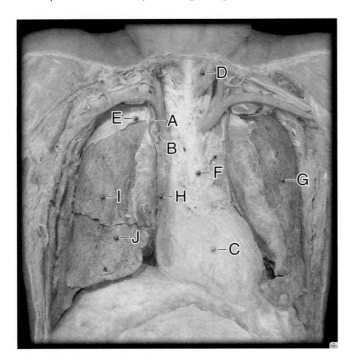

- *Mouse-over the green pins on the screen to find the information necessary to identify the following structures:*

A. _____

B. _____

C. _____

Non-cardiovascular System Structures (blue pins)

D. _____

E. _____

F. _____

G. _____

H. _____

I. _____

J. _____

Thorax, Arteries, Anterior View, cont'd

6. Name the artery that descends adjacent to the sternum within the thoracic cavity.

- *Click* **LAYER 4** *in the* **LAYER CONTROLS** *window, and you will see the following image:*

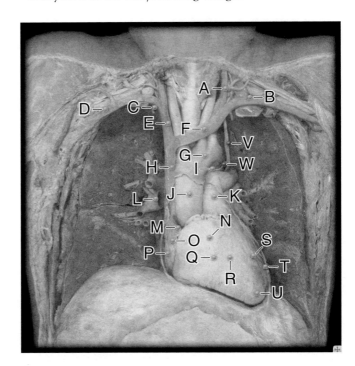

- *Mouse-over the green pins on the screen to find the information necessary to identify the following structures:*

A. _____

B. _____

C. _____

D. _____

E. _____

F. _____

G. _____

H. _____

I. _____

J. _____

K. _____

L. _____

M. _____

N. _____

O. _____

P. _____

Q. _____

R. _____

S. _____

T. _____

U. _____

Non-cardiovascular System Structures (blue pins)

V. _____

W. _____

E. _____

F. _____

G. _____

H. _____

I. _____

Non-cardiovascular System Structures (blue pins)

J. _____

K. _____

CHECK POINT:

Thorax, Arteries, Anterior View, cont'd

7. Name the large artery that originates at the aortic valve and ascends 5 mm within the pericardium.
8. Name the arched continuation of the artery in question 7.
9. Name the three branches of the artery in question 8.

- Click **LAYER 5** in the **LAYER CONTROLS** window, and you will see the following image:

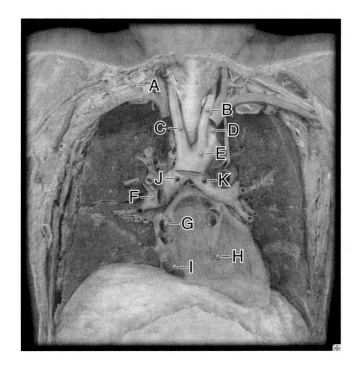

- Mouse-over the green pins on the screen to find the information necessary to identify the following structures:

A. _____

B. _____

C. _____

D. _____

CHECK POINT:

Thorax, Arteries, Anterior View, cont'd

10. Name the blood vessel also known as the innominate artery. What two vessels are found at its terminus?
11. Name the second artery to branch off of the aortic arch. What are its terminal branches?
12. Name the third artery to branch off of the aortic arch. Name the vessel that is the continuation of this artery beginning at the lateral border of rib 1.

- Click **LAYER 6** in the **LAYER CONTROLS** window, and you will see the following image:

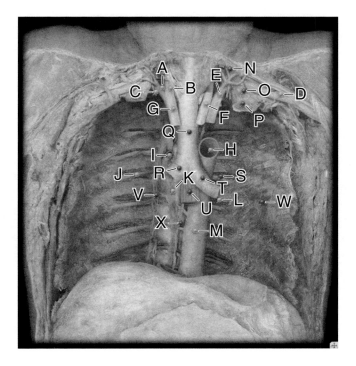

- Mouse-over the green pins on the screen to find the information necessary to identify the following structures:

A. _____

B. _____

C. _____

D. _____

E. _____

F. _____

G. _____

H. _____

I. _____

J. _____

K. _____

L. _____

M. _____

Non-cardiovascular System Structures (blue pins)

N. _____

O. _____

P. _____

Q. _____

R. _____

S. _____

T. _____

U. _____

V. _____

W. _____

X. _____

C H E C K P O I N T :

Thorax, Arteries, Anterior View, cont'd

13. Name the major artery that passes through the axilla.
14. Name the artery that ascends into the neck from the brachiocephalic trunk.
15. What are the two terminal branches of the artery in question 14?

EXERCISE 6.5:

Cardiovascular System—Thorax, Veins, Anterior View

SELECT TOPIC	SELECT VIEW	
Thorax	**Veins: Anterior**	**GO**

- *Click* **LAYER 2** *in the* **LAYER CONTROLS** *window, and you will see the following image:*

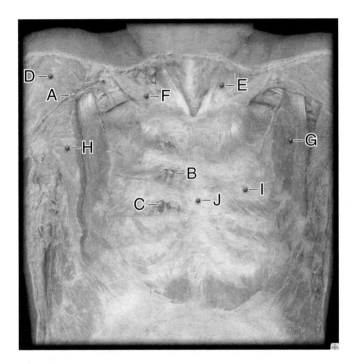

┌─ **H E A D S U P !** ──────────────
│ *We are skipping* **LAYER 1** *for this exercise because it is a repeat of Exercise 6.4. It is advisable to review this layer before you complete this exercise.*
└─────────────────────────────────

- *Mouse-over the green pins on the screen to find the information necessary to identify the following structures:*

A. _____

B. _____

C. _____

Non-cardiovascular System Structures (blue pins)

D. _____

E. _____

F. _____

G. _____

H. _____

I. _____

J. _____

CHECK POINT:

Thorax, Veins, Anterior View

1. Name the vein that ascends from the dorsum of the hand to the anterolateral forearm and arm and into the deltopectoral triangle.
2. Name the vein also known as the internal mammary vein.

- *Click* **LAYER 3** *in the* **LAYER CONTROLS** *window, and you will see the following image:*

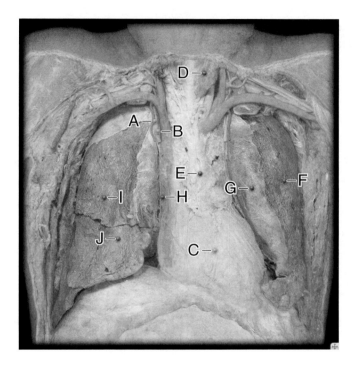

- *Mouse-over the green pins on the screen to find the information necessary to identify the following structures:*

A. _____

B. _____

C. _____

Non-cardiovascular System Structures (blue pins)

D. _____

E. _____

F. _____

G. _____

H. _____

I. _____

J. _____

CHECK POINT:

Thorax, Veins, Anterior View, cont'd

3. Name the areas drained by the internal thoracic vein.

- *Click* **LAYER 4** *in the* **LAYER CONTROLS** *window, and you will see the following image:*

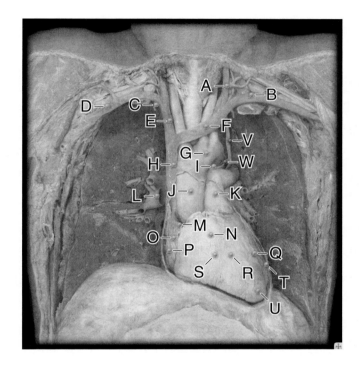

- *Mouse-over the green pins on the screen to find the information necessary to identify the following structures:*

A. _____

B. _____

C. _____

D. _____

E. _____

F. _____

G. _____

H. _____

I. _____

J. _____

K. _____

L. _____

M. _____

N. _____

O. _____

P. _____

Q. _____

R. _____

S. _____

T. _____

U. _____

Non-cardiovascular System Structures (blue pins)

V. _____

W. _____

CHECK POINT:

Thorax, Veins, Anterior View, cont'd

4. Name the areas drained by the superior vena cava. Where does it terminate?
5. What two vessels unite to form the superior vena cava?
6. Name the continuation of the sigmoid dural sinus of the cranial cavity that drains the brain, face, and neck.

• *Click* **LAYER 5** *in the* **LAYER CONTROLS** *window, and you will see the following image:*

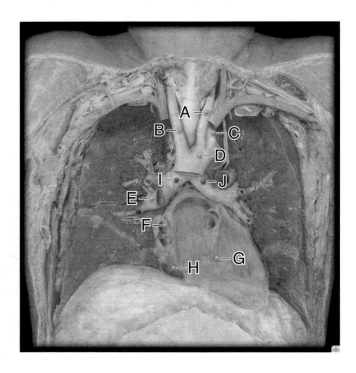

• *Mouse-over the green pins on the screen to find the information necessary to identify the following structures:*

A. _____

B. _____

C. _____

D. _____

E. _____

F. _____

G. _____

H. _____

Non-cardiovascular System Structures (blue pins)

I. _____

J. _____

CHECK POINT:

Thorax, Veins, Anterior View, cont'd

7. Name the vessels that carry oxygen-rich blood to the heart.
8. Where do they terminate?
9. Name the drainage of the inferior vena cava. Where does it terminate?

• *Click* **LAYER 6** *in the* **LAYER CONTROLS** *window, and you will see the following image:*

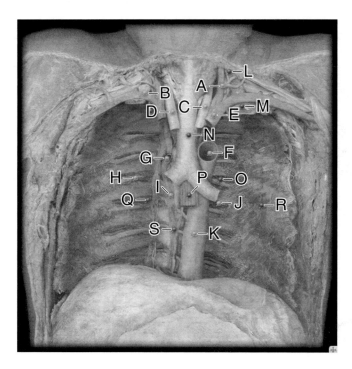

• *Mouse-over the green pins on the screen to find the information necessary to identify the following structures:*

A. _____

B. _____

C. _____

D. _____

E. _____

F. _____

G. _____

H. _____

I. _____

J. _____

K. _____

Non-cardiovascular System Structures (blue pins)

L. _____

M. _____

N. _____

O. _____

P. _____

Q. _____

R. _____

S. _____

CHECK POINT:

Thorax, Veins, Anterior View, cont'd

10. Name the venous system that forms a collateral pathway between the superior and inferior vena cavae.

Animation: Pulmonary and Systemic Circulation

SELECT ANIMATION	
Pulmonary and Systemic Circulation	**PLAY**

• *After viewing the animation, answer these questions:*

1. Name the two divisions of the cardiovascular system.

2. What are the destinations of these two circuits?

---

3. In the systemic circulation, where does gas exchange occur?

4. In the pulmonary circulation, where does gas exchange occur?

5. Name the blood vessels that carry oxygen-rich blood to the heart. How many are there? Where do they terminate?

Self Test

Take this opportunity to quiz yourself by taking the **SELF TEST.** See page 15 for a reminder on how to access the self-test for this section.

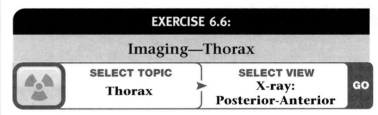

EXERCISE 6.6:
Imaging—Thorax

	SELECT TOPIC	SELECT VIEW	GO
☢	**Thorax** ▶	**X-ray: Posterior-Anterior**	

• *Click the* **TURN TAGS ON** *button, and you will see the following image:*

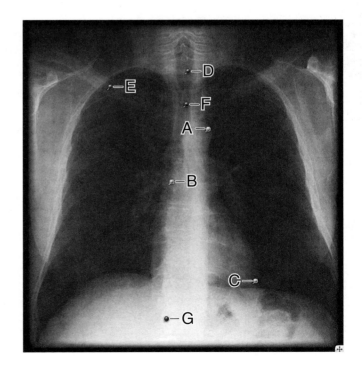

- *Mouse-over the green pins on the screen to find the information necessary to identify the following structures:*

A. _____

B. _____

C. _____

Non-cardiovascular System Structures (blue pins)

D. _____

E. _____

F. _____

G. _____

Self Test

Take this opportunity to quiz yourself by taking the **SELF TEST.** See page 15 for a reminder on how to access the self-test for this section.

IN REVIEW

What Have I Learned?

The following questions cover the material that you have just learned, the blood vessels of the thorax. Apply what you have learned to answer these questions on a separate piece of paper.

1. What is the name for the fibrous sac that encloses the heart?

2. Name the lymphatic organ that is large in children but atrophies during adolescence.

3. Name the bilobed endocrine gland located lateral to the trachea and larynx.

4. How do large arteries supply blood to body structures?

5. Name the large vessel that conveys oxygen-poor blood from the right ventricle of the heart.

6. Name the two branches of the blood vessel mentioned in question 5 that convey oxygen-poor blood to the lungs.

7. Name the blunt tip of the left ventricle.

8. What is the carotid sheath? What structures are found within it?

9. What is the serous pericardium?

10. Name the structure that served to shunt blood in the fetus from the pulmonary artery to the aorta, bypassing the lungs. What is it called in the adult?

11. Name all of the different sections of the aorta, and describe where they are found.

12. Name the two major vessels that return oxygen-poor blood to the heart. What are the drainages for each? Where do they terminate?

The Head and Neck

<table>
<tr><td colspan="3" align="center">**EXERCISE 6.7:**</td></tr>
<tr><td colspan="3" align="center">Cardiovascular System—Head and Neck, Vasculature, Lateral View</td></tr>
</table>

SELECT TOPIC	SELECT VIEW	
Head and Neck	▶ **Vasculature: Lateral**	GO

• *Click* **LAYER 1** *in the* **LAYER CONTROLS** *window, and you will see the following image:*

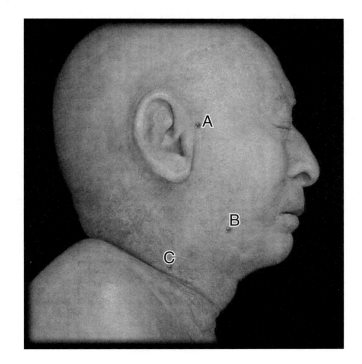

• *Mouse-over the blue pins on the screen to find the information necessary to identify the following non-cardiovascular system structures:*

A. _____

B. _____

C. _____

CHECK POINT:

Head and Neck, Vasculature, Lateral View

1. What blood vessel is responsible for the superficial temporal pulse?
2. What blood vessel is responsible for the facial pulse?
3. Between which two structures of the neck would you find the carotid pulse?

• *Click* **LAYER 2** *in the* **LAYER CONTROLS** *window, and you will see the following image:*

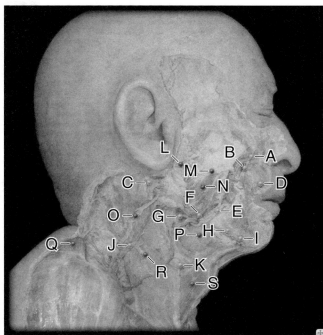

• *Mouse-over the green pins on the screen to find the information necessary to identify the following structures:*

A. _____

B. _____

C. _____

D. _____

E. _____

F. _____

G. _____

H. _____

I. _____

J. _____

K. _____

Non-cardiovascular System Structures (blue pins)

L. _____

M. _____

N. _____

O. _____

P. _____

Q. _____

R. _____

S. _____

CHECK POINT:

Head and Neck, Vasculature, Lateral View, cont'd

4. Name a subcutaneous vein with tributaries that drain the scalp and the face and terminates in the subclavian vein.
5. Name a vein, the size of which is inversely proportional to the vein in question 4.
6. Name a vein that drains the muscles, glands, and mucous membranes of the upper lip.

• *Click* **LAYER 3** *in the* **LAYER CONTROLS** *window, and you will see the following image:*

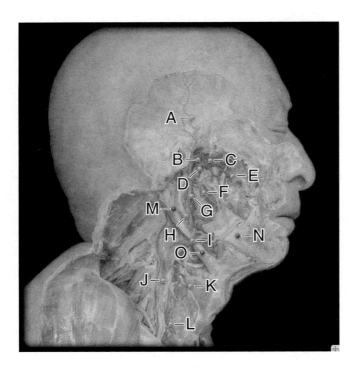

• *Mouse-over the green pins on the screen to find the information necessary to identify the following structures:*

A. _____

B. _____

C. _____

D. _____

E. _____

F. _____

G. _____

H. _____

I. _____

J. _____

K. _____

L. _____

Non-cardiovascular System Structures (blue pins)

M. _____

N. _____

O. _____

CHECK POINT:

Head and Neck, Vasculature, Lateral View, cont'd

7. Name the large vein that drains the cranial cavity, including the brain, as well as the face and neck.
8. Name a vein that drains the superior molar and premolar teeth and the mucous membrane of the maxillary sinus.
9. Name a vein that drains the deep face and temporal region and descends within the parotid gland.

• *Click* **LAYER 4** *in the* **LAYER CONTROLS** *window, and you will see the following image:*

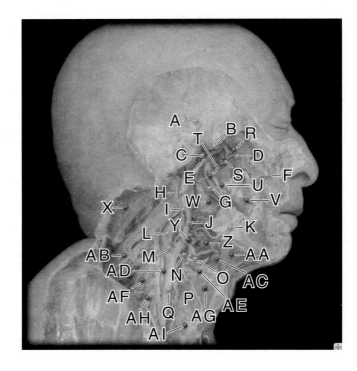

• *Mouse-over the green pins on the screen to find the information necessary to identify the following structures:*

A. _____

B. _____

C. _____

D. _____

E. _____

F. _____

G. _____

H. _____

I. _____

J. _____

K. _____

L. _____

M. _____

N. _____

O. _____

P. _____

Q. _____

**Non-cardiovascular System Structures
(blue pins)**

R. _____

S. _____

T. _____

U. _____

V. _____

W. _____

X. _____

Y. _____

Z. _____

AA. _____

AB. _____

AC. _____

AD. _____

AE. _____

AF. _____

AG. _____

AH. _____

AI. _____

• Click **LAYER 5** in the **LAYER CONTROLS** window, and you will see the following image:

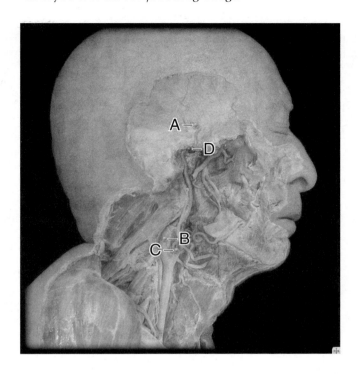

• Mouse-over the green pins on the screen to find the information necessary to identify the following structures:

A. _____

B. _____

C. _____

**Non-cardiovascular System Structure
(blue pin)**

D. _____

CHECK POINT:

Head and Neck, Vasculature, Lateral View, cont'd

13. Name the structure of the skull that contains the internal carotid artery and the carotid plexus of sympathetic nerves.

CHECK POINT:

Head and Neck, Vasculature, Lateral View, cont'd

10. Name the major arteries of the neck where the left one originates from the aortic arch and the right one originates from the brachiocephalic trunk.
11. Name the branch of the arteries in question 10 that distributes to the exterior of the head (except the orbit), the face, meninges, and neck structures.
12. Name the branch of the arteries in question 10 that enters the cranial cavity and terminates in the cerebral arterial circle.

Animation: Baroreceptor Reflex

SELECT ANIMATION
Baroreceptor Reflex

PLAY

- *After viewing the animation, answer these questions:*

1. Where are baroreceptors located?

2. What is their function?

3. How do the arteries with baroreceptors, and all arteries for that matter, respond to increased blood pressure?

4. What response do the baroreceptors have to this increased blood pressure?

5. Where are these action potentials conducted?

6. What nerves conduct these impulses?

7. What is the parasympathetic response to and result of this stimulation?

8. What is the sympathetic response and result?

9. What about sympathetic stimulation of the blood vessels?

10. What physical events combine to bring elevated blood pressure back toward normal?

Self Test
Take this opportunity to quiz yourself by taking the **SELF TEST.** See page 15 for a reminder on how to access the self-test for this section.

Animation: Chemoreceptor Reflex

SELECT ANIMATION
Chemoreceptor Reflex

PLAY

- *After viewing the animation, answer these questions:*

1. Where are chemoreceptors located?

2. What is their function?

3. Where are the impulses from the chemoreceptors conducted?

4. What nerves conduct these impulses?

5. What three events decrease parasympathetic stimulation of the heart?

6. What effect does this decreased stimulation have on the physiology of the heart?

7. What sympathetic response occurs during the three events listed in question 5?

8. What sympathetic response occurs in the blood vessels?

9. If the chemoreceptors are stimulated by decreased blood oxygen, what physical changes occur as a result of the changes in autonomic stimulation?

10. What would you deduce is the effect on blood pressure as a result of the answer to question 9?

Self Test
Take this opportunity to quiz yourself by taking the **SELF TEST.** See page 15 for a reminder on how to access the self-test for this section.

EXERCISE 6.8a:	EXERCISE 6.8b:
Imaging—Head and Neck	**Imaging—Head and Neck**

SELECT TOPIC	SELECT VIEW		SELECT TOPIC	SELECT VIEW	
Head and Neck	**Arteries: Anterior - Posterior**	**GO**	**Head and Neck**	**Arteries: Lateral**	**GO**

- *Click the* **TURN TAGS ON** *button, and you will see the following image:*

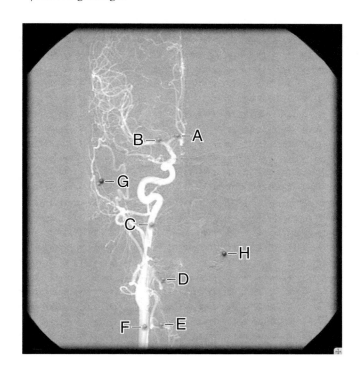

- *Click the* **TURN TAGS ON** *button, and you will see the following image:*

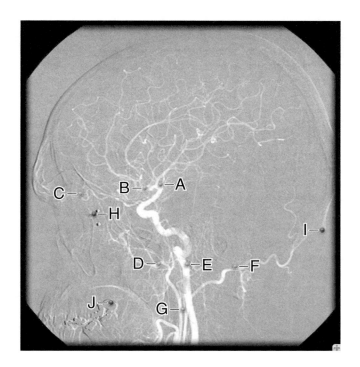

- *Mouse-over the green pins on the screen to find the information necessary to identify the following structures:*

A. _____

B. _____

C. _____

D. _____

E. _____

F. _____

Non-cardiovascular System Structures (blue pins)

G. _____

H. _____

- *Mouse-over the green pins on the screen to find the information necessary to identify the following structures:*

A. _____

B. _____

C. _____

D. _____

E. _____

F. _____

G. _____

Non-cardiovascular System Structures (blue pins)

H. _____

I. _____

J. _____

EXERCISE 6.8c:

Imaging—Head and Neck

SELECT TOPIC	SELECT VIEW	
Head and Neck	**Veins: Anterior - Posterior**	GO

- *Click the* **TURN TAGS ON** *button, and you will see the following image:*

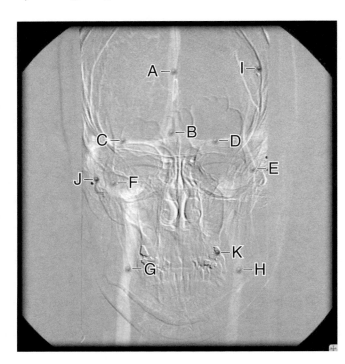

- *Mouse-over the green pins on the screen to find the information necessary to identify the following structures:*

A. _____

B. _____

C. _____

D. _____

E. _____

F. _____

G. _____

H. _____

Non-cardiovascular System Structures (blue pins)

I. _____

J. _____

K. _____

EXERCISE 6.8d:

Imaging—Head and Neck

SELECT TOPIC	SELECT VIEW	
Head and Neck	**Angiogram Veins: Lateral**	GO

- *Click the* **TURN TAGS ON** *button, and you will see the following image:*

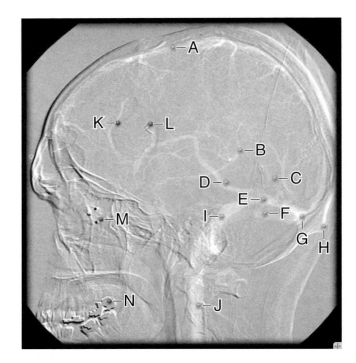

- *Mouse-over the pins on the screen to find the information necessary to identify the following structures:*

A. _____

B. _____

C. _____

D. _____

E. _____

F. _____

G. _____

H. _____

I. _____

J. _____

Non-cardiovascular System Structures (blue pins)

K. _____

L. _____

M. _____

N. _____

Self Test
Take this opportunity to quiz yourself by taking the **SELF TEST.** See page 15 for a reminder on how to access the self-test for this section.

I N   R E V I E W

What Have I Learned?

The following questions cover the material that you have just learned, the blood vessels of the head and neck. Apply what you have learned to answer these questions on a separate piece of paper.

1. Name the gland that produces 25 percent of your saliva.

2. List three structures that pass through this gland.

3. Where does the duct from this gland empty?

4. Name two veins of the head and neck that lack valves.

5. Name a superficial vein that drains the temporal region.

6. List the areas drained by the maxillary veins.

7. What is an anastomosis? Give an example of one in the head and neck region.

The Brain

EXERCISE 6.9:

Cardiovascular System—Brain, Arteries, Inferior View

SELECT TOPIC	SELECT VIEW	
Brain	**Arteries: Inferior**	GO

• Click **LAYER 1** in the **LAYER CONTROLS** window, and you will see the following image:

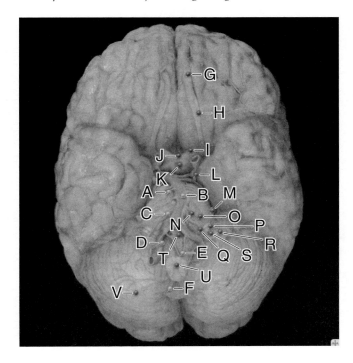

• Mouse-over the green pins on the screen to find the information necessary to identify the following structures:

A. _____

B. _____

C. _____

D. _____

E. _____

F. _____

Non-cardiovascular System Structures (blue pins)

G. _____

H. _____

I. _____

J. _____

K. _____

L. _____

M. _____

N. _____

O. _____

P. _____

Q. _____

R. _____

S. _____

T. _____

U. _____

V. _____

C H E C K   P O I N T :

Brain, Arteries, Inferior View

1. The central branches of which artery supply the anterior two-thirds of the spinal cord?
2. Name the artery that distributes to the posterior cerebellum and lateral medulla oblongata.
3. Name the artery that distributes to the anterior-inferior cerebellum, pons, and medulla oblongata.

- *Click* **LAYER 2** *in the* **LAYER CONTROLS** *window, and you will see the following image:*

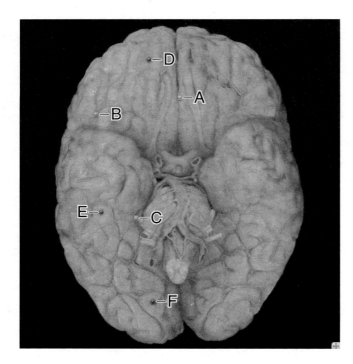

- *Mouse-over the green pins on the screen to find the information necessary to identify the following structures:*

A. _____

B. _____

C. _____

Non-cardiovascular System Structures (blue pins)

D. _____

E. _____

F. _____

C H E C K   P O I N T :

Brain, Arteries, Inferior View, cont'd

4. Name the artery that distributes to the medial aspect of the frontal lobes of the cerebral cortex.
5. Name the artery that distributes to the lateral aspects of the frontal, parietal, and temporal lobes of the cerebral cortex.
6. Name the artery that distributes to the temporal and occipital lobes of the cerebral cortex.

- *Click* **LAYER 3** *in the* **LAYER CONTROLS** *window, and you will see the following image:*

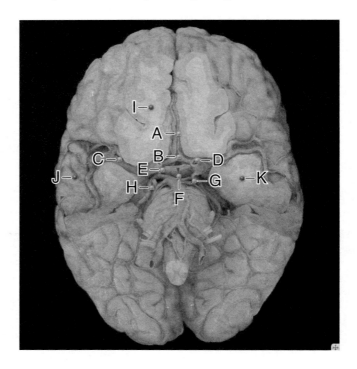

- *Mouse-over the green pins on the screen to find the information necessary to identify the following structures:*

A. _____

B. _____

C. _____

D. _____

E. _____

F. _____

G. _____

H. _____

Non-cardiovascular System Structures (blue pins)

I. _____

J. _____

K. _____

CHECK POINT:

Brain, Arteries, Inferior View, cont'd

7. Name the arterial anastomosis of the ventral surface of the brain complete in only 20 percent of individuals.
8. Name the paired arteries that course along each lateral sulcus of the cerebral hemisphere.
9. Name the cerebral lobe not visible from the surface, also known as the isle of Reil.

EXERCISE 6.10:

Cardiovascular System—Brain, Veins, Lateral View

SELECT TOPIC	SELECT VIEW	
Brain	**Veins: Lateral**	GO

- *After clicking **LAYER 2** in the **LAYER CONTROLS** window, to review the bones of the skull, click **LAYER 3** and you will see the following image:*

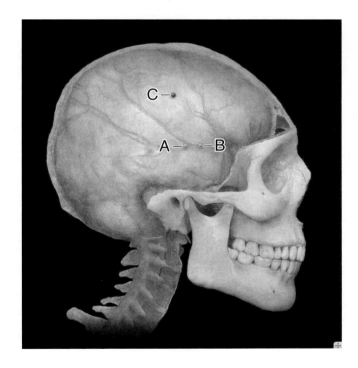

- *Mouse-over the green pins on the screen to find the information necessary to identify the following structures:*

A. _____

B. _____

Non-cardiovascular System Structure (blue pin)

C. _____

CHECK POINT:

Brain, Veins, Lateral View

1. Name the most exterior of the meninges. Where is it located? What is its function?
2. Name the artery whose distribution includes the dura mater, skull, trigeminal, and facial ganglia.
3. Name the vein that drains the dura mater and bones of the anterior and middle cranial fossae.

- *Click **LAYER 4** in the **LAYER CONTROLS** window, and you will see the following image:*

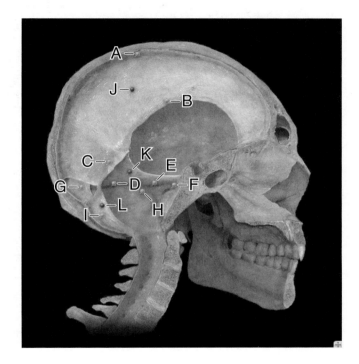

- *Mouse-over the green pins on the screen to find the information necessary to identify the following structures:*

A. _____
B. _____
C. _____
D. _____
E. _____
F. _____
G. _____
H. _____
I. _____

Non-cardiovascular System Structures (blue pins)

J. _____
K. _____
L. _____

CHECK POINT:

Brain, Veins, Lateral View, cont'd

4. What are dural venous sinuses?
5. Name the structure located in the internal occipital protuberance that drains the superior sagittal, straight, and occipital sinuses.
6. Name the sinus that drains the cerebellum.

• *Click* **LAYER 5** *in the* **LAYER CONTROLS** *window, and you will see the following image:*

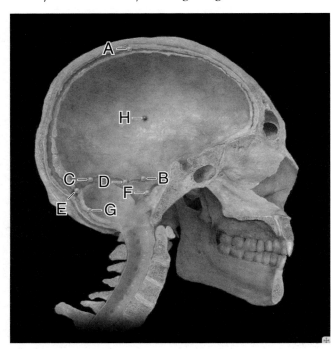

• *Mouse-over the green pins on the screen to find the information necessary to identify the following structures:*

A. _____

B. _____

C. _____

D. _____

E. _____

F. _____

G. _____

Non-cardiovascular System Structure (blue pin)

H. _____

CHECK POINT:

Brain, Veins, Lateral View, cont'd

7. Name the s-shaped continuation of the transverse sinus.
8. Name a paired dural venous sinus that terminates at the jugular foramen.
9. List the areas drained by the transverse sinus.

Animation: Blood Flow Through Brain

 SELECT ANIMATION
Blood Flow Through Brain   PLAY

• *After viewing the animation, answer these questions:*

1. Name the three arteries through which blood flows to the brain.

2. Which two arteries supply blood to 80 percent of the cerebrum? These are the terminal branches of which arteries?

3. Which arteries enter the cranial cavity through the foramen magnum? They unite to form which artery?

4. Which arteries supply the occipital and temporal lobes of the cerebrum?

5. Which arteries form an anastomosis at the base of the brain? What is the name of this anastomosis?

6. What is the function of the anastomosis?

7. Blood is drained from the brain through small veins that empty into vessel channels called . . .

8. From where does blood flow to enter the confluence of sinuses?

9. From the confluence of sinuses, where does blood flow before leaving the skull?

10. The blood leaves the skull via the . . .

Self Test

Take this opportunity to quiz yourself by taking the **SELF TEST.** See page 15 for a reminder on how to access the self-test for this section.

I N   R E V I E W

What Have I Learned?

The following questions cover the material that you have just learned, the blood vessels of the brain. Apply what you have learned to answer these questions on a separate piece of paper.

1. Name the large crescent-shaped fold of dura mater that separates the right and left cerebral hemispheres.

2. Name the small crescent-shaped fold of dura mater that separates the right and left cerebellar hemispheres.

3. Name the dural venous sinus that contains arachnoid granulations. What is the function of these granulations?

4. Name the dural venous sinus that drains the medial cerebral hemispheres.

5. Name the horizontal crescent-shaped fold of the dura mater that separates the cerebral hemispheres and the cerebellum.

The Shoulder

EXERCISE 6.11:
Cardiovascular System—The Shoulder, Arteries, Anterior View

SELECT TOPIC	SELECT VIEW	
Shoulder ▸	**Arteries: Anterior**	GO

• *Click* **LAYER 2** *in the* **LAYER CONTROLS** *window, and you will see the following image:*

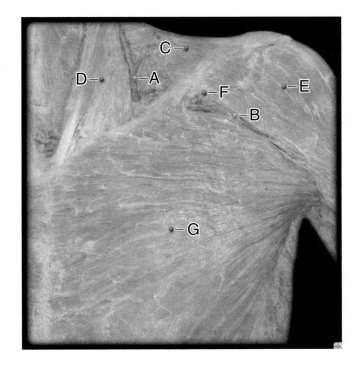

• *Mouse-over the green pins on the screen to find the information necessary to identify the following structures:*

A. _____

B. _____

Non-cardiovascular System Structures (blue pins)

C. _____

D. _____

E. _____

F. _____

G. _____

C H E C K   P O I N T :

The Shoulder, Arteries, Anterior View

1. Name the large vein of the neck that drains the face and scalp.
2. What are the major tributaries of this vein?
3. List the structures drained by the cephalic vein.

- *Click* **LAYER 3** *in the* **LAYER CONTROLS** *window, and you will see the following image:*

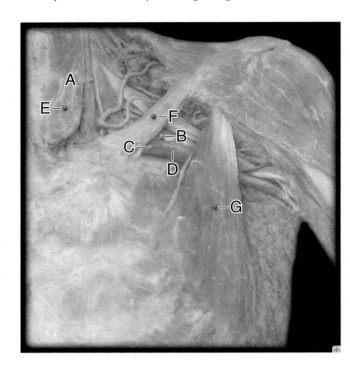

- *Mouse-over the green pins on the screen to find the information necessary to identify the following structures:*

A. _____

B. _____

C. _____

D. _____

Non-cardiovascular System Structures (blue pins)

E. _____

F. _____

G. _____

- *Click* **LAYER 4** *in the* **LAYER CONTROLS** *window, and you will see the following image:*

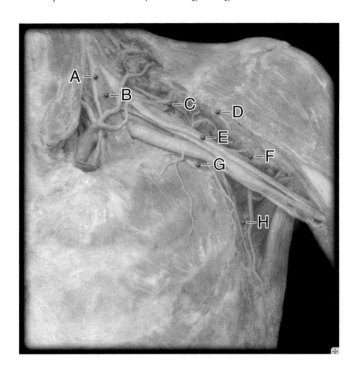

- *Mouse-over the blue pins on the screen to find the information necessary to identify the following non-cardiovascular system structures:*

A. _____

B. _____

C. _____

D. _____

E. _____

F. _____

G. _____

H. _____

CHECK POINT:

The Shoulder, Arteries, Anterior View, cont'd

4. Name the large vein within the carotid sheath that is a continuation of the sigmoid dural sinus of the cranial cavity.
5. Which two veins unite to form the brachiocephalic vein?
6. Where does the axillary vein become the subclavian vein?

• *Click* **LAYER 5** *in the* **LAYER CONTROLS** *window, and you will see the following image:*

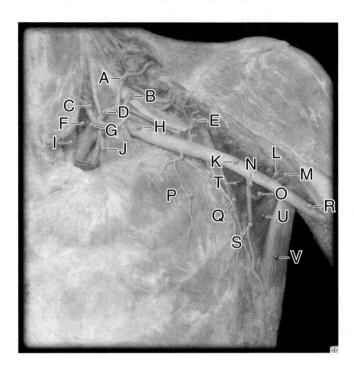

• *Mouse-over the green pins on the screen to find the information necessary to identify the following structures:*

A. _____

B. _____

C. _____

D. _____

E. _____

F. _____

G. _____

H. _____

I. _____

J. _____

K. _____

L. _____

M. _____

N. _____

O. _____

P. _____

Q. _____

R. _____

S. _____

Non-cardiovascular System Structures (blue pins)

T. _____

U. _____

V. _____

CHECK POINT:

The Shoulder, Arteries, Anterior View, cont'd

7. What is the definition of a "trunk"? (This one will require inference from the information available in the **STRUCTURE INFORMATION** window.)
8. Name a trunk that ascends in the lower neck.
9. Where does the subclavian artery become the axial artery? Where does the axillary artery become the brachial artery?

EXERCISE 6.12:

Cardiovascular System—Shoulder, Veins, Anterior View

SELECT TOPIC	SELECT VIEW	
Shoulder	**Veins: Anterior**	GO

• *Click* **LAYER 4** *in the* **LAYER CONTROLS** *window, and you will see the following image:*

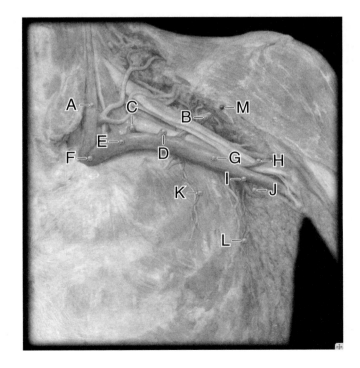

• *Mouse-over the green pins on the screen to find the information necessary to identify the following structures:*

A. _____

B. _____

C. _____

D. _____

E. _____

F. _____

G. _____

H. _____

I. _____

J. _____

K. _____

L. _____

Non-cardiovascular System Structure (blue pin)

M. _____

CHECK POINT:

Shoulder, Veins, Anterior View

1. List the areas drained by the left brachiocephalic vein. What is the other name for this vein in Latin and in English?
2. Where does the left brachiocephalic vein terminate?
3. List the areas drained by the axillary vein. What are its tributaries?
4. Does the left brachiocephalic vein contain valves?

• *Click **LAYER 5** in the **LAYER CONTROLS** window, and you will see the following image:*

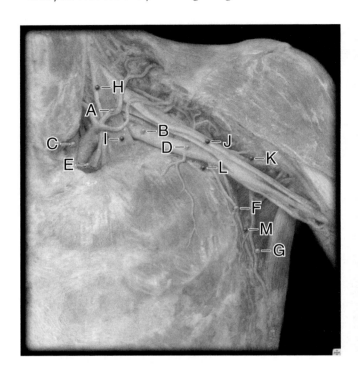

• *Mouse-over the green pins on the screen to find the information necessary to identify the following structures:*

A. _____

B. _____

C. _____

D. _____

E. _____

F. _____

G. _____

Non-cardiovascular System Structures (blue pins)

H. _____

I. _____

J. _____

K. _____

L. _____

M. _____

Self Test

Take this opportunity to quiz yourself by taking the **SELF TEST**. See page 15 for a reminder on how to access the self-test for this section.

I N   R E V I E W

What Have I Learned?

The following questions cover the material that you have just learned, the blood vessels of the shoulder. Apply what you have learned to answer these questions on a separate piece of paper.

1. What is the deltopectoral triangle? What is another name for it?

2. Name three arteries that distribute to the muscles of the scapula.

3. Name the paired arteries that ascend through the neck via the transverse foramina of the cervical vertebrae.

4. Where do these paired arteries enter the cranial cavity?

The Shoulder and Arm

EXERCISE 6.13:
Cardiovascular System—Shoulder and Arm, Arteries, Anterior View

SELECT TOPIC	SELECT VIEW	
Shoulder and Arm ▶	**Arteries: Anterior**	GO

- *Click* **LAYER 1** *in the* **LAYER CONTROLS** *window, and you will see the following image:*

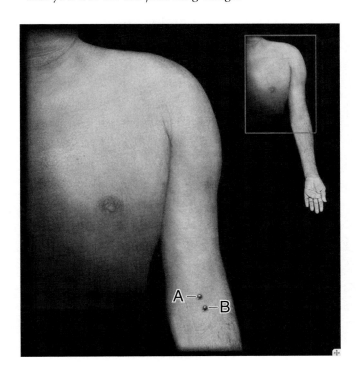

- *Mouse-over the blue pins on the screen to find the information necessary to identify the following non-cardiovascular system structures:*

A. _____

B. _____

CHECK POINT:

Shoulder and Arm, Arteries, Anterior View

1. What is the cubital fossa?
2. What structures are found in the cubital fossa?
3. Where is the location for the stethoscope when taking blood pressure with a pressure cuff?

- *Click* **LAYER 2** *in the* **LAYER CONTROLS** *window, and you will see the following image:*

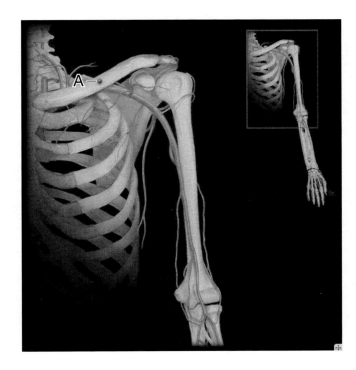

- *Mouse-over the blue pin on the screen to find the information necessary to identify the following non-cardiovascular system structure:*

A. _____

- *Click* **LAYER 3** *in the* **LAYER CONTROLS** *window, and you will see the following image:*

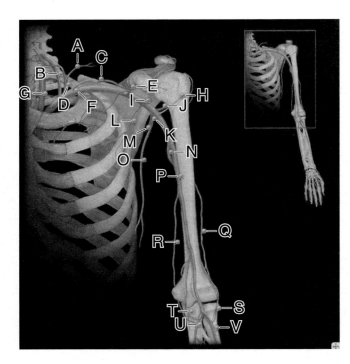

- *Mouse-over the green pins on the screen to find the information necessary to identify the following structures:*

A. _____

B. _____

C. _____

D. _____

E. _____

F. _____

G. _____

H. _____

I. _____

J. _____

K. _____

L. _____

M. _____

N. _____

O. _____

P. _____

Q. _____

R. _____

S. _____

T. _____

U. _____

V. _____

- *Click* **LAYER 4** *in the* **LAYER CONTROLS** *window, and you will see the following image:*

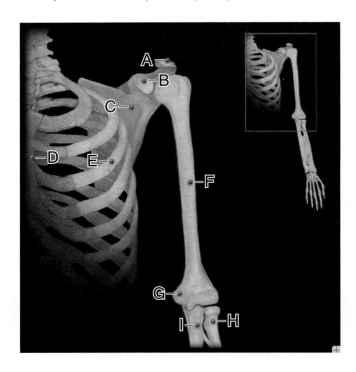

- *Mouse-over the blue pins on the screen to find the information necessary to identify the following non-cardiovascular system structures:*

A. _____

B. _____

C. _____

D. _____

E. _____

F. _____

G. _____

H. _____

I. _____

CHECK POINT:

Shoulder and Arm, Arteries, Anterior View, cont'd

4. Name the major artery of the arm.
5. What two arteries branch from this artery at the anterior elbow?
6. What artery is also known as the *profunda brachii* artery? How does *profunda brachii* translate?

EXERCISE 6.14:

Cardiovascular System—Shoulder and Arm, Veins, Anterior View

SELECT TOPIC	SELECT VIEW	
Shoulder and Arm ➤	**Veins: Anterior**	**GO**

• *Click* **LAYER 1** *in the* **LAYER CONTROLS** *window, and you will see the following image:*

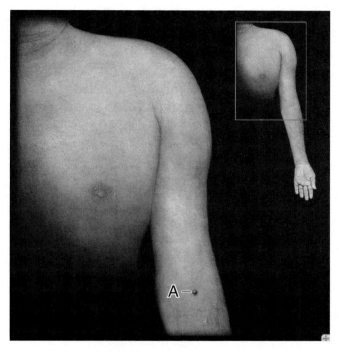

• *Mouse-over the blue pin on the screen to find the information necessary to identify the following non-cardiovascular system structure:*

A. _____

• *Click* **LAYER 2** *in the* **LAYER CONTROLS** *window, and you will see the following image:*

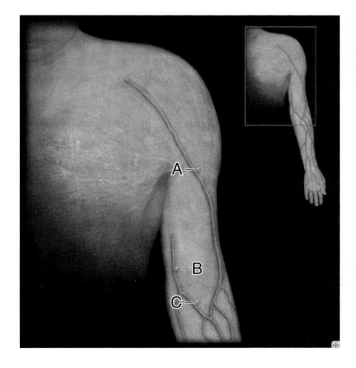

• *Mouse-over the pins on the screen to find the information necessary to identify the following structures:*

A. _____

B. _____

C. _____

CHECK POINT:

Shoulder and Arm, Veins, Anterior View

1. Which vein of the arm is frequently used for venipuncture? What is venipuncture?
2. What structures are drained by the basilica vein?
3. What structures are drained by the cephalic vein?

• *Click* **LAYER 3** *in the* **LAYER CONTROLS** *window, and you will see the following image:*

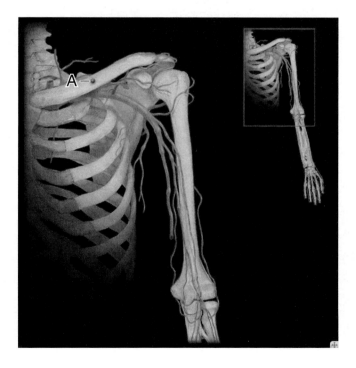

• *Mouse-over the blue pin on the screen to find the information necessary to identify the following non-cardiovascular system structure:*

A. _____

- *Click* **LAYER 4** *in the* **LAYER CONTROLS** *window, and you will see the following image:*

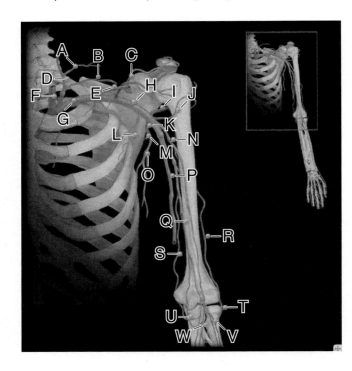

- *Mouse-over the pins on the screen to find the information necessary to identify the following structures:*

A. _____

B. _____

C. _____

D. _____

E. _____

F. _____

G. _____

H. _____

I. _____

J. _____

K. _____

L. _____

M. _____

N. _____

O. _____

P. _____

Q. _____

R. _____

S. _____

T. _____

U. _____

V. _____

W. _____

CHECK POINT:

Shoulder and Arm, Veins, Anterior View, cont'd

4. Name the paired veins that ascend along the brachial artery.
5. Name the structures drained by the axillary vein. What are its major tributaries?
6. Name the large vein that terminates in the axillary vein.

- *Click* **LAYER 5** *in the* **LAYER CONTROLS** *window, and you will see the following image:*

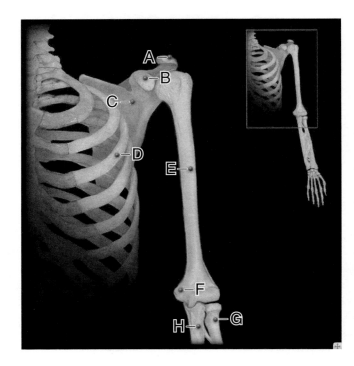

- *Mouse-over the blue pins on the screen to find the information necessary to identify the following non-cardiovascular system structures:*

A. _____

B. _____

C. _____

D. _____

E. _____

F. _____

G. _____

H. _____

Self Test

Take this opportunity to quiz yourself by taking the **SELF TEST.** See page 15 for a reminder on how to access the self-test for this section.

I N   R E V I E W

What Have I Learned?

The following questions cover the material that you have just learned, the blood vessles of the shoulder and arm. Apply what you have learned to answer these questions on a separate piece of paper.

1. Name the two arteries that arise from the brachiocephalic trunk.

2. Which of the two ascends into the neck?

3. Name the two branches at the terminus of this artery.

4. Name the arteries that form an anastomosis around the elbow.

5. Name two arteries that form an anastomosis around the scapula.

6. Name the structures drained by the external jugular vein. What are its major tributaries?

7. Name the structures drained by the internal jugular vein. What are its major tributaries?

8. Name two locations where venous anastomoses occur in the shoulder and arm.

The Forearm and Hand

EXERCISE 6.15:
Cardiovascular System—Forearm and Hand, Arteries, Anterior View

  SELECT TOPIC   **SELECT VIEW**   **GO**
Forearm and Hand ▶ **Arteries: Anterior**

- *Click **LAYER 1** in the **LAYER CONTROLS** window, and you will see the following image:*

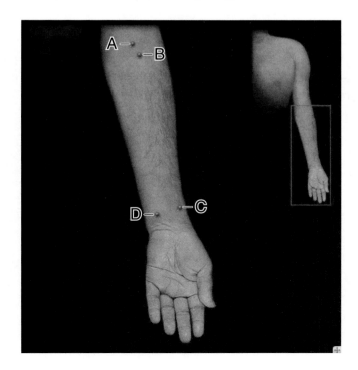

- *Mouse-over the blue pins on the screen to find the information necessary to identify the following non-cardiovascular system structures:*

A. _____

B. _____

C. _____

D. _____

C H E C K   P O I N T :

Forearm and Hand, Arteries, Anterior View

1. Name the location commonly used to take a pulse at the wrist.
2. Name another location for taking a pulse at the wrist.
3. Name a location for taking a pulse at the elbow.

- *Click* **LAYER 2** *in the* **LAYER CONTROLS** *window, and you will see the following image:*

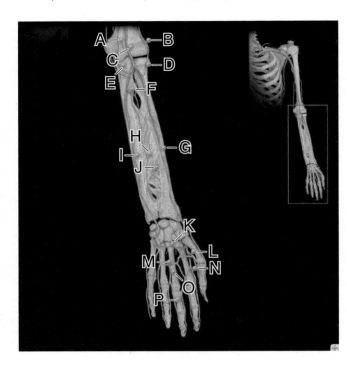

- *Mouse-over the pins on the screen to find the information necessary to identify the following structures:*

A. _____

B. _____

C. _____

D. _____

E. _____

F. _____

G. _____

H. _____

I. _____

J. _____

K. _____

L. _____

M. _____

N. _____

O. _____

P. _____

- *Click* **LAYER 3** *in the* **LAYER CONTROLS** *window, and you will see the following image:*

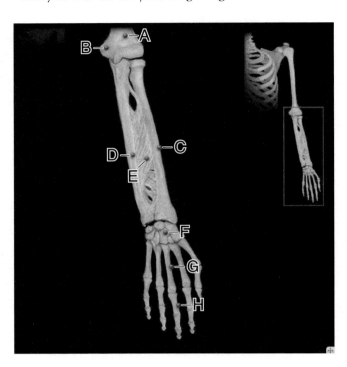

- *Mouse-over the blue pins on the screen to find the information necessary to identify the following non-cardiovascular system structures:*

A. _____

B. _____

C. _____

D. _____

E. _____

F. _____

G. _____

H. _____

CHECK POINT:

Forearm and Hand, Arteries, Anterior View, cont'd

4. Name the two arteries that form an anastomosis through the superficial and deep palmar arch.
5. Name the artery that descends along the anterior aspect of the interosseous membrane of the forearm.
6. Name the artery commonly used for taking a pulse at the wrist.

EXERCISE 6.16:

Cardiovascular System—Forearm and Hand, Veins, Anterior View

SELECT TOPIC	SELECT VIEW	
Forearm and Hand ➤	**Veins: Anterior**	**GO**

• *Click* **LAYER 1** *in the* **LAYER CONTROLS** *window, and you will see the following image:*

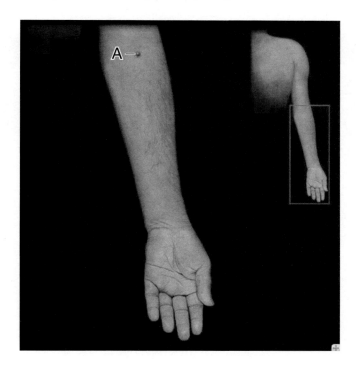

• *Mouse-over the blue pin on the screen to find the information necessary to identify the following non-cardiovascular system structure:*

A. _____

• *Click* **LAYER 2** *in the* **LAYER CONTROLS** *window, and you will see the following image:*

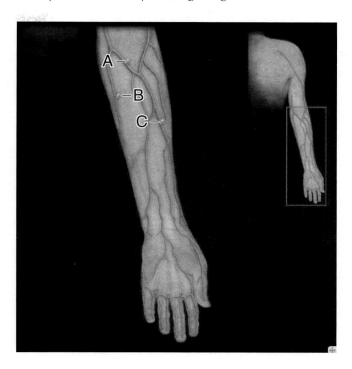

• *Mouse-over the pins on the screen to find the information necessary to identify the following structures:*

A. _____

B. _____

C. _____

CHECK POINT:

Forearm and Hand, Veins, Anterior View

1. Name the two veins that extend from the dorsal hand to the axillary vein.
2. Name the vein that has an oblique subcutaneous path between these two veins over the cubital fossa.

• *Click* **LAYER 3** *in the* **LAYER CONTROLS** *window, and you will see the following image:*

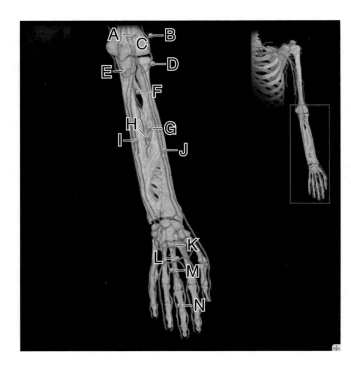

• *Mouse-over the pins on the screen to find the information necessary to identify the following structures:*

A. _____

B. _____

C. _____

D. _____

E. _____

F. _____

G. _____

H. _____

I. _____

J. _____

K. _____

L. _____

M. _____

N. _____

CHECK POINT:

Forearm and Hand, Veins, Anterior View, cont'd

3. Name the paired veins that ascend along the lateral aspect of the forearm.
4. Name the paired veins that ascend along the medial aspect of the forearm.
5. Name a vein of the forearm that is not always present.

EXERCISE 6.17:

Imaging—Hand

SELECT TOPIC	SELECT VIEW	
Hand	▶**Angiogram Arteries: Anterior -Posterior**	GO

• *Click the* **TURN TAGS ON** *button, and you will see the following image:*

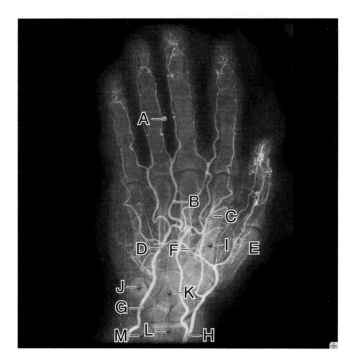

• *Mouse-over the green pins on the screen to find the information necessary to identify the following structures:*

A. _____

B. _____

C. _____

D. _____

E. _____

F. _____

G. _____

H. _____

Non-Cardiovascular System Structures (blue pins)

I. _____

J. _____

K. _____

L. _____

M. _____

Self Test

Take this opportunity to quiz yourself by taking the **SELF TEST.** See page 15 for a reminder on how to access the self-test for this section.

See page 15 for a reminder on how to access the self-test for this section.

I N   R E V I E W

What Have I Learned?

The following questions cover the material that you have just learned, the blood vessels of the forearm and hand. Apply what you have learned to answer these questions on a separate piece of paper.

1. Describe the pathway of arterial blood flow from the arm through the elbow and lateral forearm to the palm of the hand.

2. Describe the pathway of arterial blood flow from the arm through the elbow and medial forearm to the palm of the hand.

3. Describe the pathway of arterial blood flow from the palm of the hand to the middle finger.

4. Describe the *superficial* pathway of venous blood drainage from the anterior forearm to the elbow.

5. Describe the *deep* pathway of venous blood drainage from the middle finger to the anterior elbow and forearm.

The Abdomen

• *Click* **LAYER 2** in the **LAYER CONTROLS** *window, and you will see the following image:*

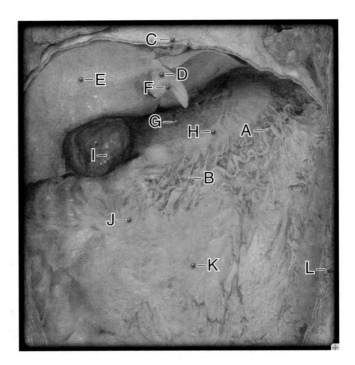

EXERCISE 6.18:

Cardiovascular System—Abdomen, Celiac Trunk, Anterior View

SELECT TOPIC	SELECT VIEW	
Abdomen	**Celiac Trunk: Anterior**	**GO**

• *Click* **LAYER 1** *in the* **LAYER CONTROLS** *window, and you will see the following image:*

• *Mouse-over the green pins on the screen to find the information necessary to identify the following structures:*

A. _____

B. _____

Non-cardiovascular System Structures (blue pins)

C. _____

D. _____

E. _____

F. _____

G. _____

H. _____

I. _____

J. _____

K. _____

L. _____

• *Mouse-over the blue pins on the screen to find the information necessary to identify the following non-cardiovascular system structures:*

A. _____

B. _____

C. _____

D. _____

E. _____

F. _____

G. _____

H. _____

CHECK POINT:

Abdomen, Celiac Trunk, Anterior View, cont'd

4. Name the structure that is the remnant of the umbilical vein of the fetus.
5. Name the anastomosis that has distribution to the stomach and the greater omentum.
6. Once again, what is an anastomosis?

CHECK POINT:

Abdomen, Celiac Trunk, Anterior View

1. Name the upper median abdominal region.
2. Name the two abdominal regions lateral to question 1.
3. Name the median abdominal region.

- *Click* **LAYER 3** *in the* **LAYER CONTROLS** *window, and you will see the following image:*

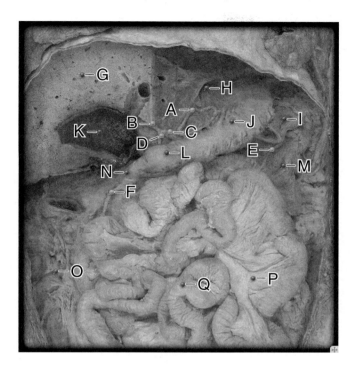

- *Mouse-over the green pins on the screen to find the information necessary to identify the following structures:*

A. _____

B. _____

C. _____

D. _____

E. _____

F. _____

Non-cardiovascular System Structures (blue pins)

G. _____

H. _____

I. _____

J. _____

K. _____

L. _____

M. _____

N. _____

O. _____

P. _____

Q. _____

CHECK POINT:

Abdomen, Celiac Trunk, Anterior View, cont'd

7. Name two arteries that originate at the celiac artery.
8. What is the smallest branch of the celiac artery? What is its distribution?
9. Name the large, highly vascular, accessory digestive organ. What are its four lobes?

- *Click* **LAYER 4** *in the* **LAYER CONTROLS** *window, and you will see the following image:*

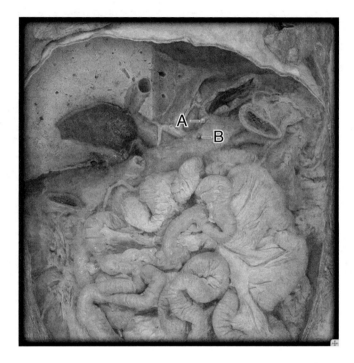

- *Mouse-over the green pin on the screen to find the information necessary to identify the following structure:*

A. _____

Non-cardiovascular System Structure (blue pin)

B. _____

CHECK POINT:

Abdomen, Celiac Trunk, Anterior View, cont'd

10. Name the artery that is the largest branch of the celiac trunk.
11. List the branches of this artery.
12. What is meant by serpentine?

- *Click **LAYER 5** in the **LAYER CONTROLS** window, and you will see the following image:*

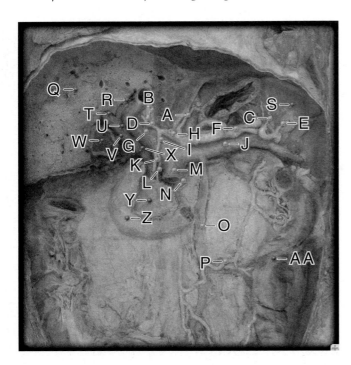

- *Mouse-over the green pins on the screen to find the information necessary to identify the following structures:*

A. _____

B. _____

C. _____

D. _____

E. _____

F. _____

G. _____

H. _____

I. _____

J. _____

K. _____

L. _____

M. _____

N. _____

O. _____

P. _____

Non-cardiovascular System Structures (blue pins)

Q. _____

R. _____

S. _____

T. _____

U. _____

V. _____

W. _____

X. _____

Y. _____

Z. _____

AA. _____

CHECK POINT:

Abdomen, Celiac Trunk, Anterior View, cont'd

13. Name the unpaired anterior artery immediately inferior to the celiac trunk.
14. What is the distribution of this artery?
15. Name the arteries that originate at the celiac trunk.

- *Click **LAYER 6** in the **LAYER CONTROLS** window, and you will see the following image:*

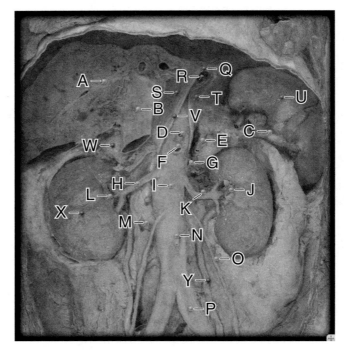

- *Mouse-over the green pins on the screen to find the information necessary to identify the following structures:*

A. _____

B. _____

C. _____

D. _____

E. _____

F. _____

G. _____

H. _____

I. _____

J. _____

K. _____

L. _____

M. _____

N. _____

O. _____

P. _____

Non-cardiovascular System Structures (blue pins)

Q. _____

R. _____

S. _____

T. _____

U. _____

V. _____

W. _____

X. _____

Y. _____

CHECK POINT:

Abdomen, Celiac Trunk, Anterior View, cont'd

16. Name the three arteries that branch off of the anterior abdominal aorta, from superior to inferior.
17. Name the large paired arteries that branch off from the abdominal aorta laterally to the kidneys.
18. Name the small paired arteries that branch off from the abdominal aorta inferiolaterally in the vicinity of the kidneys. What is their name in the male? In the female?

EXERCISE 6.19:

Cardiovascular System—Abdomen, Mesenteric Arteries, Anterior View

SELECT TOPIC	SELECT VIEW	
Abdomen	▶ **Mesenteric Arteries: Anterior**	**GO**

- *Click* **LAYER 1** *in the* **LAYER CONTROLS** *window, and you will see the following image:*

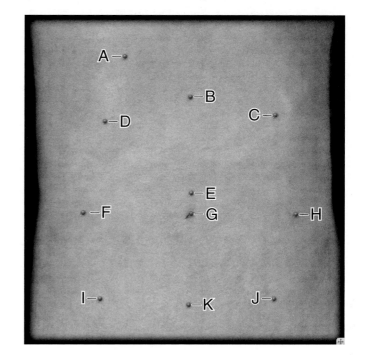

- *Mouse-over the blue pins on the screen to find the information necessary to identify the following non-cardiovascular system structures:*

A. _____

B. _____

C. _____

D. _____

E. _____

F. _____

G. _____

H. _____

I. _____

J. _____

K. _____

CHECK POINT:

Abdomen, Mesenteric Arteries, Anterior View

1. Name the lower medial abdominal region.
2. What regions flank this lower medial region?
3. What regions flank the umbilical region?

- *Click* **LAYER 2** *in the* **LAYER CONTROLS** *window, and you will see the following image:*

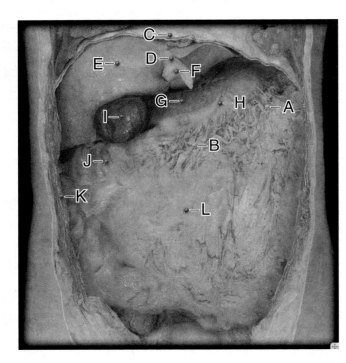

- *Mouse-over the green pins on the screen to find the information necessary to identify the following structures:*

A. _____

B. _____

Non-cardiovascular System Structures (blue pins)

C. _____

D. _____

E. _____

F. _____

G. _____

H. _____

I. _____

J. _____

K. _____

L. _____

CHECK POINT:

Abdomen, Mesenteric Arteries, Anterior View, cont'd

4. Name two arteries that distribute to the stomach and the greater omentum.
5. Describe the greater omentum.
6. Name the largest visceral organ.

- *Click* **LAYER 3** *in the* **LAYER CONTROLS** *window, and you will see the following image:*

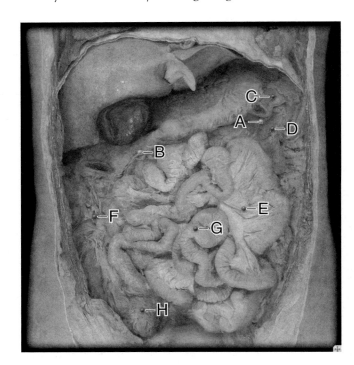

- *Mouse-over the green pins on the screen to find the information necessary to identify the following structures:*

A. _____

B. _____

Non-cardiovascular System Structures (blue pins)

C. _____

D. _____

E. _____

F. _____

G. _____

H. _____

CHECK POINT:

Abdomen, Mesenteric Arteries, Anterior View, cont'd

7. What is the origin of the middle colic artery?
8. What is the origin of the left colic artery?
9. An anastomosis is formed between these two colic arteries and what other artery?

• *Click* **LAYER 4** *in the* **LAYER CONTROLS** *window, and you will see the following image:*

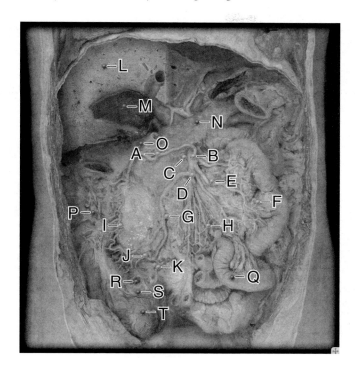

• *Mouse-over the green pins on the screen to find the information necessary to identify the following structures:*

A. _____

B. _____

C. _____

D. _____

E. _____

F. _____

G. _____

H. _____

I. _____

J. _____

K. _____

**Non-cardiovascular System Structures
(blue pins)**

L. _____

M. _____

N. _____

O. _____

P. _____

Q. _____

R. _____

S. _____

T. _____

CHECK POINT:
_____
Abdomen, Mesenteric Arteries, Anterior View, cont'd

10. Name the abdominal artery that courses inferiorly to enter the mesentery of the small intestine.
11. What is the termination of this artery?
12. What arteries are also known as *vasa recta?* What is their distribution?

• *Click* **LAYER 5** *in the* **LAYER CONTROLS** *window, and you will see the following image:*

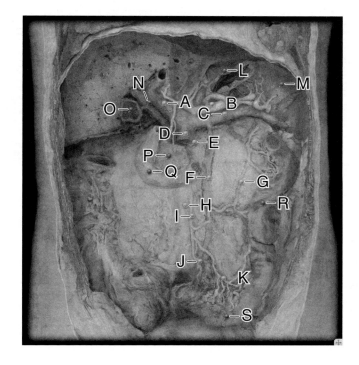

• *Mouse-over the green pins on the screen to find the information necessary to identify the following structures:*

A. _____

B. _____

C. _____

D. _____

E. _____

F. _____

G. _____

H. _____

I. _____

J. _____

K. _____

Non-cardiovascular System Structures (blue pins)

L. _____

M. _____

N. _____

O. _____

P. _____

Q. _____

R. _____

S. _____

CHECK POINT:

Abdomen, Mesenteric Arteries, Anterior View, cont'd

13. What is the distribution of the inferior mesenteric artery?
14. Name the artery with the distribution to the lower part of the descending colon and the sigmoid colon.
15. Name the artery that is the direct continuation of the inferior mesenteric artery and distributes to the rectum.

- *Click* **LAYER 6** *in the* **LAYER CONTROLS** *window, and you will see the following image:*

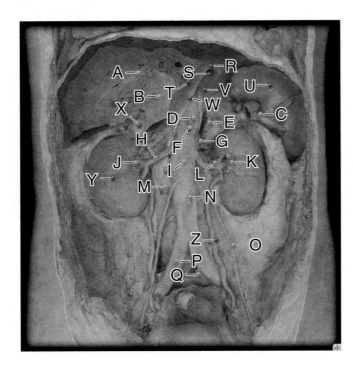

- *Mouse-over the green pins on the screen to find the information necessary to identify the following structures:*

A. _____

B. _____

C. _____

D. _____

E. _____

F. _____

G. _____

H. _____

I. _____

J. _____

K. _____

L. _____

M. _____

N. _____

O. _____

P. _____

Q. _____

Non-cardiovascular System Structures (blue pins)

R. _____

S. _____

T. _____

U. _____

V. _____

W. _____

X. _____

Y. _____

Z. _____

CHECK POINT:

Abdomen, Mesenteric Arteries, Anterior View, cont'd

16. List the drainages of the common iliac veins. Where do they terminate?
17. What is the origin and termination of the common iliac arteries?
18. Name the vein that drains the spleen, pancreas, and the fundus and greater curvature of the stomach.

Animation: Hepatic Portal System

| | SELECT ANIMATION
Hepatic Portal System | PLAY |

- *After viewing the animation, answer these questions:*

1. What is the hepatic portal system?

2. What does this do for the liver?

3. The hepatic portal system originates from which organs?

4. Which vein is the largest in the hepatic portal system?

5. Which two veins unite to form it?

6. Which areas are drained by the splenic vein?

7. Which areas are drained by the superior mesenteric vein?

8. Blood from which structures drains directly into the hepatic portal vein?

9. What happens to the hepatic portal vein upon entering the liver?

10. What are hepatic sinusoids?

11. What occurs there?

12. What occurs in central veins of the liver?

13. What veins are formed by the union of thousands of these central veins?

14. Upon exiting the liver, these veins . . .

Self Test
Take this opportunity to quiz yourself by taking the **SELF TEST.** See page 15 for a reminder on how to access the self-test for this section.

EXERCISE 6.20:

Cardiovascular System—Abdomen, Veins, Anterior View

| SELECT TOPIC
Abdomen | SELECT VIEW
Veins: Anterior | GO |

- *After reviewing the structures of* **LAYER 1,** *click* **LAYER 2** *in the* **LAYER CONTROLS** *window, and you will see the following image:*

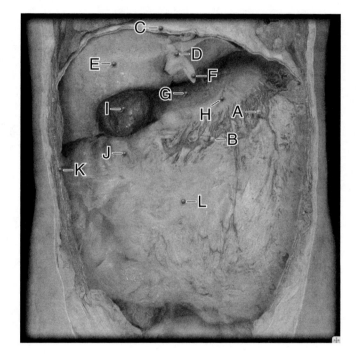

- *Mouse-over the green pins on the screen to find the information necessary to identify the following structures:*

A. _____

B. _____

Non-cardiovascular System Structures (blue pins)

C. _____

D. _____

E. _____

F. _____

G. _____

H. _____

I. _____

J. _____

K. _____

L. _____

CHECK POINT:

Abdomen, Veins, Anterior View

1. Name the veins that drain the stomach and greater omentum.

- *Click* **LAYER 3** *in the* **LAYER CONTROLS** *window, and you will see the following image:*

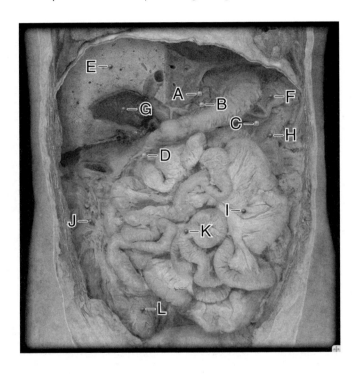

- *Mouse-over the green pins on the screen to find the information necessary to identify the following structures:*

A. _____

B. _____

C. _____

D. _____

Non-cardiovascular System Structures (blue pins)

E. _____

F. _____

G. _____

H. _____

I. _____

J. _____

K. _____

L. _____

CHECK POINT:

Abdomen, Veins, Anterior View, cont'd

2. Name a vein that drains the stomach and the inferior thoracic and abdominal parts of the esophagus.
3. Name a vein that drains only the stomach.
4. Where do these two veins terminate?

- *Click* **LAYER 4** *in the* **LAYER CONTROLS** *window, and you will see the following image:*

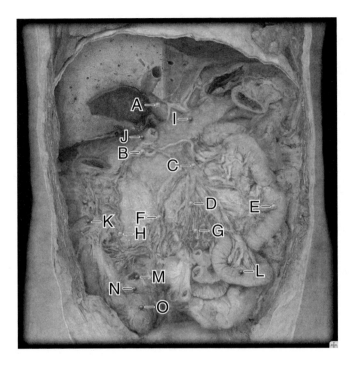

- *Mouse-over the green pins on the screen to find the information necessary to identify the following structures:*

A. _____

B. _____

C. _____

D. _____

E. _____

F. _____

G. _____

H. _____

Non-cardiovascular System Structures (blue pins)

I. _____

J. _____

K. _____

L. _____

M. _____

N. _____

O. _____

CHECK POINT:

Abdomen, Veins, Anterior View, cont'd

5. List the structures drained by the hepatic portal vein.
6. What do marginal venous arcades represent?
7. Where is the termination of the intestinal venous arcades?

- *Click **LAYER 5** in the **LAYER CONTROLS** window, and you will see the following image:*

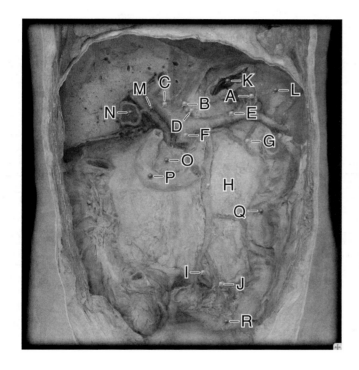

- *Mouse-over the green pins on the screen to find the information necessary to identify the following structures:*

A. _____

B. _____

C. _____

D. _____

E. _____

F. _____

G. _____

H. _____

I. _____

J. _____

Non-cardiovascular System Structures (blue pins)

K. _____

L. _____

M. _____

N. _____

O. _____

P. _____

Q. _____

R. _____

CHECK POINT:

Abdomen, Veins, Anterior View, cont'd

8. Name the vein that ascends from the true pelvis to become the inferior mesenteric vein in the false pelvis.

- *Click **LAYER 6** in the **LAYER CONTROLS** window, and you will see the following image:*

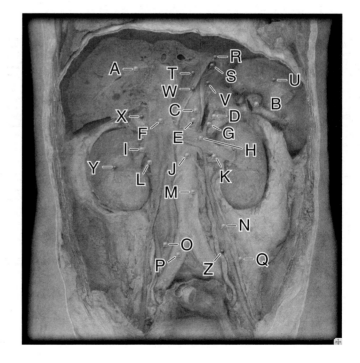

- *Mouse-over the green pins on the screen to find the information necessary to identify the following structures:*

A. _____

B. _____

C. _____

D. _____

E. _____

F. _____

G. _____

H. _____

I. _____

J. _____

K. _____

L. _____

M. _____

N. _____

O. _____

P. _____

Q. _____

Non-cardiovascular System Structures (blue pins)

R. _____

S. _____

T. _____

U. _____

V. _____

W. _____

X. _____

Y. _____

Z. _____

CHECK POINT:

Abdomen, Veins, Anterior View, cont'd

9. What is normally on the right side of the body, the abdominal aorta or the inferior vena cava?
10. Which are normally anterior to the others, the renal arteries or the renal veins?
11. What are the two terminations of the gonadal veins? (Be sure to note which one terminates where.)

Self Test

Take this opportunity to quiz yourself by taking the **SELF TEST**. See page 15 for a reminder on how to access the self-test for this section.

IN REVIEW

What Have I Learned?

The following questions cover the material that you have just learned, blood vessels of the abdomen. Apply what you have learned to answer these questions on a separate piece of paper.

1. Using the following figure, label the nine abdominal regions.

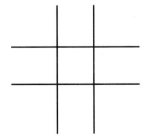

2. What is the umbilicus? (Not the region, the structure.)

3. Name the insertion point for the diaphragm.

4. Name the vein that carries absorbed products of digestion to the liver.

5. What two veins unite to form the hepatic portal vein?

6. List the organs drained by the splenic vein.

7. Name the opening that transmits the aorta from the thoracic to the abdominal cavity.

8. Name the opening that transmits the esophagus from the thoracic to the abdominal cavity.

9. Name the paired arteries at the terminus of the abdominal aorta.

10. Name the vein that drains the diaphragm and the left suprarenal gland.

The Pelvis

EXERCISE 6.21:

Cardiovascular System—Pelvis—Female, Vasculature, Anterior View

SELECT TOPIC	SELECT VIEW	
Pelvis—Female ▶	**Vasculature: Anterior**	**GO**

• *Click* **LAYER 1** *in the* **LAYER CONTROLS** *window, and you will see the following image:*

• *Click* **LAYER 2** *in the* **LAYER CONTROLS** *window, and you will see the following image:*

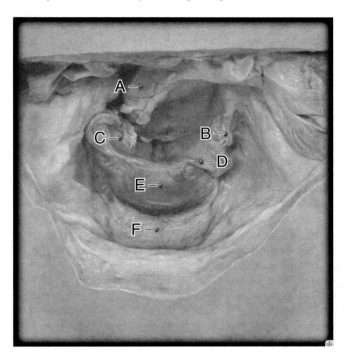

• *Mouse-over the blue pins on the screen to find the information necessary to identify the following non-cardiovascular system structures:*

A. _____

B. _____

C. _____

D. _____

E. _____

F. _____

• *Mouse-over the blue pins on the screen to find the information necessary to identify the following non-cardiovascular system structures:*

A. _____

B. _____

C. _____

• *Click* **LAYER 3** *in the* **LAYER CONTROLS** *window, and you will see the following image:*

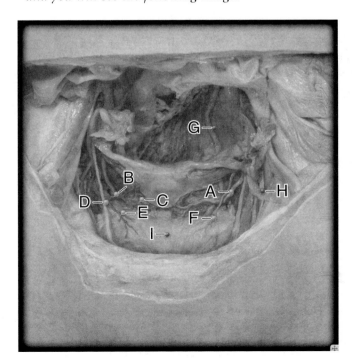

- *Mouse-over the green pins on the screen to find the information necessary to identify the following structures:*

A. _____

B. _____

C. _____

D. _____

E. _____

F. _____

Non-cardiovascular System Structures (blue pins)

G. _____

H. _____

I. _____

CHECK POINT:

Pelvis—Female, Vasculature, Anterior View

1. Name the artery that carries blood between the placenta and fetus.
2. Name the artery that reflects along the side of the uterus, distributing to the uterus and vagina.
3. Name the vein that drains the uterus.

- *Click **LAYER 4** in the **LAYER CONTROLS** window, and you will see the following image:*

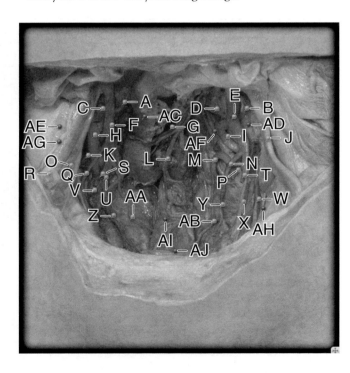

- *Mouse-over the green pins on the screen to find the information necessary to identify the following structures:*

A. _____

B. _____

C. _____

D. _____

E. _____

F. _____

G. _____

H. _____

I. _____

J. _____

K. _____

L. _____

M. _____

N. _____

O. _____

P. _____

Q. _____

R. _____

S. _____

T. _____

U. _____

V. _____

W. _____

X. _____

Y. _____

Z. _____

AA. _____

AB. _____

Non-cardiovascular System Structures (blue pins)

AC. _____

AD. _____

AE. _____

AF. _____

AG. _____

AH. _____

AI. _____

AJ. _____

CHECK POINT:

Pelvis—Female, Vasculature, Anterior View, cont'd

4. Name the vein that drains the pelvis and gluteal region and terminates in the common iliac vein.
5. Name two veins that are tributaries to the vein in question 4.
6. Name the artery that originates at the common iliac artery and continues as the femoral artery.

Self Test

Take this opportunity to quiz yourself by taking the **SELF TEST.** See page 15 for a reminder on how to access the self-test for this section.

EXERCISE 6.22:

Cardiovascular System—Pelvis—Male, Vasculature, Anterior View

SELECT TOPIC	SELECT VIEW	
Pelvis—Male	➤ Vasculature: Anterior	GO

• Click **LAYER 1** in the **LAYER CONTROLS** window, and you will see the following image:

• Mouse-over the blue pins on the screen to find the information necessary to identify the following non-cardiovascular system structures:

A. _____

B. _____

C. _____

D. _____

• Click **LAYER 2** in the **LAYER CONTROLS** window, and you will see the following image:

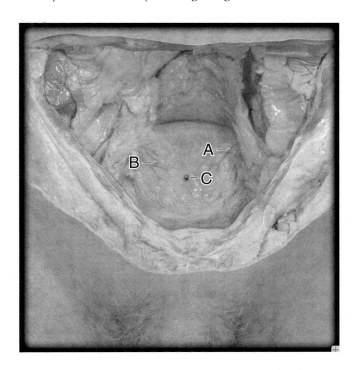

• Mouse-over the green pins on the screen to find the information necessary to identify the following structures:

A. _____

B. _____

Non-cardiovascular System Structure (blue pin)

C. _____

CHECK POINT:

Pelvis—Male, Vasculature, Anterior View

1. Name an artery that distributes to the urinary bladder.
2. Name a vein that drains the urinary bladder.
3. The volume of which organ effects the position of the surrounding organs?

- *Click* **LAYER 3** *in the* **LAYER CONTROLS** *window, and you will see the following image:*

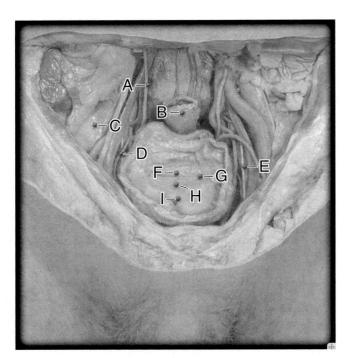

- *Mouse-over the blue pins on the screen to find the information necessary to identify the following non-cardiovascular system structures:*

A. _____

B. _____

C. _____

D. _____

E. _____

F. _____

G. _____

H. _____

I. _____

- *Click* **LAYER 4** *in the* **LAYER CONTROLS** *window, and you will see the following image:*

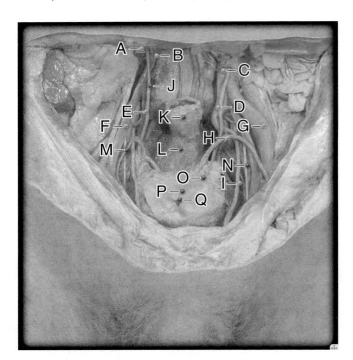

- *Mouse-over the green pins on the screen to find the information necessary to identify the following structures:*

A. _____

B. _____

C. _____

D. _____

E. _____

F. _____

G. _____

H. _____

I. _____

Non-cardiovascular System Structures (blue pins)

J. _____

K. _____

L. _____

M. _____

N. _____

O. _____

P. _____

Q. _____

CHECK POINT:

Pelvis—Male, Vasculature, Anterior View, cont'd

4. Name an artery that passes medially to reach the superior surface of the urinary bladder.

• *Click* **LAYER 5** *in the* **LAYER CONTROLS** *window, and you will see the following image:*

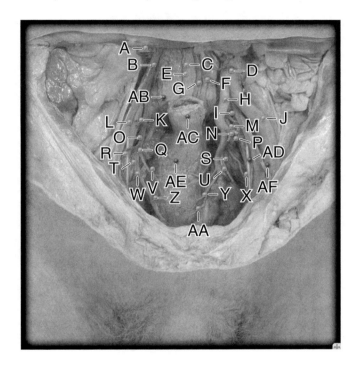

• *Mouse-over the green pins on the screen to find the information necessary to identify the following structures:*

A. _____

B. _____

C. _____

D. _____

E. _____

F. _____

G. _____

H. _____

I. _____

J. _____

K. _____

L. _____

M. _____

N. _____

O. _____

P. _____

Q. _____

R. _____

S. _____

T. _____

U. _____

V. _____

W. _____

X. _____

Y. _____

Z. _____

AA. _____

Non-cardiovascular System Structures (blue pins)

AB. _____

AC. _____

AD. _____

AE. _____

AF. _____

CHECK POINT:

Pelvis—Male, Vasculature, Anterior View, cont'd

5. Name a vein that passes laterally from the inferior surface of the urinary bladder and drains the urinary bladder and the prostate gland. What is this vein known as in the female?
6. List the structures drained by the median sacral vein.
7. List the structures distributed by the iliolumbar artery.

Self Test

Take this opportunity to quiz yourself by taking the **SELF TEST.** See page 15 for a reminder on how to access the self-test for this section.

IN REVIEW

What Have I Learned?

The following questions cover the material that you have just learned, the vasculature of the female and male pelvis. Apply what you have learned to answer these questions on a separate piece of paper.

1. Name the thick-walled, pear-shaped hollow muscular organ that is the site of implantation of the blastocyst.

2. The position of the uterus varies with . . .

3. Name the two parts of the uterus.

4. Name the paired female gonads.

5. What is their function?

6. Name the specific site of fertilization of the egg.

7. Name the artery that descends into the pelvis and passes posteriorly toward the superior margin of the greater sciatic notch.

8. Name an artery that distributes to the anal canal and external genitalia of both sexes.

9. Name an artery whose distribution includes the rectum and vagina in females and the rectum, prostate, and seminal vesicle in males.

10. Name a pelvic vein that drains the muscles and skin of the medial thigh and terminates at the internal iliac vein.

The Hip and Thigh

Cardiovascular System—Hip and Thigh, Arteries, Anterior View

SELECT TOPIC	SELECT VIEW	
Hip and Thigh ▶	**Arteries: Anterior**	**GO**

• *Click **LAYER 1** in the **LAYER CONTROLS** window, and you will see the following image:*

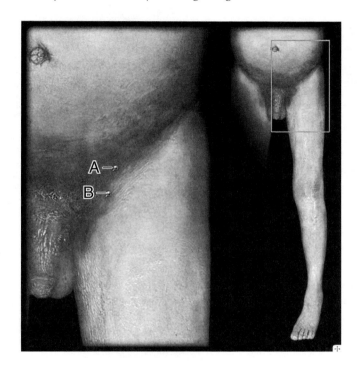

• *Mouse-over the blue pins on the screen to find the information necessary to identify the following non-cardiovascular system structures:*

A. _____

B. _____

CHECK POINT:

Hip and Thigh, Arteries, Anterior View

1. Name the location for checking the pulse at the groin.
2. What emergency function is applicable at this location?
3. What structures are contained within the femoral triangle?

• *Click **LAYER 2** in the **LAYER CONTROLS** window, and you will see the following image:*

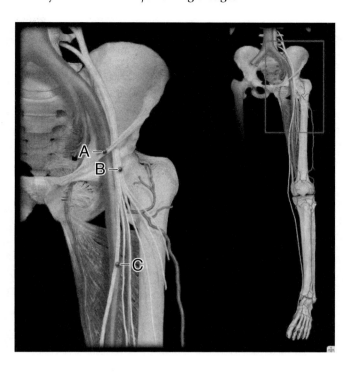

• *Mouse-over the blue pins on the screen to find the information necessary to identify the following non-cardiovascular system structures:*

A. _____

B. _____

C. _____

CHECK POINT:

Hip and Thigh, Arteries, Anterior View, cont'd

4. What structure forms the floor of the inguinal canal?
5. Name the nerve and its branches that enter the thigh posterior to the inguinal ligament.
6. Which branch is the largest?

- *Click* **LAYER 3** *in the* **LAYER CONTROLS** *window, and you will see the following image:*

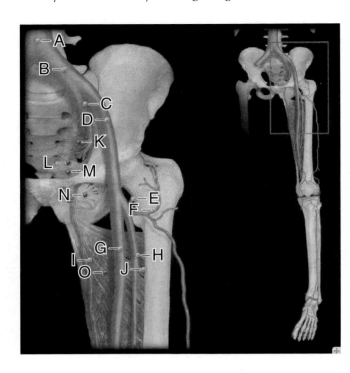

- *Mouse-over the green pins on the screen to find the information necessary to identify the following structures:*

A. _____

B. _____

C. _____

D. _____

E. _____

F. _____

G. _____

H. _____

I. _____

J. _____

Non-cardiovascular System Structures (blue pins)

K. _____

L. _____

M. _____

N. _____

O. _____

CHECK POINT:

Hip and Thigh, Arteries, Anterior View, cont'd

7. List the sequence of the continuous major arteries from the abdominal aorta through the thigh.
8. Name the largest branch of the femoral artery. What is its distribution? What are two other names for this artery?
9. List the distribution of the obturator artery.

EXERCISE 6.24:

Cardiovascular System—Hip and Thigh, Arteries, Posterior View

SELECT TOPIC	SELECT VIEW	
Hip and Thigh	**Arteries: Posterior**	GO

- *After viewing the structures in* **LAYER 1,** *click* **LAYER 2** *in the* **LAYER CONTROLS** *window, and you will see the following image:*

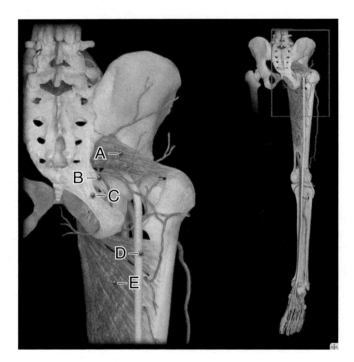

- *Mouse-over the blue pins on the screen to find the information necessary to identify the following non-cardiovascular system structures:*

A. _____

B. _____

C. _____

D. _____

E. _____

CHECK POINT:

Hip and Thigh, Arteries, Posterior View

1. Name the largest nerve in the body.
2. Name the two nerves that form that nerve.
3. List the sensory and motor innervations of this nerve.

• *Click* **LAYER 3** *in the* **LAYER CONTROLS** *window, and you will see the following image:*

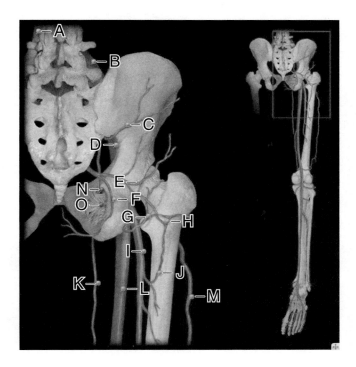

• *Mouse-over the green pins on the screen to find the information necessary to identify the following structures:*

A. _____

B. _____

C. _____

D. _____

E. _____

F. _____

G. _____

H. _____

I. _____

J. _____

K. _____

L. _____

M. _____

Non-cardiovascular System Structures (blue pins)

N. _____

O. _____

CHECK POINT:

Hip and Thigh, Arteries, Posterior View, cont'd

4. Name the largest branch of the internal iliac artery. What is its distribution?
5. List the distribution of the abdominal aorta.
6. What artery is usually crossed by the ureter and gonadal vessels at its origin?

EXERCISE 6.25:

Cardiovascular System—Hip and Thigh, Veins, Anterior View

SELECT TOPIC	SELECT VIEW	
Hip and Thigh	**Veins: Anterior**	GO

• *After viewing the structures in* **LAYER 1,** *click* **LAYER 2** *in the* **LAYER CONTROLS** *window, and you will see the following image:*

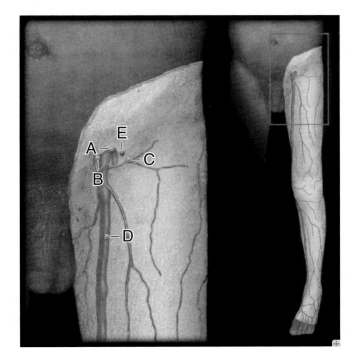

• *Mouse-over the green pins on the screen to find the information necessary to identify the following structures:*

A. _____

B. _____

C. _____

D. _____

Non-cardiovascular System Structure (blue pin)

E. _____

CHECK POINT:

Hip and Thigh, Veins, Anterior View

1. Name the common source of vessel tissue for coronary bypass surgery.
2. Defective valves in this vein may cause . . .
3. What is another name for this vein?

• *Click **LAYER 4** in the **LAYER CONTROLS** window, and you will see the following image:*

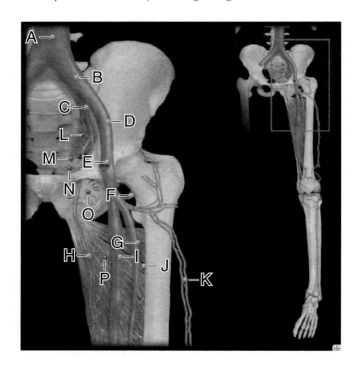

• *Mouse-over the green pins on the screen to find the information necessary to identify the following structures:*

A. _____

B. _____

C. _____

D. _____

E. _____

F. _____

G. _____

H. _____

I. _____

J. _____

K. _____

Non-cardiovascular System Structures (blue pins)

L. _____

M. _____

N. _____

O. _____

P. _____

CHECK POINT:

Hip and Thigh, Veins, Anterior View, cont'd

4. Name a vein of the thigh that has the popliteal vein as a tributary.
5. What is the drainage for this vein?
6. List the sequence of this vein and those that continue with it superiorly to the heart.

EXERCISE 6.26:

Cardiovascular System—Hip and Thigh, Veins, Posterior View

SELECT TOPIC	SELECT VIEW	
Hip and Thigh	**Veins: Posterior**	GO

• *After viewing the structures in **LAYER 1**, click **LAYER 2** in the **LAYER CONTROLS** window, and you will see the following image:*

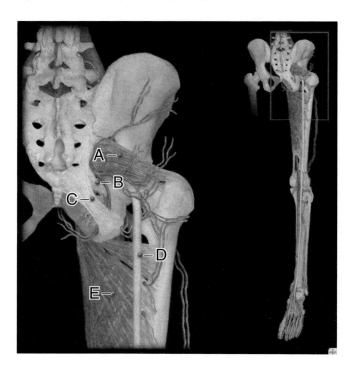

• *Mouse-over the blue pins on the screen to find the information necessary to identify the following non-cardiovascular system structures:*

A. _____

B. _____

C. _____

D. _____

E. _____

• *Click* **LAYER 3** *in the* **LAYER CONTROLS** *window, and you will see the following image:*

• *Mouse-over the green pins on the screen to find the information necessary to identify the following structures:*

A. _____

B. _____

C. _____

D. _____

E. _____

F. _____

G. _____

H. _____

I. _____

J. _____

K. _____

L. _____

Non-cardiovascular System Structures (blue pins)

M. _____

N. _____

CHECK POINT:

Hip and Thigh, Veins, Posterior View

1. Name the medial vein of the leg that spans from the anterior medial malleolus to the femoral vein just inferior to the inguinal ligament.
2. List the drainages of the deep vein of the thigh.
3. Name the largest vein in the body.

Self Test

Take this opportunity to quiz yourself by taking the **SELF TEST.** See page 15 for a reminder on how to access the self-test for this section.

I N   R E V I E W

What Have I Learned?

The following questions cover the material that you have just learned, the blood vessels of the hip and thigh. Apply what you have learned to answer these questions on a separate piece of paper.

1. Name the four arteries that contribute to the cruciate anastomosis of the hip joint.

2. Name the three veins that contribute to the anastomosis of the hip joint.

3. What is the largest vein in the body? Did you know that it is also the thickest *vessel* in your body? Which do you suppose is the longest vein in your body?

4. Name the triangular region of the groin that contains the femoral nerve, artery, and vein.

5. Name the opening in the deep fascia through which the great saphenous vein passes.

6. Name the subcutaneous vein that runs parallel to the inguinal ligament.

7. Name the deep vertical space that separates the left and right buttock.

8. What structure represents the superior limit of the posterior thigh?

The Knee

EXERCISE 6.27:

Cardiovascular System—Knee, Arteries, Anterior View

SELECT TOPIC	SELECT VIEW	
Knee	▶ **Arteries: Anterior**	**GO**

• *Click* **LAYER 1** *in the* **LAYER CONTROLS** *window, and you will see the following image:*

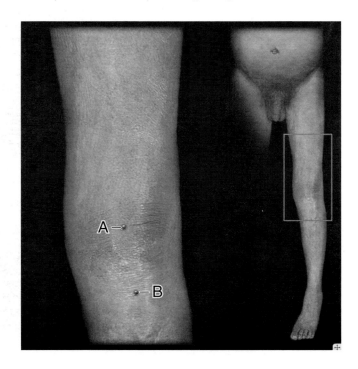

• *Mouse-over the blue pins on the screen to find the information necessary to identify the following non-cardiovascular system structures:*

A. _____

B. _____

C H E C K   P O I N T :

Knee, Arteries, Anterior View

1. Name the sesamoid bone embedded in the tendon of the quadriceps femoris muscles.
2. Name the bony elevation on the proximal shaft of the tibia.

- *Click* **LAYER 2** *in the* **LAYER CONTROLS** *window, and you will see the following image:*

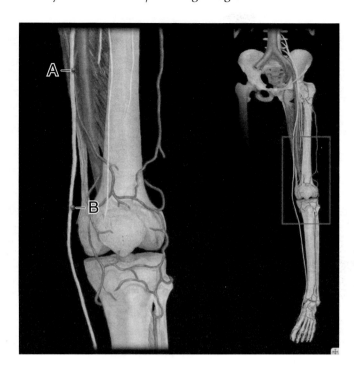

- *Mouse-over the blue pins on the screen to find the information necessary to identify the following non-cardiovascular system structures:*

A. _____

B. _____

- *Click* **LAYER 3** *in the* **LAYER CONTROLS** *window, and you will see the following image:*

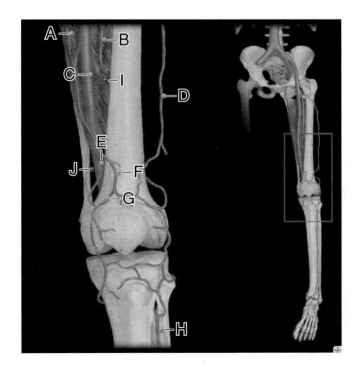

- *Mouse-over the green pins on the screen to find the information necessary to identify the following structures:*

A. _____

B. _____

C. _____

D. _____

E. _____

F. _____

G. _____

H. _____

Non-cardiovascular System Structures (blue pins)

I. _____

J. _____

CHECK POINT:

Knee, Arteries, Anterior View, cont'd

3. List the areas distributed by the femoral artery.
4. Name a deep artery of the thigh that distributes to the hip joint and the quadriceps femoris muscle.
5. Name an artery that distributes to the muscles and skin of the medial thigh.

EXERCISE 6.28:

**Cardiovascular System—Knee,
Arteries, Posterior View**

SELECT TOPIC	SELECT VIEW	
Knee	▸ **Arteries: Posterior**	GO

- *Click* **LAYER 1** *in the* **LAYER CONTROLS** *window, and you will see the following image:*

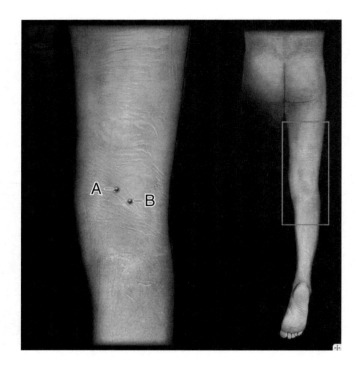

- *Mouse-over the blue pins on the screen to find the information necessary to identify the following non-cardiovascular system structures:*

A. _____

B. _____

CHECK POINT:

Knees, Arteries, Posterior View

1. Name the diamond-shaped area of the posterior knee.
2. Name the artery that provides a pulse in this area.
3. What special requirements must be met to appreciate this pulse?

- *Click* **LAYER 2** *in the* **LAYER CONTROLS** *window, and you will see the following image:*

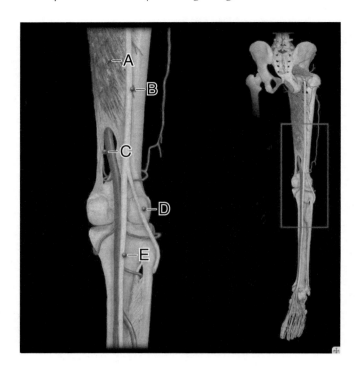

- *Mouse-over the blue pins on the screen to find the information necessary to identify the following non-cardiovascular system structures:*

A. _____

B. _____

C. _____

D. _____

E. _____

CHECK POINT:

Knee, Arteries, Posterior View, cont'd

4. Name the large nerve of the posterior hip and thigh. Name the two nerves that merge to form this nerve.
5. What effect will a herniated intervertebral disk in the lower lumbar region have along the distribution of this nerve?
6. What is the adductor hiatus?

- *Click* **LAYER 3** *in the* **LAYER CONTROLS** *window, and you will see the following image:*

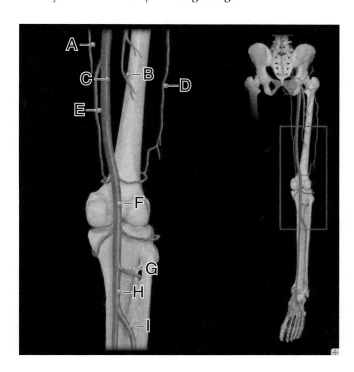

- *Mouse-over the pins on the screen to find the information necessary to identify the following structures:*

A. _____

B. _____

C. _____

D. _____

E. _____

F. _____

G. _____

H. _____

I. _____

CHECK POINT:

Knee, Arteries, Posterior View, cont'd

7. What artery does the femoral artery become in the vicinity of the knee?
8. Where exactly does this name change occur?
9. Name an artery that distributes to the anterior (extensor) compartment of the leg.

EXERCISE 6.29:

Cardiovascular System—Knee, Veins, Anterior View

SELECT TOPIC	SELECT VIEW	
Knee	**Veins: Anterior**	**GO**

- *Click* **LAYER 2** *in the* **LAYER CONTROLS** *window, and you will see the following image:*

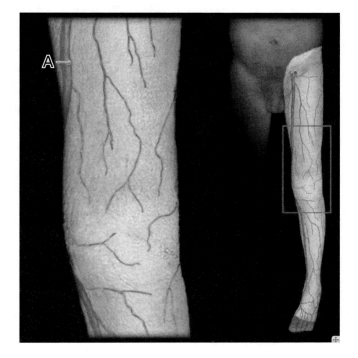

- *Mouse-over the pin on the screen to find the information necessary to identify the following structure:*

A. _____

CHECK POINT:

Knee, Veins, Anterior View

1. Name the large subcutaneous vein of the median leg and thigh.
2. What is its drainage?

- *Click* **LAYER 3** *in the* **LAYER CONTROLS** *window, and you will see the following image:*

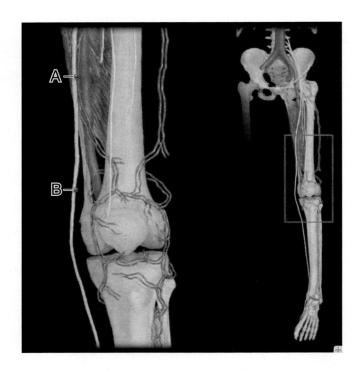

- *Mouse-over the blue pins on the screen to find the information necessary to identify the following non-cardiovascular system structures:*

A. _____

B. _____

- *Click* **LAYER 4** *in the* **LAYER CONTROLS** *window, and you will see the following image:*

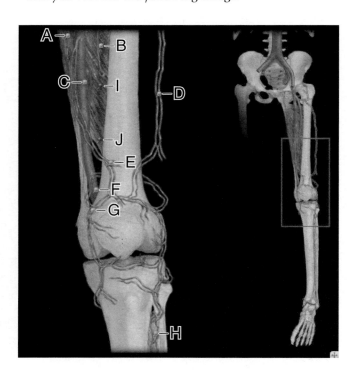

- *Mouse-over the green pins on the screen to find the information necessary to identify the following structures:*

A. _____

B. _____

C. _____

D. _____

E. _____

F. _____

G. _____

H. _____

Non-cardiovascular System Structures (blue pins)

I. _____

J. _____

CHECK POINT:

Knee, Veins, Anterior View, cont'd

3. List the structures drained by the femoral vein.
4. Name the veins that drain the anterior (extensor) compartment of the leg.
5. Name the venous anastomosis that forms a network around the knee. What veins contribute to this anastomosis?

EXERCISE 6.30:

Cardiovascular System—Knee, Veins, Posterior View

SELECT TOPIC	SELECT VIEW	
Knee	▶ **Veins: Posterior**	**GO**

- *Click* **LAYER 2** *in the* **LAYER CONTROLS** *window, and you will see the following image:*

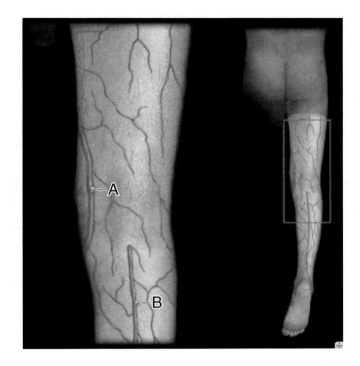

- *Mouse-over the pins on the screen to find the information necessary to identify the following structures:*

A. _____

B. _____

CHECK POINT:

Knee, Veins, Posterior View

1. Name the vein providing drainage for the dorsum of the foot and subcutaneous posterior leg.
2. Name the vein providing drainage for the dorsum of the foot and subcutaneous medial leg and thigh.

- *Click* **LAYER 4** *in the* **LAYER CONTROLS** *window, and you will see the following image:*

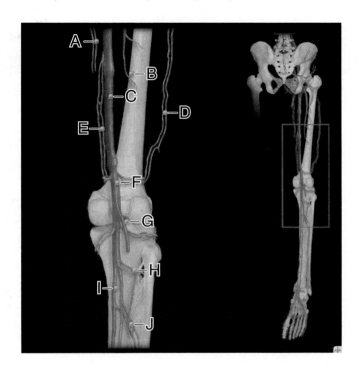

- *Mouse-over the pins on the screen to find the information necessary to identify the following structures:*

A. _____

B. _____

C. _____

D. _____

E. _____

F. _____

G. _____

H. _____

I. _____

J. _____

CHECK POINT:

Knee, Veins, Posterior View, cont'd

3. Name the vein that drains the knee joint and the surrounding structures.
4. Name the paired veins that drain the posterior muscles of the leg.
5. List the drainage of the fibular veins.

EXERCISE 6.31:

Imaging—Knee

SELECT TOPIC	SELECT VIEW	
Knee	▶ Angiogram Arteries: Anterior-Posterior	GO

- Click the **TURN TAGS ON** button, and you will see the following image:

- Mouse-over the green pins on the screen to find the information necessary to identify the following structures:

A. _____

B. _____

C. _____

D. _____

E. _____

Non-cardiovascular System Structures (blue pins)

F. _____

G. _____

H. _____

I. _____

Self Test

Take this opportunity to quiz yourself by taking the **SELF TEST.** See page 15 for a reminder on how to access the self-test for this section.

IN REVIEW

What Have I Learned?

The following questions cover the material that you have just learned, the blood vessels of the knee. Apply what you have learned to answer these questions on a separate piece of paper.

1. Name two blood vessels that change their names at the posterior knee.

2. Where exactly does this name change occur?

3. What new names do these vessels acquire?

4. The terminal branches of which artery supplies everything inferior to the knee?

5. Which artery is the primary blood supply for the posterior leg muscles?

6. List the distribution of the fibular artery.

7. Name the artery with the distribution to the knee joint and the structures around it.

8. Many arteries of the limbs are accompanied by . . .

9. What is the meaning of the Latin word *genu?*

10. Name the paired veins that drain the quadriceps femoris muscles and the hip joint.

The Leg and Foot

EXERCISE 6.32:

Cardiovascular System—Leg and Foot, Arteries, Anterior View

SELECT TOPIC	SELECT VIEW	
Leg and Foot	► **Arteries: Anterior**	**GO**

• *Click* **LAYER 1** *in the* **LAYER CONTROLS** *window, and you will see the following image:*

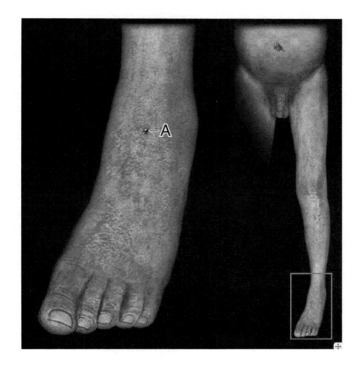

• *Mouse-over the blue pin on the screen to find the information necessary to identify the following non-cardiovascular system structure:*

A. _____

CHECK POINT:

Leg and Foot, Arteries, Anterior View

1. Name the artery used for a pedal pulse.
2. Where is this pulse located?
3. What is this pulse used to assess?

• *Click* **LAYER 2** *in the* **LAYER CONTROLS** *window, and you will see the following image:*

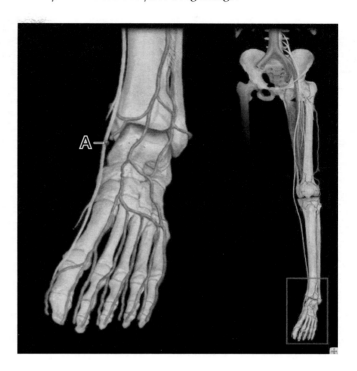

• *Mouse-over the blue pin on the screen to find the information necessary to identify the following non-cardiovascular system structure:*

A. _____

CHECK POINT:

Leg and Foot Arteries, Anterior View, cont'd

4. Name the distal branch of the femoral nerve.
5. Where is it located?
6. Where is the general sensory innervation associated with this nerve?

• Click **LAYER 3** in the **LAYER CONTROLS** window, and you will see the following image:

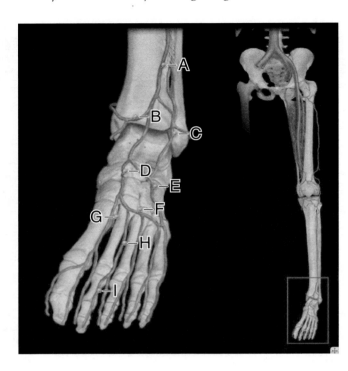

• Mouse-over the pins on the screen to find the information necessary to identify the following structures:

A. _____

B. _____

C. _____

D. _____

E. _____

F. _____

G. _____

H. _____

I. _____

CHECK POINT:

Leg and Foot, Arteries, Anterior View, cont'd

7. Name the terminal branch of the popliteal artery that continues on the dorsum of the foot.

8. What is the name for this artery on the dorsum of the foot?

9. Name the arteries that distribute to the toes and their joints.

EXERCISE 6.33:

Cardiovascular System—Leg and Foot, Arteries, Posterior View

SELECT TOPIC	SELECT VIEW	
Leg and Foot	▶ **Arteries: Posterior**	**GO**

• Click **LAYER 2** in the **LAYER CONTROLS** window, and you will see the following image:

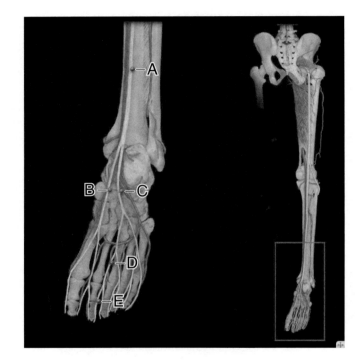

• Mouse-over the blue pins on the screen to find the information necessary to identify the following non-cardiovascular system structures:

A. _____

B. _____

C. _____

D. _____

E. _____

CHECK POINT:

Leg and Foot, Arteries, Posterior View

1. Name the nerve that is both motor and general sensory and innervates the muscles of the posterior leg.

2. Name the lateral terminal branch of this nerve. Where is its motor and general sensory innervation?

3. Name the medial terminal branch of this nerve. Where is its motor and general sensory innervation?

• *Click* **LAYER 3** *in the* **LAYER CONTROLS** *window, and you will see the following image:*

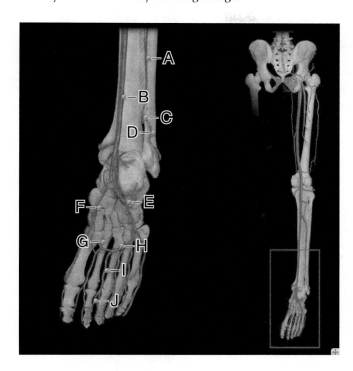

• *Mouse-over the pins on the screen to find the information necessary to identify the following structures:*

A. _____

B. _____

C. _____

D. _____

E. _____

F. _____

G. _____

H. _____

I. _____

J. _____

CHECK POINT:

Leg and Foot, Arteries, Posterior View, cont'd

4. Name the arterial vessel that arches across the plantar surface at the bases of metatarsals II–IV.
5. Name the artery that distributes to the muscles and joints of the medial plantar foot.
6. Name the artery that distributes to the muscles and joints of the lateral plantar foot.

Cardiovascular System—Leg and Foot, Veins, Anterior View

SELECT TOPIC	SELECT VIEW	
Leg and Foot	Veins: Anterior	GO

• *Click* **LAYER 2** *in the* **LAYER CONTROLS** *window, and you will see the following image:*

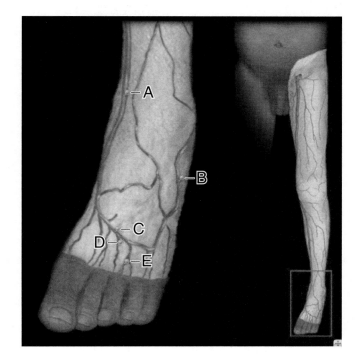

• *Mouse-over the pins on the screen to find the information necessary to identify the following structures:*

A. _____

B. _____

C. _____

D. _____

E. _____

CHECK POINT:

Leg and Foot, Veins, Anterior View

1. Name the venous vessel that curves medial to lateral over the proximal ends of the metatarsals.
2. Name the veins that drain the toes, plus the distal dorsum and joints of the foot.
3. Name the veins that course along the medial and lateral sides of the digits and drain the toes.

• Click **LAYER 4** in the **LAYER CONTROLS** window, and you will see the following image:

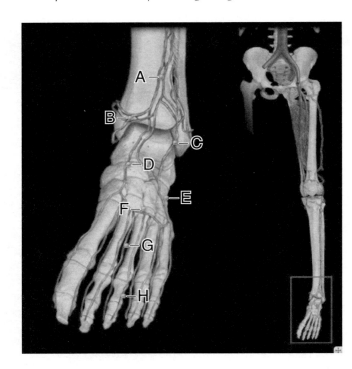

• Mouse-over the pins on the screen to find the information necessary to identify the following structures:

A. _____

B. _____

C. _____

D. _____

E. _____

F. _____

G. _____

H. _____

CHECK POINT:

Leg and Foot, Veins, Anterior View, cont'd

4. Name the vein that drains the lateral tarsal bones.
5. Name the vein that drains the dorsum of the foot, including the toes.
6. Name the veins that drain the toes and metatarsal region of the foot.

EXERCISE 6.35:

Cardiovascular System—Leg and Foot, Veins, Posterior View

SELECT TOPIC	SELECT VIEW	
Leg and Foot	**Veins: Posterior**	GO

• Click **LAYER 2** in the **LAYER CONTROLS** window, and you will see the following image:

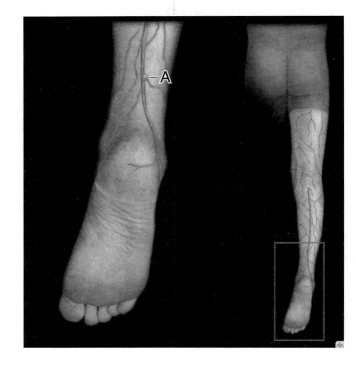

• Mouse-over the pin on the screen to find the information necessary to identify the following structure:

A. _____

CHECK POINT:

Leg and Foot, Veins, Posterior View

1. List the drainage and the course of the small saphenous vein.

- *Click* **LAYER 4** *in the* **LAYER CONTROLS** *window, and you will see the following image:*

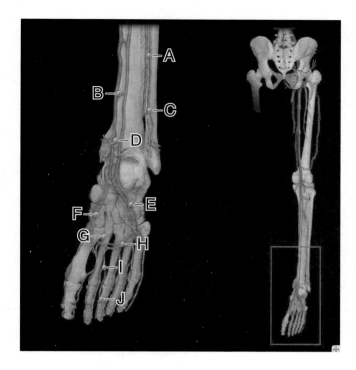

- *Mouse-over the pins on the screen to find the information necessary to identify the following structures:*

A. _____

B. _____

C. _____

D. _____

E. _____

F. _____

G. _____

H. _____

I. _____

J. _____

CHECK POINT:
_____

Leg and Foot, Veins, Posterior View, cont'd

2. Name the venous vessels that arch across the plantar surface at the bases of metatarsals II–IV.
3. List the drainages of the medial plantar veins.
4. List the drainages of the lateral plantar veins.

Self Test

Take this opportunity to quiz yourself by taking the **SELF TEST.** See page 15 for a reminder on how to access the self-test for this section.

IN REVIEW

What Have I Learned?

The following questions cover the material that you have just learned, the blood vessels of the leg and foot. Apply what you have learned to answer these questions on a separate piece of paper.

1. Name the arteries that pass distally along the metatarsals and distribute to the dorsum of the foot and its joints and toes.

2. Name the artery that distributes to the tarsal bones of the lateral foot.

3. Name the artery used for pedal pulse.

4. Name the artery that distributes to the ankle joint and the dorsum of the foot.

5. Name the artery that distributes to the lateral and medial aspects of the ankle joint.

6. Name the nerve located along the plantar aspect of the metatarsals supplying motor innervation to the intrinsic foot muscles. Where is its general sensory innervation?

7. Name the nerve located on the medial and lateral sides of the digits and supplying general sensory innervation to the skin and joints of the toes.

8. List the distribution of the plantar metatarsal arteries.

9. Name the arteries that course along the medial and lateral sides of the digits, distributing to the toes and their joints.

10. Name the vein that drains the posterior leg muscles as it ascends from the ankle through the posterior leg.

Elastic Artery—The Aorta

EXERCISES 6.36:

Cardiovascular System—Elastic Artery (Aorta), Histology

SELECT TOPIC	SELECT VIEW	
Elastic Artery (Aorta) ▶		**GO**

• *Click the* **TURN TAGS ON** *button, and you will see the following image:*

• *Mouse-over the pins on the screen to find the information necessary to identify the following structures:*

A. _____

B. _____

C. _____

D. _____

E. _____

C H E C K P O I N T :

Elastic Artery (Aorta), Histology

1. What is the name for the outermost layer of the vessel wall?
2. What is the name for the middle layer of the vessel wall?
3. What is the name for the innermost layer of the vessel wall?

I N   R E V I E W

What Have I Learned?

The following questions cover the material that you have just learned, the histology of an elastic anterior. Apply what you have learned to answer these questions on a separate piece of paper.

1. In arteries, which tunic or layer is the thickest?

2. In veins, which tunic or layer is the thickest?

3. What is the function of the tunica externa of elastic arteries?

4. What is the function of the tunica media of elastic arteries?

5. What is the function of the tunica intima of elastic arteries?

6. Which of the three tunics contains the endothelium? Did you know that the endothelium of the blood vessels is continuous with the endocardium of the heart?

7. Which structures in the walls of medium and large arteries provide the elastic properties of those blood vessel walls?

8. Name the small blood vessels in the tunica externa.

9. How are the other two tunics supplied with oxygen and nutrients?

Large Vein—The Inferior Vena Cava

EXERCISE 6.37:
Cardiovascular System—Large Vein (Inferior Vena Cava), Histology

 SELECT TOPIC
Large Vein
(Inferior Vena Cava) ▸ **SELECT VIEW** **GO**

• *Click the* **TURN TAGS ON** *button, and you will see the following image:*

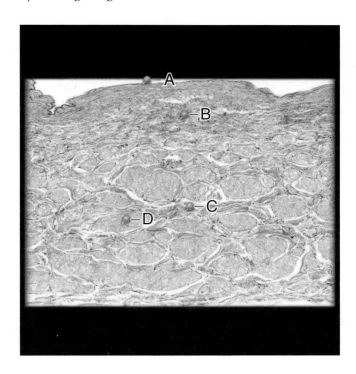

• *Mouse-over the pins on the screen to find the information necessary to identify the following structures:*

A. _____

B. _____

C. _____

D. _____

CHECK POINT:

Large Vein (Inferior Vena Cava), Histology

1. What is the name for the outermost layer of the vessel wall?
2. What is the name for the middle layer of the vessel wall?
3. What is the name for the innermost layer of the vessel wall?

IN REVIEW

What Have I Learned?

The following questions cover the material that you have just learned, the histology of a large vein. Apply what you have learned to answer these questions on a separate piece of paper.

1. Where are the smooth muscle bundles located in veins?

2. What is their function?

3. How does the tunica intima of the inferior vena cava compare with that of the aorta?

4. How does the tunica media of the inferior vena cava compare with that of the aorta?

5. How does the tunica externa of the inferior vena cava compare with that of the aorta?

Muscular Artery and Medium-Sized Vein

EXERCISE 6.38:

Cardiovascular System—Muscular Artery and Medium-Sized Vein, Histology

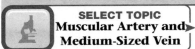

SELECT TOPIC	SELECT VIEW	
Muscular Artery and Medium-Sized Vein		GO

- *Click the **TURN TAGS ON** button, and you will see the following image:*

- *Mouse-over the pins on the screen to find the information necessary to identify the following structures:*

A. _____

B. _____

C. _____

D. _____

E. _____

F. _____

G. _____

H. _____

CHECK POINT:

Muscular Artery and Medium-Sized Vein, Histology

1. Where are muscular arteries located?
2. Where are medium-sized veins located?
3. What is the function of muscular arteries?

IN REVIEW

What Have I Learned?

The following questions cover the material that you have just learned, the histology of muscular arteries and medium-sized veins. Apply what you have learned to answer these questions on a separate piece of paper.

1. What tissue predominates the tunica media of a muscular artery?

2. Give three examples of muscular arteries.

3. What is the function of medium-sized veins?

4. What tissues predominate the tunica media of these veins?

5. Give two examples of medium-sized veins.

Arteriole and Venule

EXERCISE 6.39:
Cardiovascular System—Arteriole and Venule, Histology

SELECT TOPIC
Elastic Artery (Aorta)

SELECT VIEW

GO

• Click the **TURN TAGS ON** button, and you will see the following image:

• Mouse-over the pins on the screen to find the information necessary to identify the following structures:

A. _____

B. _____

C. _____

D. _____

E. _____

F. _____

G. _____

H. _____

I. _____

CHECK POINT:

Arteriole and Venule, Histology

1. What is the primary point for control of blood flow to organs?
2. Where are these vessels located?
3. Where are venules located?

IN REVIEW

What Have I Learned?

The following questions cover the material that you have just learned, the histology of arterioles and venules. Apply what you have learned to answer these questions on a separate piece of paper.

1. What is the diameter of an arteriole?

2. Describe the tunica media of an arteriole.

3. What is the function of an arteriole?

4. What is the diameter of a venule?

5. What is the function of venules?

Neurovascular Bundle

EXERCISE 6.40:

Cardiovascular System—Neurovascular Bundle, Histology

SELECT TOPIC **Neurovascular Bundle**	▶	SELECT VIEW	
			GO

- *Click the* **TURN TAGS ON** *button, and you will see the following image:*

- *Mouse-over the pins on the screen to find the information necessary to identify the following structures:*

A. _____

B. _____

C. _____

D. _____

E. _____

F. _____

CHECK POINT:

Neurovascular Bundle, Histology

1. Where are peripheral nerves located?
2. Describe peripheral nerves.
3. Give at least two examples of peripheral nerves.

IN REVIEW

What Have I Learned?

The following questions cover the material that you have just learned, the histology of a neurovascular bundle. Apply what you have learned to answer these questions on a separate piece of paper.

1. Describe adipose connective tissue.

2. What is its function?

3. Name five structures, other than connective tissue, found in neurovascular bundles.

Animation: Capillary Exchange

| SELECT ANIMATION **Capillary Exchange** | **PLAY** |

- *After viewing the animation, answer these questions:*

1. Describe the structure of capillaries.

2. What is the function of capillaries? What term is used to describe this function?

3. What specific characteristics of capillaries make them optimal for their function?

4. Fluid moves out of the capillaries at the _____ end.

5. What happens to most of that fluid?

6. Name the forces that drive fluids and their dissolved contents into and out of the capillaries.

7. What is net filtration pressure? What does net filtration pressure regulate?

8. What force favors movement of fluid from the blood into the interstitial spaces? Where does this occur?

9. What force causes a shift from filtration to reabsorption in the capillary? Where does this occur?

10. What percent of the fluid that leaves the blood capillary at its arterial end will re-enter the capillary at its venous end?

11. What happens to the remaining fluid that does not re-enter the blood capillary?

12. Using the information from this animation, fill in the blank spaces in the Fluid Exchange Summary table:

Fluid Exchange Summary

ARTERIAL END		VENOUS END
_____	Net hydrostatic pressure	_____
_____	Oncotic pressure	_____
_____	Net filtration pressure	_____

Self Test

Take this opportunity to quiz yourself by taking the **SELF TEST**. See page 15 for a reminder on how to access the self-test for this section.

Cardiac Muscle

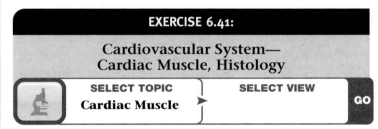

EXERCISE 6.41:

Cardiovascular System— Cardiac Muscle, Histology

| SELECT TOPIC **Cardiac Muscle** ▶ | SELECT VIEW | **GO** |

- *Click the* **TURN TAGS ON** *button, and you will see the following image:*

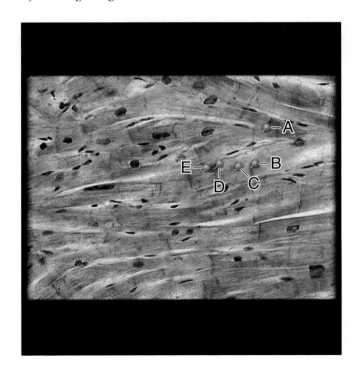

- *Mouse-over the pins on the screen to find the information necessary to identify the following structures:*

A. _____

B. _____

C. _____

D. _____

E. _____

CHECK POINT:

Cardiac Muscle, Histology

1. Name three locations where cardiac muscle is found.
2. What is the function of cardiac muscle tissue?
3. This function may be modulated by . . .

IN REVIEW

What Have I Learned?

The following questions cover the material that you have just learned, the histology of cardiac muscle. Apply what you have learned to answer these questions on a separate piece of paper.

1. Unlike skeletal muscle, cardiac muscle fibers
   _____.

2. Name one way skeletal muscle tissue is similar to cardiac muscle tissue.

3. Describe intercalated disks.

4. What is their function?

5. What is sarcoplasm?

Animation: Hemopoiesis

SELECT ANIMATION	
Hemopoiesis	PLAY

- *After viewing the animation, answer these questions:*

1. Hemopoiesis is the process of _____, which occurs primarily _____.

2. Hemopoiesis begins with undifferentiated cells called _____.

3. These _____ give rise to _____.

4. What types of hormones influence the differentiation of the blood stem cells?

5. What are the two lines of cells that differentiate from the hemocytoblast?

6. Which groups of cells arise from each of these lines?

7. Which cells are produced by the lymphoid cell line?

8. Erythropoiesis produces . . .

9. List the steps for red blood cell (erythrocyte) production.

10. Thrombopoiesis produces _____.

11. What is the name for the committed progenitor cell in thrombopoiesis?

12. In response to the hormone _____, the _____ differentiates into a _____.

13. Platelets are formed when . . .

14. Leukopoiesis is the . . .

15. The myeloid cell line gives rise to . . .

16. The lymphoid cell line produces . . .

17. Eosinophilic myelocytes differentiate into _____.

18. Basophilic myelocytes differentiate into _____.

19. Neutrophilic myelocytes develop into _____.

20. Monocytes are derived from _____.

21. Lymphocytes are derived from _____.

22. The two types of lymphocytes that develop are . . . .

Self Test

Take this opportunity to quiz yourself by taking the **SELF TEST.** See page 15 for a reminder on how to access the self-test for this section.

Animation: Hemoglobin Breakdown

SELECT ANIMATION
Hemoglobin Breakdown
PLAY

- *After viewing the animation, answer these questions:*

1. What happens to the hemoglobin released by the rupture of old red blood cells?

2. The globin chains . . .

3. What is released from the heme?

4. The remaining structure of the heme goes through a two-step process, being converted to the following two products sequentially . . .

5. What plasma protein transports the iron?

6. Where is iron transported for storage (two locations)?

7. Where is iron transported to make new hemoglobin?

8. What plasma protein transports free bilirubin?

9. Where is it transported?

10. Liver cells make _____, excreted as part of _____ into the _____.

11. _____ convert _____ into _____, which contribute to the _____.

12. Some of the _____ are absorbed into the blood and excreted from the _____ in the _____. (This product is urochrome, which gives urine its yellow color.)

Self Test
Take this opportunity to quiz yourself by taking the **SELF TEST.** See page 15 for a reminder on how to access the self-test for this section.

I N   R E V I E W

What Have I Learned?

The following questions cover the material that you have just learned, hemopoiesis and hemoglobin breakdown. Apply what you have learned to answer these questions on a separate piece of paper.

1. Hemopoiesis is the process of _____, which occurs primarily . . .

2. What types of hormones influence the differentiation of the blood stem cells?

3. Erythropoiesis produces _____.

4. Describe the production of platelets.

5. Name the two types of lymphocytes. Where does each type mature?

6. What is hemopoiesis?

7. Name the oxygen-carrying molecule in red blood cells.

8. What two components of this molecule are released by hemopoiesis?

9. Describe how these components are recycled.

10. What products of hemoglobin breakdown end up in the feces? In the urine?

EXERCISE 6.42:

Coloring Exercise

Look up the structures of the heart and then color them with colored pens or pencils.

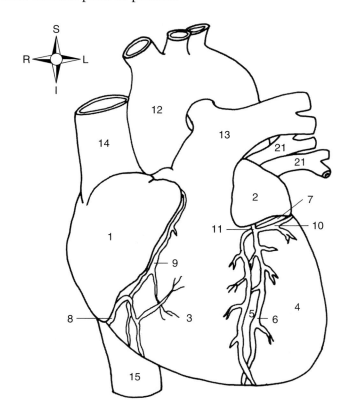

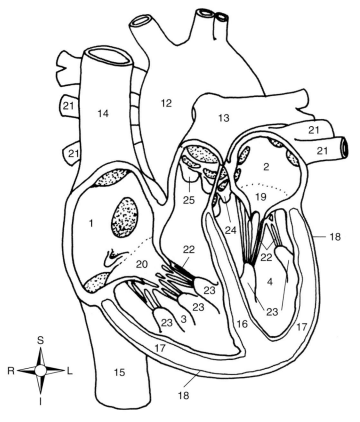

□ 1. _____

□ 2. _____

□ 3. _____

□ 4. _____

□ 5. _____

□ 6. _____

□ 7. _____

□ 8. _____

□ 9. _____

□ 10. _____

□ 11. _____

□ 12. _____

□ 13. _____

□ 14. _____

□ 15. _____

□ 16. _____

□ 17. _____

□ 18. _____

□ 19. _____

□ 20. _____

□ 21. _____

□ 22. _____

□ 23. _____

□ 24. _____

□ 25. _____

EXERCISE 6.43:

Labeling Exercise

Major arteries of the body. Write the names of the arteries indicated on the label lines.

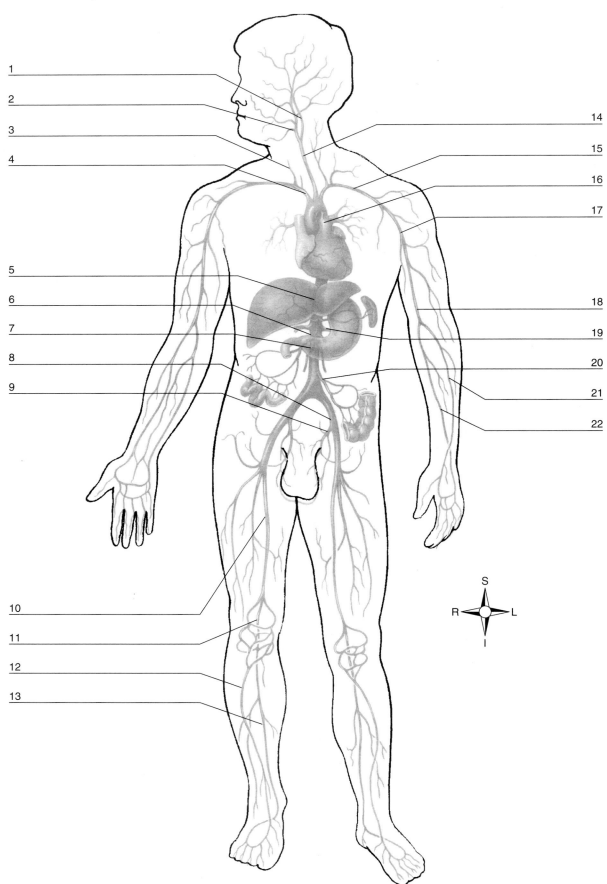

1 _____

2 _____

3 _____

4 _____

5 _____

6 _____

7 _____

8 _____

9 _____

10 _____

11 _____

12 _____

13 _____

14 _____

15 _____

16 _____

17 _____

18 _____

19 _____

20 _____

21 _____

22 _____

EXERCISE 6.44:

Labeling Exercise

Major veins of the body. Write the names of the veins indicated on the label lines.

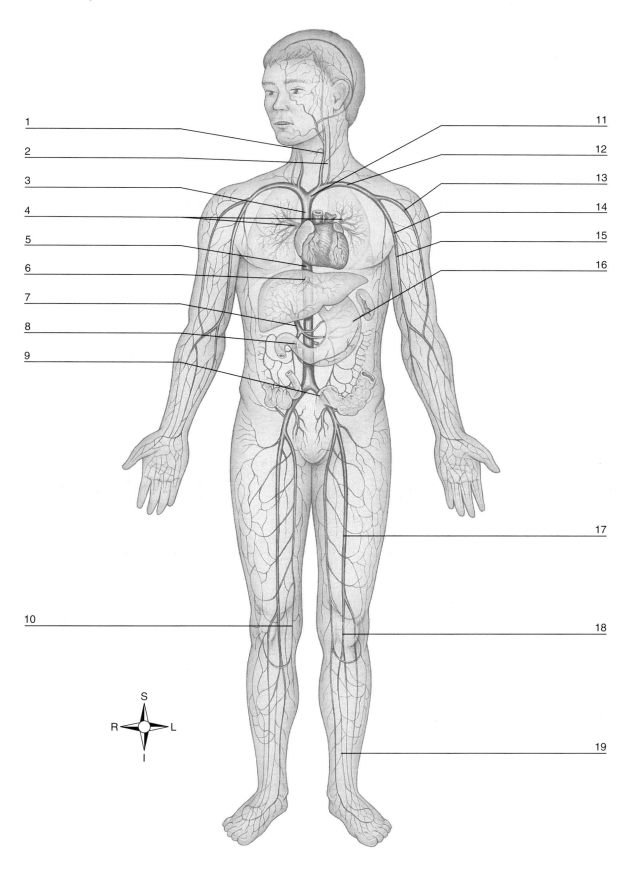

1 _____

2 _____

3 _____

4 _____

5 _____

6 _____

7 _____

8 _____

9 _____

10 _____

11 _____

12 _____

13 _____

14 _____

15 _____

16 _____

17 _____

18 _____

19 _____

S
R — ○ — L
I

CHAPTER 7

The Lymphatic System

Overview: Lymphatic System

You are probably not familiar with the lymphatic system, yet it is vital to your survival. It's not one of those systems that you think about every day—unless you are sick! What does this system do for us? Plenty! It recovers fluid lost from the blood capillaries, is intricately involved with the immune system, and absorbs dietary lipids through lacteals located in the small intestine.

Let's begin our exploration of this mostly unknown organ system by taking an overview of the entire lymphatic system, and then looking at the specific structures in more detail.

- *From the* **Home screen,** *click the drop-down box on the* **Select system** *menu.*
- *From the systems listed, click on* **Lymphatic,** *and you will see the image to the right:*
- *This is the opening screen for the* **Lymphatic System.**

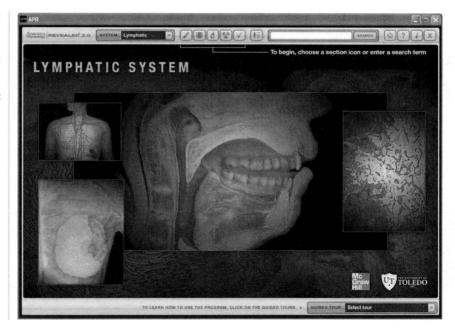

Animation: Lymphatic System

SELECT ANIMATION
Lymphatic System Overview

PLAY

- *After viewing the animation, answer these questions:*

1. What is the lymphatic system?

2. Name the fluid involved in this system.

3. As a system, what is its function?

4. Name the fluid that seeps from the blood capillaries throughout your body.

5. What percentage of this fluid becomes lymph?

6. What are lymphatic capillaries?

7. What do they form when they converge?

8. What do these vessels form when they merge?

9. Where do these vessels drain?

10. From where does the right lymphatic duct receive lymph?

11. Where does the right lymphatic duct empty?

12. From where does the thoracic duct receive lymph?

13. Where does it empty?

14. Which lymphatic duct is larger, the right lymphatic duct or the thoracic duct?

15. Lymphatic tissues include . . .

16. What are lymphatic nodules?

17. What do they contain?

18. Where are clusters of lymphatic nodules associated?

19. Name the large groups of lymphatic nodules found in the walls of the nasal and oral cavities.

20. _____ tonsils are located in the nasopharynx. When they are inflamed, they are known as _____.

21. Where are the palatine tonsils located? The lingual tonsils?

22. Name the lymphatic organs. What is their structure?

23. What are lymph nodes? Where are prominent clusters of lymph nodes located?

24. What are the primary functions of lymph nodes?

25. What is the basic structure of a lymph node?

26. Describe the passage of lymph into, through, and out of the lymph node.

27. Where is the thymus located? What is its function?

28. When is the thymus most active?

29. What becomes of the thymus, beginning at adolescence?

30. Name the body's largest lymph organ. Where is it located?

31. What is its function?

32. How does it act like a lymph node?

Self Test
Take this opportunity to quiz yourself by taking the **SELF TEST.** See page 15 for a reminder on how to access the self-test for this section.

The Thorax

> **EXERCISE 7.1:**
>
> ## Lymphatic System—Thorax, Anterior View
>
> | SELECT TOPIC | SELECT VIEW | |
> |---|---|---|
> | **Thorax** ▶ | **Anterior** | GO |

- *After clicking* **LAYERS 2** *and* **3** *to orient yourself to our location, click* **LAYER 4** *in the* **LAYER CONTROLS** *window, and you will see the following image:*

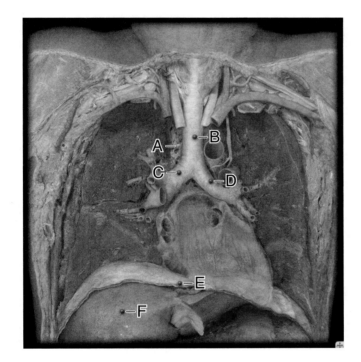

- *Mouse-over the green pin on the screen to find the information necessary to identify the following structure:*

A. _____

Non-lymphatic System Structures (blue pins)

B. _____

C. _____

D. _____

E. _____

F. _____

C H E C K P O I N T:

Thorax, Anterior View

1. What are lymph nodes?
2. The mediastinal lymph nodes are clusters found along what structures?
3. Where are lymph nodes typically found?

- *Click* **LAYER 5** *in the* **LAYER CONTROLS** *window, and you will see the following image:*

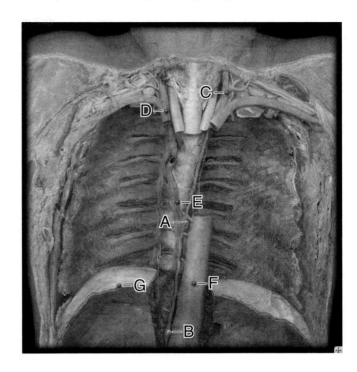

- *Mouse-over the green pins on the screen to find the information necessary to identify the following structures:*

A. _____

B. _____

Non-lymphatic System Structures (blue pins)

C. _____

D. _____

E. _____

F. _____

G. _____

C H E C K P O I N T:

Thorax, Anterior View, cont'd

4. Name the small irregular-shaped lymph sac found in the abdomen.
5. From where does it receive lymph?
6. This structure in question 4 forms the origin of which lymph duct?
7. List the areas drained by the thoracic duct.

I N   R E V I E W

What Have I Learned?

The following questions cover the material that you have just learned, the lymphatic system of the thorax. Apply what you have learned to answer these questions on a separate piece of paper.

1. List the areas drained by the thoracic duct.

2. Lymph nodes are _____.

3. Lymph nodes are typically found in _____.

4. Where are these clusters typically found?

5. Name the structure that forms the origin of the thoracic duct.

Lymph Nodes

Animation: Antigen Processing

SELECT ANIMATION	
Antigen Processing	PLAY

- *After viewing the animation, answer these questions:*

1. What is an antigen?

2. After antigens are produced, where are they transported?

3. What molecules combine with the antigens there? This combination is then transported to the _____ and from there to the _____.

4. What then happens to foreign antigens? To self-antigens?

5. When an antigen originates from outside of the cell, how do the particles enter the cell?

6. What happens to the particles inside the cell?

7. The vesicle containing the foreign particles . . .

8. What then happens to the MHC class II/antigen complex?

Animation: Cytotoxic T Cells

SELECT ANIMATION	
Cytotoxic T Cells	PLAY

- *After viewing the animation, answer these questions:*

1. When a virus infects a cell, what does it produce?

2. What happens to some of these proteins?

3. What molecule complexes with these fragments?

4. Where are they displayed?

5. How do the cytotoxic T cells interact with the virus-infected cells?

6. What substances are released by the cytotoxic T cells?

7. The release of these substances results in . . .

8. What is the result for self-proteins?

9. What then becomes of the cytotoxic T cells?

Animation: Helper T Cells

| SELECT ANIMATION **Helper T Cells** | PLAY |

- *After viewing the animation, answer these questions:*

1. Proteins (antigens) require the cooperation of helper T cells for what purpose?

2. These antigens are therefore said to be . . .

3. What does an antigen presenting cell do in the presence of the antigen?

4. The antigen is then moved _____ on a _____.

5. How does the helper T cell become activated?

6. What is the activated T cell capable of doing?

7. The antigen reacts with an _____ on the surface of the B cell and is then _____.

8. How does the B cell interact with the activated T cell?

9. The helper T cell produces _____, which stimulate the B cell to _____.

Animation: IgE Mediated Hypersensitivity

| SELECT ANIMATION **IgE Mediated Hypersensitivity** | PLAY |

- *After viewing the animation, answer these questions:*

1. Another name for an allergic reaction is . . .

2. This is mediated by . . .

3. How does sensitization occur?

4. Which tissues are rich in B cells committed to IgE production?

5. IgE producing cells are more abundant in . . .

6. What do the helper T cells produce, and what is the effect on B cells?

7. Where and how do IgE molecules attach?

8. What are mast cells?

9. When an antigen-sensitive person is exposed a second time to the antigen, where does the antigen bind?

10. What is required to trigger a response?

11. Within seconds, what chemicals are released from the mast cells? What do they trigger?

12. What are some of those symptoms?

Self Test

Take this opportunity to quiz yourself by taking the **SELF TEST.** See page 15 for a reminder on how to access the self-test for this section.

• *Mouse-over the pins on the screen to find the information necessary to identify the following structures:*

A. _____

B. _____

C. _____

D. _____

E. _____

F. _____

G. _____

H. _____

I. _____

CHECK POINT:

Lymph Node, Histology (Low Magnification)

1. Name the dense irregular connective tissue that covers the outer surface of a lymph node.
2. Name the outer zone of the lymph node.
3. What is its function?

EXERCISE 7.2:

Lymphatic System—Lymph Node, Histology (Low Magnification)

SELECT TOPIC	SELECT VIEW	
Lymph Node ▶	Low Magnification	GO

• *Click the* **TURN TAGS ON** *button, and you will see the following image:*

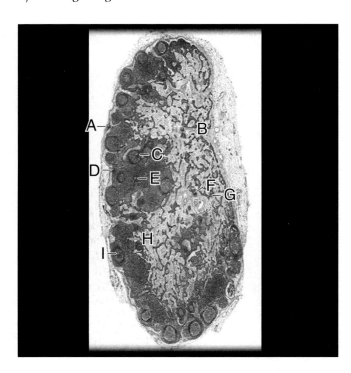

EXERCISE 7.3:

Lymphatic System—Lymph Node, Histology (High Magnification)

SELECT TOPIC	SELECT VIEW	
Lymph Node ▶	High Magnification	GO

• *Click the* **TURN TAGS ON** *button, and you will see the following image:*

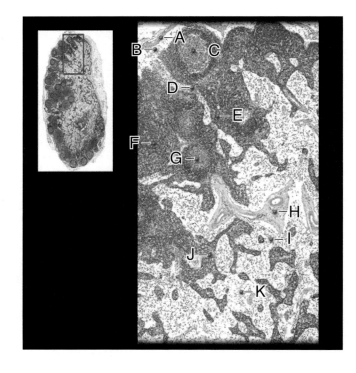

• *Mouse-over the pins on the screen to find the information necessary to identify the following structures:*

A. _____

B. _____

C. _____

D. _____

E. _____

F. _____

G. _____

H. _____

I. _____

J. _____

K. _____

CHECK POINT:

Lymph Node, Histology (High Magnification)

1. Name three structures that are part of a channel system that allows lymph to filter through the node.
2. Name the structure that provides the main structural support for the lymph node.
3. Name the lymph-filled space between the medullary cords.

I N   R E V I E W

What Have I Learned?

The following questions cover the material that you have just learned, the histology of a lymph node. Apply what you have learned to answer these questions on a separate piece of paper.

1. Name the inner zone of the lymph node.

2. What is its function?

3. What percent of the lymphocytes remain in the medulla while the rest leave the node via the lymph?

4. Name the location for memory B lymphocyte and plasma cell formation.

5. Name the site of B lymphocyte localization.

6. Name the structure of the lymph node that forms in response to antigenic challenge. What is its function?

7. Name the site of antibody production. What cells are responsible for this production?

EXERCISE 7.4:

Lymphatic System—Breast and Axillary Nodes—Female

SELECT TOPIC	SELECT VIEW	
Breast and Axillary ➤ Nodes—Female	Anterior	GO

- *After clicking* **LAYERS 1** *through* **3** *to orient yourself to our location, click* **LAYER 4** *in the* **LAYER CONTROLS** *window and you will see the following image:*

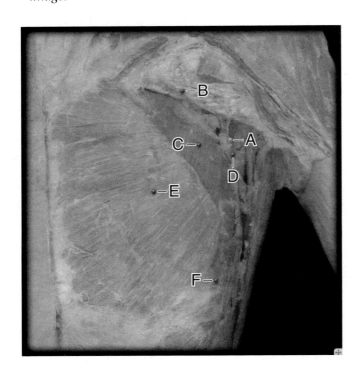

- *Mouse-over the green pin on the screen to find the information necessary to identify the following structure:*

A. _____

Non-lymphatic System Structures (blue pins)

B. _____

C. _____

D. _____

E. _____

F. _____

CHECK POINT:

Breast and Axillary Nodes—Female

1. Describe the axillary lymph nodes.
2. Where are they located?
3. What is their function?

EXERCISE 7.5:

Lymphatic System—Tonsils

SELECT TOPIC	SELECT VIEW	
Tonsils	Lateral	GO

- *Click* **LAYER 2** *in the* **LAYER CONTROLS** *window, and you will see the following image:*

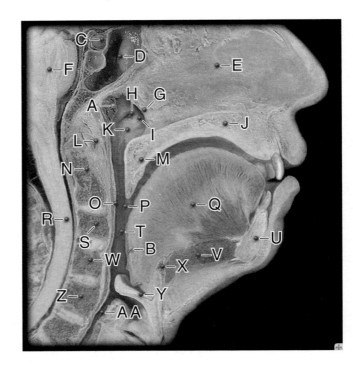

- *Mouse-over the green pins on the screen to find the information necessary to identify the following structures:*

A. _____

B. _____

Non-lymphatic System Structures (blue pins)

C. _____

D. _____

E. _____

F. _____

G. _____

H. _____

I. _____

J. _____

K. _____

L. _____

M. _____

N. _____

O. _____

P. _____

Q. _____

R. _____

S. _____

T. _____

U. _____

V. _____

W. _____

X. _____

Y. _____

Z. _____

AA. _____

CHECK POINT:

Tonsils

1. Describe the pharyngeal tonsil.
2. Where is it located?
3. What is it known as when infected or inflamed?

- *Click* **LAYER 3** *in the* **LAYER CONTROLS** *window, and you will see the following image:*

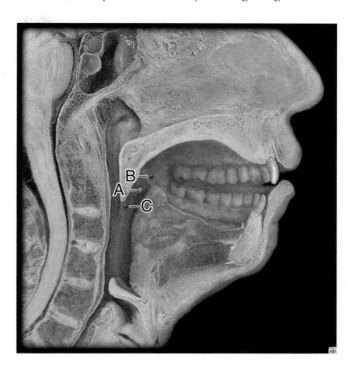

- *Mouse-over the green pin on the screen to find the information necessary to identify the following structure:*

A. _____

Non-lymphatic System Structures (blue pins)

B. _____

C. _____

CHECK POINT:

Tonsils, cont'd

4. Describe the palatine tonsil.
5. Where is it located?
6. "Tonsils" most commonly refers to which tonsils?

I N   R E V I E W

What Have I Learned?

The following questions cover the material that you have just learned, the axillary lymph nodes and tonsils. Apply what you have learned to answer these questions on a separate piece of paper.

1. Describe the axillary lymph nodes.

2. Where are they located?

3. What is their function?

4. Describe the pharyngeal tonsil.

5. Where is it located?

6. What is it known as when infected or inflamed?

7. Describe the lingual tonsil.

8. Where is it located?

9. Describe the palatine tonsil.

10. Where is it located?

11. Which tonsils are visible through the open mouth?

12. "Tonsils" most commonly refers to which tonsils?

13. Which tonsil is the largest during childhood? What occurs to this tonsil by middle age?

Palatine Tonsil

EXERCISE 7.6:

Lymphatic System—Palatine Tonsil, Histology (Low Magnification)

SELECT TOPIC **Palatine Tonsil** ▸ SELECT VIEW **Low Magnification** **GO**

- *Click the* **TURN TAGS ON** *button, and you will see the following image:*

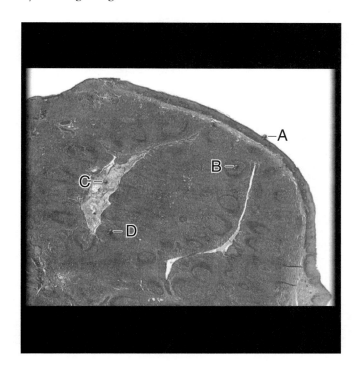

- *Mouse-over the pins on the screen to find the information necessary to identify the following structures:*

A. _____

B. _____

C. _____

D. _____

CHECK POINT:

Palatine Tonsil, Histology (Low Magnification)

1. Name the superficial structure of the palatine tonsils.
2. What is its function?
3. What is the function of the germinal center?

EXERCISE 7.7:

Lymphatic System—Palatine Tonsil, Histology (High Magnification)

SELECT TOPIC **Palatine Tonsil** ▸ SELECT VIEW **High Magnification** **GO**

- *Click the* **TURN TAGS ON** *button, and you will see the following image:*

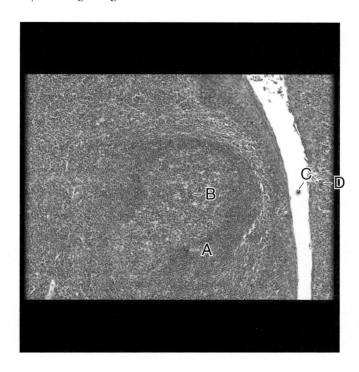

- *Mouse-over the pins on the screen to find the information necessary to identify the following structures:*

A. _____

B. _____

C. _____

D. _____

I N   R E V I E W

What Have I Learned?

The following questions cover the material that you have just learned, the histology of a palatine tonsil. Apply what you have learned to answer these questions on a separate piece of paper.

1. What is the function of the lymphatic nodules?

2. Name the structures that contain discarded epithelial cells, dead white blood cells, bacteria, and debris from the oral cavity.

3. What is the function of these structures? How many are found on each tonsil?

4. Name the structure of the palatine tonsils continuous with the epithelium of the oral cavity and oropharynx.

Peyer's Patch

EXERCISE 7.8:
Lymphatic System—Peyer's Patch, Histology (Low Magnification)

SELECT TOPIC	SELECT VIEW	
Peyer's Patch ▶	**Low Magnification**	**GO**

- *Click the* **TURN TAGS ON** *button, and you will see the following image:*

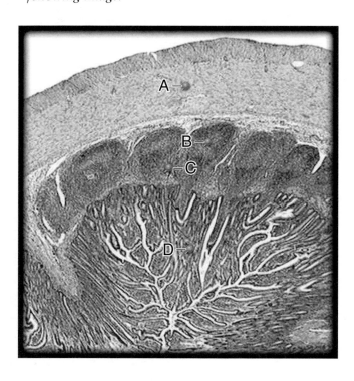

- *Mouse-over the pins on the screen to find the information necessary to identify the following structures:*

A. _____

B. _____

C. _____

D. _____

C H E C K   P O I N T :

Peyer's Patch, Histology (Low Magnification)

1. Specifically, where are the Peyer's patches found?
2. What is their function?

EXERCISE 7.9:

Lymphatic System—Peyer's Patch, Histology (High Magnification)

| SELECT TOPIC
Peyer's Patch | > | SELECT VIEW
**High
Magnification** | **GO** |

- *Click the* **TURN TAGS ON** *button, and you will see the following image:*

- *Mouse-over the pins on the screen to find the information necessary to identify the following structures:*

A. _____

B. _____

C. _____

I N   R E V I E W

What Have I Learned?

The following questions cover the material that you have just learned, the histology of a Peyer's Patch. Apply what you have learned to answer these questions on a separate piece of paper.

1. Where are Peyer's patches located in the body?

2. What is the function of the lymphatic nodules? Why do they form?

3. What is the function of the ileum?

The Spleen

EXERCISE 7.10:

Lymphatic System—Spleen, Anterior View

SELECT TOPIC	SELECT VIEW	
Spleen	**Anterior**	GO

- *Click* **LAYER 1** *in the* **LAYER CONTROLS** *window, and you will see the following image:*

- *Mouse-over the blue pins on the screen to find the information necessary to identify the following non-lymphatic system structures:*

A. _____

B. _____

C. _____

D. _____

E. _____

- *Click* **LAYER 2** *in the* **LAYER CONTROLS** *window, and you will see the following image:*

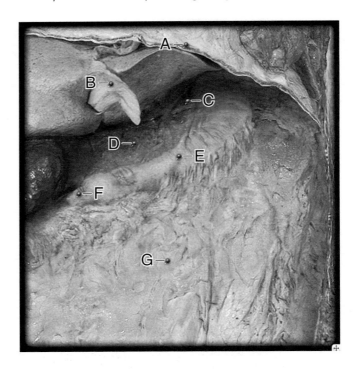

- *Mouse-over the blue pins on the screen to find the information necessary to identify the following non-lymphatic system structures:*

A. _____

B. _____

C. _____

D. _____

E. _____

F. _____

G. _____

- Click **LAYER 3** in the **LAYER CONTROLS** window, and you will see the following image:

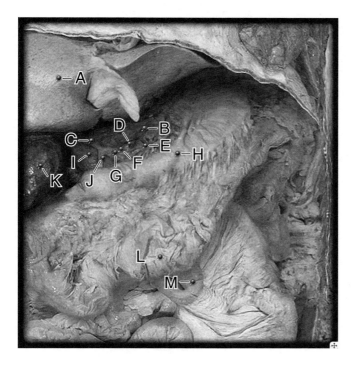

- Mouse-over the blue pins on the screen to find the information necessary to identify the following non-lymphatic system structures:

A. _____

B. _____

C. _____

D. _____

E. _____

F. _____

G. _____

H. _____

I. _____

J. _____

K. _____

L. _____

M. _____

- Click **LAYER 4** in the **LAYER CONTROLS** window, and you will see the following image:

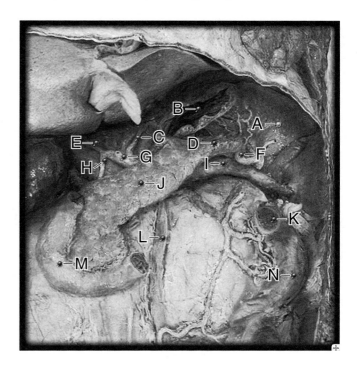

- Mouse-over the green pin on the screen to find the information necessary to identify the following structure:

A. _____

Non-lymphatic System Structures (blue pins)

B. _____

C. _____

D. _____

E. _____

F. _____

G. _____

H. _____

I. _____

J. _____

K. _____

L. _____

M. _____

N. _____

CHECK POINT:

Spleen, Anterior View

1. Name the location of the spleen.
2. What structure of the respiratory system does it contact?
3. Describe the spleen.

• Click **LAYER 5** in the **LAYER CONTROLS** window, and you will see the following image:

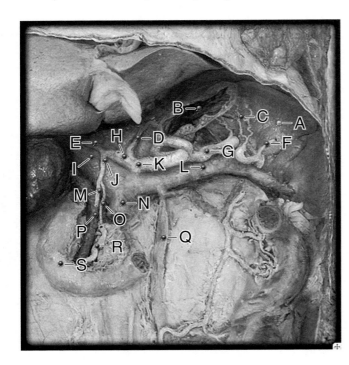

• Mouse-over the green pin on the screen to find the information necessary to identify the following structure:

A. _____

Non-lymphatic System Structures (blue pins)

B. _____

C. _____

D. _____

E. _____

F. _____

G. _____

H. _____

I. _____

J. _____

K. _____

L. _____

M. _____

N. _____

O. _____

P. _____

Q. _____

R. _____

S. _____

CHECK POINT:

Spleen, Anterior View, cont'd

Name four functions of the spleen:

4.

5.

6.

7.

• Click **LAYER 6** in the **LAYER CONTROLS** window, and you will see the following image:

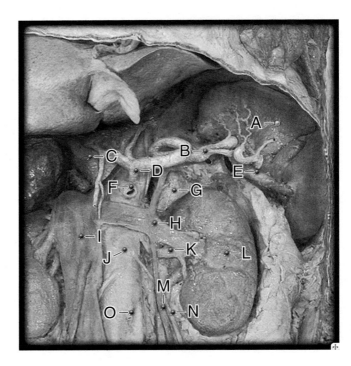

• Mouse-over the green pin on the screen to find the information necessary to identify the following structure:

A. _____

Non-lymphatic System Structures (blue pins)

B. _____

C. _____

D. _____

E. _____

F. _____

G. _____

H. _____

I. _____

J. _____

K. _____

L. _____

M. _____

N. _____

O. _____

C H E C K   P O I N T :

Spleen, Anterior View, cont'd

8. Name the structure that protects the spleen posteriorly.

C H E C K   P O I N T :

Spleen, Histology (Low Magnification)

1. Name the outer covering of the spleen. This structure consists of what tissues?
2. What structure provides the main structural support for the spleen?
3. Name the functions of the red pulp.

EXERCISE 7.11:

Lymphatic System—Spleen, Histology (Low Magnification)

SELECT TOPIC	SELECT VIEW	
Spleen	Low Magnification	GO

- *Click the* **TURN TAGS ON** *button, and you will see the following image:*

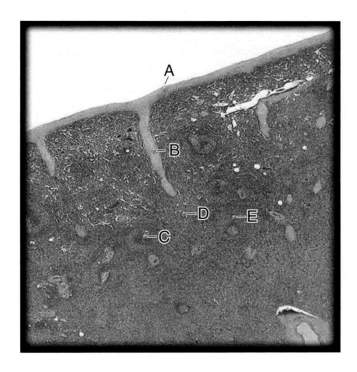

- *Mouse-over the pins on the screen to find the information necessary to identify the following structures:*

A. _____

B. _____

C. _____

D. _____

E. _____

EXERCISE 7.12:

Lymphatic System—Spleen, Histology (High Magnification)

SELECT TOPIC	SELECT VIEW	
Spleen	High Magnification	GO

- *Click the* **TURN TAGS ON** *button, and you will see the following image:*

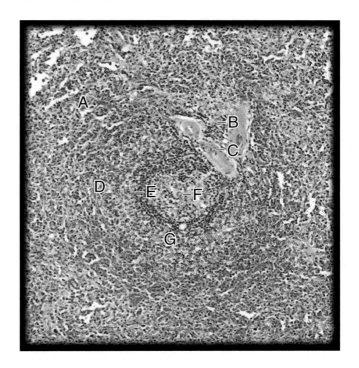

- *Mouse-over the pins on the screen to find the information necessary to identify the following structures:*

A. _____

B. _____

C. _____

D. _____

E. _____

F. _____

G. _____

CHECK POINT:

Spleen, Histology (High Magnification)

1. Name the source of arterial blood to the white pulp. These are a small branch of the _____ artery from the _____ artery.
2. What is the function of the marginal zone?
3. What is the structure of the marginal zone?

IN REVIEW

What Have I Learned?

The following questions cover the material that you have just learned, the spleen. Apply what you have learned to answer these questions on a separate piece of paper.

1. Name the largest lymphatic organ.

2. Name the location of this organ.

3. What structure of the respiratory system does it contact?

4. Describe the largest lymphatic organ.

5. Name four functions of this organ:

6. Name the structure that protects the spleen posteriorly.

7. What is the function of the white pulp? What percent of the total mass of the spleen is white pulp?

8. Name the four structures that make up the white pulp.

9. Name the location for memory B lymphocyte and plasma cell formation. How are they affected by age?

10. What tissue/structure makes up the bulk of the spleen?

11. This tissue/structure is composed of _____.

The Thymus

EXERCISE 7.13:

Lymphatic System—Thymus—Adult, Anterior View

| | SELECT TOPIC
Thymus—Adult > | SELECT VIEW
Anterior | GO |

- *Click* **LAYER 1** *in the* **LAYER CONTROLS** *window, and you will see the following image:*

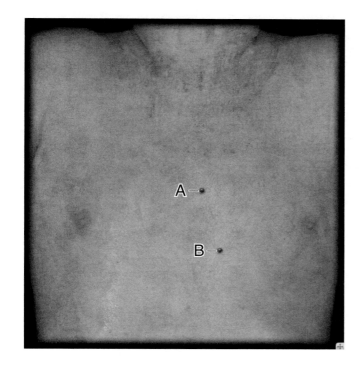

- *Mouse-over the blue pins on the screen to find the information necessary to identify the following non-lymphatic system structures:*

A. _____

B. _____

- *Click* **LAYER 2** *in the* **LAYER CONTROLS** *window, and you will see the following image:*

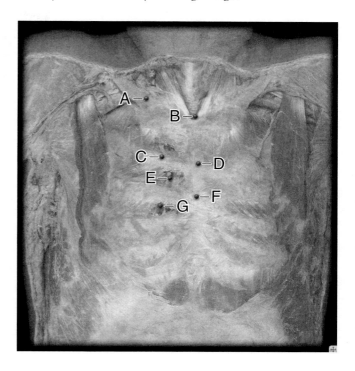

- *Mouse-over the blue pins on the screen to find the information necessary to identify the following non-lymphatic system structures:*

A. _____

B. _____

C. _____

D. _____

E. _____

F. _____

G. _____

- *Click* **LAYER 3** *in the* **LAYER CONTROLS** *window and you will see the following image:*

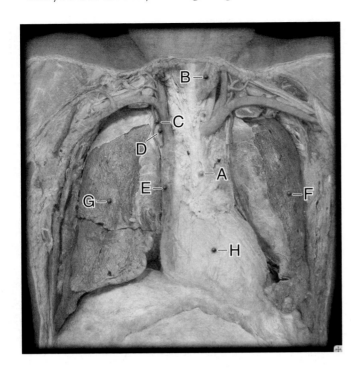

- *Mouse-over the green pin on the screen to find the information necessary to identify the following structure:*

A. _____

Non-lymphatic System Structures (blue pins)

B. _____

C. _____

D. _____

E. _____

F. _____

G. _____

H. _____

C H E C K   P O I N T:
_____

Thymus—Adult, Anterior View

1. Where is the thymus located?
2. Name two functions of the adult thymus:

EXERCISE 7.14:

Lymphatic System—Thymus—Fetus, Anterior View

SELECT TOPIC	SELECT VIEW	
Thymus—Fetus ▶	**Anterior**	GO

- *Click* **LAYER 2** *in the* **LAYER CONTROLS** *window, and you will see the following image:*

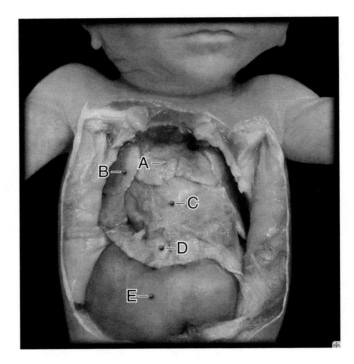

- *Mouse-over the green pin on the screen to find the information necessary to identify the following structure:*

A. _____

Non-lymphatic System Structures (blue pins)

B. _____

C. _____

D. _____

E. _____

C H E C K   P O I N T:
_____

Thymus—Fetus, Anterior View

1. What is the function of thymopoietin and thymosins?
2. What function of the thymus occurs primarily in young individuals?
3. What occurs to the thymus during adolescence?

EXERCISE 7.15:

Lymphatic System—Thymus, Histology (Low Magnification)

SELECT TOPIC
Thymus

SELECT VIEW
Low Magnification

GO

- *Click the **TURN TAGS ON** button, and you will see the following image:*

- *Mouse-over the pins on the screen to find the information necessary to identify the following structures:*

A. _____

B. _____

C. _____

D. _____

E. _____

CHECK POINT:

Thymus, Histology (Low Magnification)

1. Where is the cortex of the thymus located? What is its function?
2. Where is the medulla of the thymus located? What is its function?
3. What is the main subdivision of the thymus?

EXERCISE 7.16:

Lymphatic System—Thymus, Histology (High Magnification)

SELECT TOPIC
Thymus

SELECT VIEW
High Magnification

GO

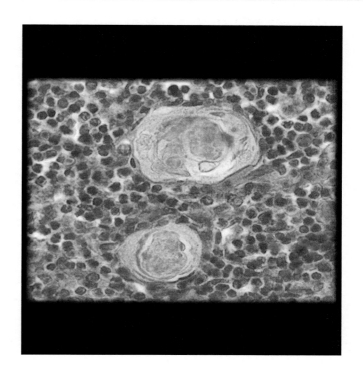

- *There are no pins on the screen, so find the information necessary to identify the highlighted structures:*

A. _____

CHECK POINT:

Thymus, Histology (High Magnification)

1. Name the structures of the thymus that increase in number with age.
2. What is their function?
3. Where in the thymus are these structures located?

Self Test

Take this opportunity to quiz yourself by taking the **SELF TEST**. See page 15 for a reminder on how to access the self-test for this section.

I N   R E V I E W

What Have I Learned?

The following questions cover the material that you have just learned, the histology of the thymus. Apply what you have learned to answer these questions on a separate piece of paper.

1. What is the thymus? Where is it located? What is its function?

2. How does it change from birth to adulthood?

3. Name the subdivisions of the thymic lobe. What is its function?

4. What is the function of the septa/trabeculae?

CHAPTER **8**

The Respiratory System

Overview: The Respiratory System

Take a deep breath. Now exhale. What has just occurred? You have taken in much-needed oxygen and breathed out toxic carbon dioxide. This is foundational to all of your body processes as you maintain homeostasis. Your cells require oxygen for metabolic reactions and produce carbon dioxide as a waste product of that metabolism. Without the respiratory system, metabolism, and thus homeostasis, would be impossible.

But, the function of the respiratory system doesn't stop there. It is also involved in regulating the body's acid–base balance, blood gases, and other homeostatic controls of the circulatory system. Like so many of our organ systems, the respiratory system doesn't exist in a vacuum, but works in coordination with the circulatory and urinary systems.

We will begin with an overview of the respiratory system. Next, we will take a regional view of the respiratory structures, followed by a detailed look at these structures.

- *From the* **Home screen,** *click the drop-down box on the* **Select system** *menu.*

- *From the systems listed, click on* **Respiratory,** *and you will see the image to the right:*

- *This is the opening screen for the* **Respiratory System.**

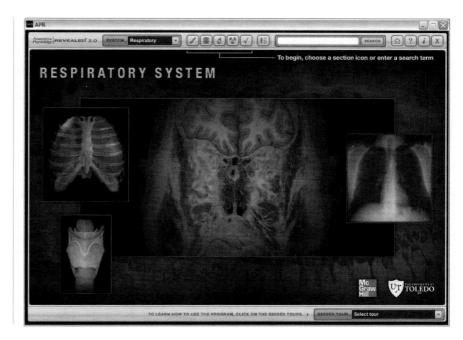

Animation: Respiratory System Overview

	SELECT ANIMATION	
	Respiratory System Overview	PLAY

- *After viewing the animation, answer these questions:*

1. What are the two functions of the respiratory system?

2. Name the structures of the upper respiratory tract.

3. Name the structures of the lower respiratory tract.

4. What effect does the nasal cavity have on inhaled air?

5. Name the structure shared by both the respiratory and digestive systems.

6. Name the structure that contains the vestibular and vocal folds. What protective function do these folds have?

7. The vocal folds are also known as the _____. Why?

8. Name the organ that maintains an open passageway to and from the lungs. What particular structure helps to keep this passageway open?

9. This passageway divides into two _____, which upon entering the lungs, continue to divide into smaller _____ until they ultimately divide into _____.

10. Each _____ divides repetitively to form _____, _____, and _____.

11. What structure serves as the site for gas exchange? These rounded structures are surrounded by . . .

12. How do oxygen and carbon dioxide move between the blood and the alveoli?

Self Test

Take this opportunity to quiz yourself by taking the **SELF TEST**. See page 15 for a reminder on how to access the self-test for this section.

The Upper Respiratory System

EXERCISE 8.1:
Respiratory System—Upper Respiratory, Lateral View

SELECT TOPIC	SELECT VIEW	
Upper Respiratory ➤	Lateral	**GO**

• *Click* **LAYER 1** *in the* **LAYER CONTROLS** *window, and you will see the following image:*

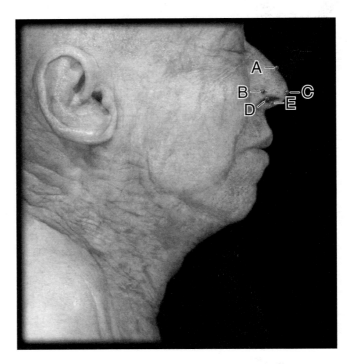

• *Mouse-over the pins on the screen to find the information necessary to identify the following structures:*

A. _____

B. _____

C. _____

D. _____

E. _____

C H E C K   P O I N T :

Upper Respiratory, Lateral View

1. Name the external opening of the nose.
2. Name the structure that separates both of the openings of the nose.
3. Name the midline dorsal aspect of the nose.

• *Click* **LAYER 2** *in the* **LAYER CONTROLS** *window, and you will see the following image:*

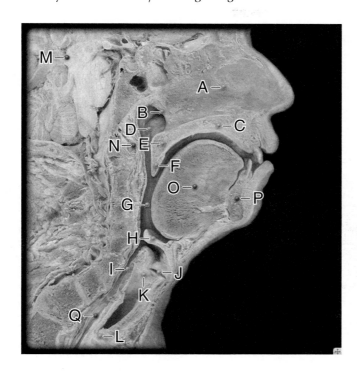

• *Mouse-over the green pins on the screen to find the information necessary to identify the following structures:*

A. _____

B. _____

C. _____

D. _____

E. _____

F. _____

G. _____

H. _____

I. _____

J. _____

K. _____

L. _____

Non-respiratory System Structures (blue pins)

M. _____

N. _____

O. _____

P. _____

Q. _____

C H E C K   P O I N T :

Upper Respiratory, Lateral View, cont'd

4. Name the paired posterior nasal openings.
5. Name two structures contained in the nasopharynx.
6. Name the structure that closes over the laryngeal inlet when swallowing.

• *Click* **LAYER 3** *in the* **LAYER CONTROLS** *window, and you will see the following image:*

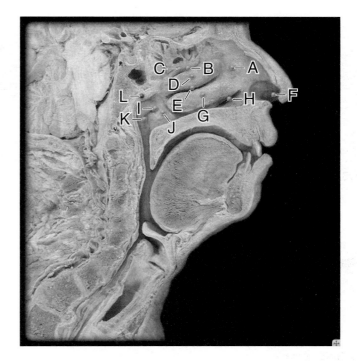

• *Mouse-over the green pins on the screen to find the information necessary to identify the following structures:*

A. _____

B. _____

C. _____

D. _____

E. _____

F. _____

G. _____

H. _____

I. _____

J. _____

K. _____

Non-respiratory System Structure (blue pin)

L. _____

CHECK POINT:

Upper Respiratory, Lateral View, cont'd

7. Name the expanded area inside of the nares. What does the Latin term mean?
8. The mucosa of what structure has both respiratory and olfactory parts?
9. Name the two shelflike projections of bone covered with mucosa and part of the ethmoid bone.

CHECK POINT:

Upper Respiratory, Lateral View, cont'd

10. Name the paranasal sinuses.
11. Name the narrow curved gap that contains the openings of the frontal and maxillary sinuses, plus the anterior ethmoid air cells.
12. Name the mucous membrane-lined cavity in the body of the sphenoid bone. Where does it drain?

• *Click **LAYER 4** in the **LAYER CONTROLS** window, and you will see the following image:*

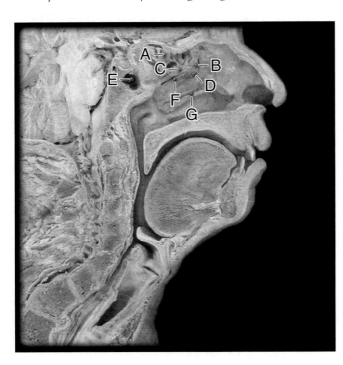

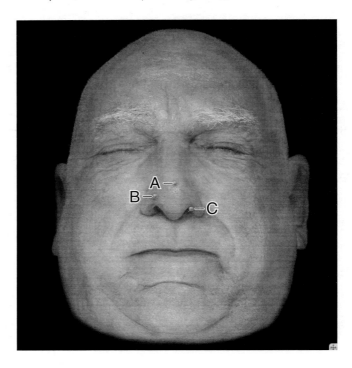

EXERCISE 8.2:

Respiratory System—Nasal Cavity, Coronal View

SELECT TOPIC	SELECT VIEW	
Nasal Cavity ▶	Coronal	GO

• *Click **LAYER 1** in the **LAYER CONTROLS** window, and you will see the following image:*

• *Mouse-over the pins on the screen to find the information necessary to identify the following structures:*

A. _____

B. _____

C. _____

D. _____

E. _____

F. _____

G. _____

• *Mouse-over the pins on the screen to find the information necessary to identify the following structures:*

A. _____

B. _____

C. _____

CHECK POINT:

Nasal Cavity, Coronal View

1. Name the lateral wall of the nostril.
2. What tissue composes this structure?
3. Define ala.

• *Click* **LAYER 2** *in the* **LAYER CONTROLS** *window, and you will see the following image:*

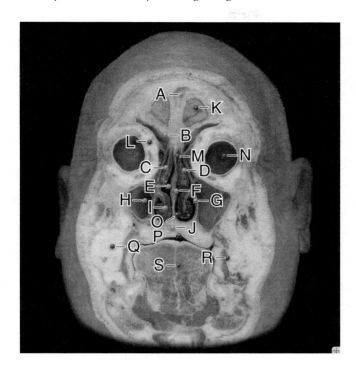

• *Mouse-over the green pins on the screen to find the information necessary to identify the following structures:*

A. _____

B. _____

C. _____

D. _____

E. _____

F. _____

G. _____

H. _____

I. _____

J. _____

Non-respiratory System Structures (blue pins)

K. _____

L. _____

M. _____

N. _____

O. _____

P. _____

Q. _____

R. _____

S. _____

CHECK POINT:

Nasal Cavity, Coronal View, cont'd

4. Name a structure commonly deviated from the midline, impacting airflow.
5. Name the structure containing the nasolacrimal duct.
6. Name the superior opening in the frontal sinus.

• *Click* **LAYER 3** *in the* **LAYER CONTROLS** *window, and you will see the following image:*

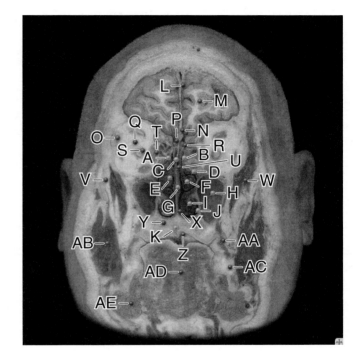

• *Mouse-over the green pins on the screen to find the information necessary to identify the following structures:*

A. _____

B. _____

C. _____

D. _____

E. _____

F. _____

G. _____

H. _____

I. _____

J. _____

K. _____

Non-respiratory System Structures (blue pins)

L. _____

M. _____

N. _____

O. _____

P. _____

Q. _____

R. _____

S. _____

T. _____

U. _____

V. _____

W. _____

X. _____

Y. _____

Z. _____

AA. _____

AB. _____

AC. _____

AD. _____

AE. _____

CHECK POINT:

Nasal Cavity, Coronal View, cont'd

7. Name the largest paranasal sinus.
8. This sinus is a common site of . . .
9. Where is this sinus located?

- *Click* **LAYER 4** *in the* **LAYER CONTROLS** *window, and you will see the following image:*

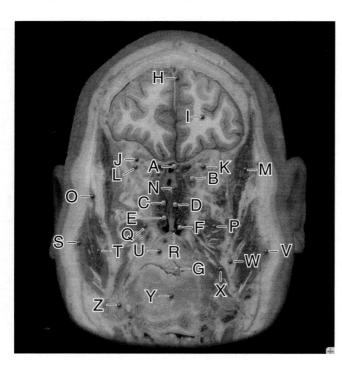

- *Mouse-over the green pins on the screen to find the information necessary to identify the following structures:*

A. _____

B. _____

C. _____

D. _____

E. _____

F. _____

G. _____

Non-respiratory System Structures (blue pins)

H. _____

I. _____

J. _____

K. _____

L. _____

M. _____

N. _____

O. _____

P. _____

Q. _____

R. _____

S. _____

T. _____

U. _____

V. _____

W. _____

X. _____

Y. _____

Z. _____

CHECK POINT:

Nasal Cavity, Coronal View, cont'd

10. Name a shelflike projection of the ethmoid bone.
11. What is its function?
12. Name a shelflike projection that is not a part of, but does articulate with, the ethmoid bone.

- Click **LAYER 5** in the **LAYER CONTROLS** window, and you will see the following image:

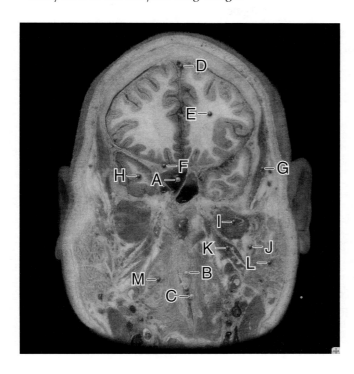

- Mouse-over the green pins on the screen to find the information necessary to identify the following structures:

A. _____

B. _____

C. _____

Non-respiratory System Structures (blue pins)

D. _____

E. _____

F. _____

G. _____

H. _____

I. _____

J. _____

K. _____

L. _____

M. _____

CHECK POINT:

Nasal Cavity, Coronal View, cont'd

13. Describe the uvula.
14. Where is it located?
15. Name the subdivision of the pharynx that is part of both the respiratory and digestive tracts.

EXERCISE 8.3a:

Imaging—Upper Respiratory System

SELECT TOPIC	SELECT VIEW	
Upper Respiratory ➤	**Coronal**	GO

- Click the **TURN TAGS ON** button, and you will see the following image:

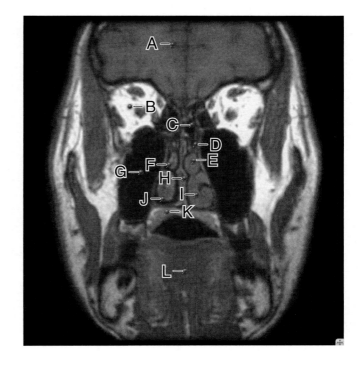

- Mouse-over the blue pins on the screen to find the information necessary to identify the following non-respiratory system structures:

A. _____

B. _____

C. _____

D. _____

E. _____

F. _____

G. _____

H. _____

I. _____

J. _____

K. _____

L. _____

EXERCISE 8.3b:

Imaging—Upper Respiratory System

SELECT TOPIC	SELECT VIEW	
Upper Respiratory ▶	**Sagittal**	**GO**

- *Click the* **TURN TAGS ON** *button, and you will see the following image:*

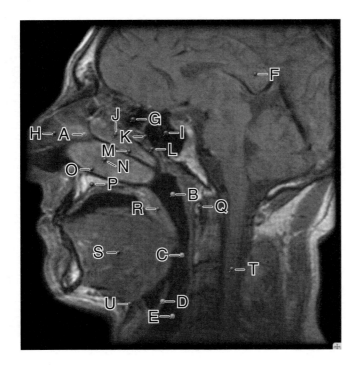

- *Mouse-over the green pins on the screen to find the information necessary to identify the following structures:*

A. _____

B. _____

C. _____

D. _____

E. _____

Non-respiratory System Structures (blue pins)

F. _____

G. _____

H. _____

I. _____

J. _____

K. _____

L. _____

M. _____

N. _____

O. _____

P. _____

Q. _____

R. _____

S. _____

T. _____

U. _____

Self Test

Take this opportunity to quiz yourself by taking the **SELF TEST.** See page 15 for a reminder on how to access the self-test for this section.

I N   R E V I E W

What Have I Learned?

The following questions cover the material that you have just learned, the upper respiratory system. Apply what you have learned to answer these questions on a separate piece of paper.

1. What is the Latin term for the vestibule? What does this word mean?

2. Name the three midline structures that make up the nasal septum.

3. Name the largest paranasal sinus.

4. Name the portion of the throat posterior to the oral cavity. What lymphatic organs are located there?

5. Name the structure of the soft palate that elevates during swallowing to prevent food from entering the nasopharynx.

6. Name a structure commonly deviated from the midline, impacting airflow.

7. Name the shelflike projection of mucosa-covered bone that is a separate bone.

8. Name the narrow space that contains the opening of the nasolacrimal duct.

9. What is the three-fold purpose for increasing the surface area of the nasal cavity with conchae and meatus?

10. Name the collection of lymphatic tissue in the nasopharynx.

11. Name the structure of the nasopharynx that represents the anterior end of the narrow duct between the middle ear and the pharynx. Name the structure of the nasopharynx that surrounds it.

The Lower Respiratory System

EXERCISE 8.4:

Respiratory System—Lower Respiratory, Anterior View

SELECT TOPIC	SELECT VIEW	
Lower Respiratory ▶	**Anterior**	**GO**

- *Click* **LAYER 1** *in the* **LAYER CONTROLS** *window, and you will see the following image:*

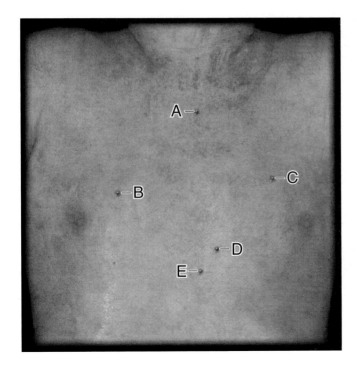

- *Mouse-over the blue pins on the screen to find the information necessary to identify the following non-respiratory system structures:*

A. _____

B. _____

C. _____

D. _____

E. _____

- *Click* **LAYER 2** *in the* **LAYER CONTROLS** *window, and you will see the following image:*

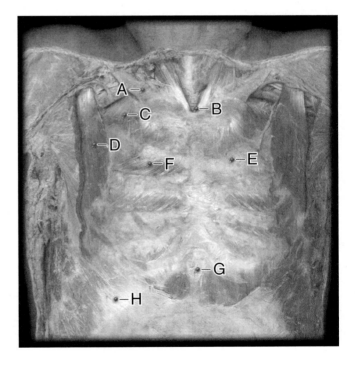

- *Mouse-over the blue pins on the screen to find the information necessary to identify the following non-respiratory system structures:*

A. _____

B. _____

C. _____

D. _____

E. _____

F. _____

G. _____

H. _____

- *Click* **LAYER 3** *in the* **LAYER CONTROLS** *window, and you will see the following image:*

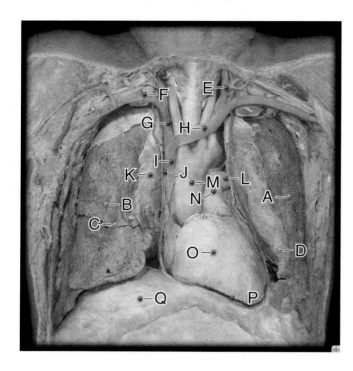

- *Mouse-over the green pins on the screen to find the information necessary to identify the following structures:*

A. _____

B. _____

C. _____

D. _____

Non-respiratory System Structures (blue pins)

E. _____

F. _____

G. _____

H. _____

I. _____

J. _____

K. _____

L. _____

M. _____

N. _____

O. _____

P. _____

Q. _____

CHECK POINT:

Lower Respiratory, Anterior View

1. How many lobes are in the left lung? What are they?
2. How many lobes are in the right lung? What are they?
3. Why does the left lung have one fewer lobe than the right lung?

- *Click* **LAYER 4** *in the* **LAYER CONTROLS** *window, and you will see the following image:*

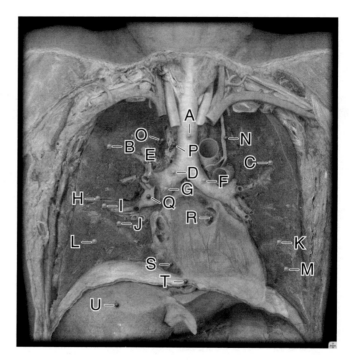

- *Mouse-over the green pins on the screen to find the information necessary to identify the following structures:*

A. _____

B. _____

C. _____

D. _____

E. _____

F. _____

G. _____

H. _____

I. _____

J. _____

K. _____

L. _____

M. _____

Non-respiratory System Structures (blue pins)

N. _____

O. _____

P. _____

Q. _____

R. _____

S. _____

T. _____

U. _____

C H E C K   P O I N T :

Lower Respiratory, Anterior View, cont'd

4. Name the structures formed by the bifurcation of the trachea. Which of the two is where foreign bodies that enter the trachea tend to pass?
5. Name the branches eminating from the main bronchi. How many are found in each lung?
6. Name the branches eminating from the bronchi in question 5.

• Click **LAYER 5** in the **LAYER CONTROLS** window, and you will see the following image:

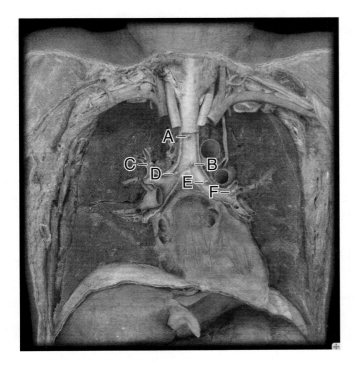

• Mouse-over the green pins on the screen to find the information necessary to identify the following structures:

A. _____

B. _____

C. _____

D. _____

E. _____

F. _____

Animation: Thoracic Cavity Dimensional Changes

> **SELECT ANIMATION**
> **Thoracic Cavity Dimensional Changes**   **PLAY**

• *After viewing the animation, answer these questions:*

1. Another term for breathing is _____.

2. Both _____ and _____ result from changes in _____ in the _____.

3. What structures drive these changes?

4. During inspiration, the thoracic cavity increases in _____. Why?

5. What regulates the length of the thoracic cavity?

6. How is the length of the thoracic cavity increased during inspiration?

7. What changes occur with the diaphragm and the thoracic cavity during expiration?

8. What regulates the depth and width of the thoracic cavity? How?

9. How does the elevation of the ribs affect the thoracic cavity width? This motion is similar to _____.

10. What effect does this elevation of the ribs have on the sternum? What effect does this have on the thoracic cavity depth?

11. How does the movement of the sternum and ribs facilitate inspiration?

Self Test

Take this opportunity to quiz yourself by taking the **SELF TEST.** See page 15 for a reminder on how to access the self-test for this section.

Animation: Partial Pressure

SELECT ANIMATION
Partial Pressure   PLAY

- *After viewing the animation, answer these questions:*

1. What is the partial pressure of oxygen in fresh air entering the lungs?

2. What effect does moisture in the lungs have on this number?

3. What is the partial pressure of carbon dioxide in fresh air entering the lungs?

4. What effect does carbon dioxide delivered to the lungs from the blood have on this number?

5. Describe the direction of diffusion of oxygen and carbon dioxide in the alveoli.

6. This occurs because of _____.

7. This occurs at the _____ ends of the _____ _____.

8. What occurs as a result of diffusion at the venous ends of the pulmonary capillaries?

9. With no differences in partial pressure, . . .

10. How do oxygen and carbon dioxide diffuse into/out of the tissue capillaries?

11. This occurs because of . . .

12. What occurs at the venous ends of the tissue capillaries?

13. The blood now carries the _____ to the _____.

14. In the body, all of these exchanges occur . . .

Self Test

Take this opportunity to quiz yourself by taking the **SELF TEST.** See page 15 for a reminder on how to access the self-test for this section.

Animation: Alveolar Pressure Changes

SELECT ANIMATION
Alveolar Pressure Changes   PLAY

- *After viewing the animation, answer these questions:*

1. At the end of expiration, . . .

2. Therefore, . . .

3. Inspiration begins with _____ of the _____ to _____.

4. This results in . . .

5. The increased alveolar pressure causes a _____ in _____ below _____ and _____ flows _____.

6. At the end of inspiration, . . .

7. Air flow into the lungs causes . . .

8. The pressure becomes equal, so . . .

9. During expiration, the _____ of the _____ _____ as the _____ _____, and the _____ and the _____ _____.

10. This results in a _____ in _____ and an _____ in _____.

11. The _____ is now _____ than _____, so air flows _____ of the lungs.

12. Air continues to flow out of the lungs until . . .

Self Test
Take this opportunity to quiz yourself by taking the **SELF TEST.** See page 15 for a reminder on how to access the self-test for this section.

Animation: Diffusion Across Respiratory Membrane

	SELECT ANIMATION	
	Diffusion Across Respiratory Membrane	PLAY

• *After viewing the animation, answer these questions:*

1. In the lungs, gas exchange takes place . . .

2. What are alveoli? Where are they located?

3. What is the diameter of an alveolus? How many are in each lung?

4. What are the two types of specialized cells in the wall of the alveoli?

5. The cells that form 90 percent of the alveolar wall are . . .

6. Describe these cells.

7. Name and describe the second type of cells.

8. What do these cells secrete? What is the function of this secretion?

9. _____ form a network around each alveolus.

10. Name the thin structure that separates the capillary blood from the air in the alveolus.

11. What is this thin structure comprised of?

12. This structure has a thickness of only _____,
    which facilitates _____.

13. What causes the diffusion of gases across this
    membrane? Explain it for both oxygen and carbon
    dioxide.

Self Test

Take this opportunity to quiz yourself by taking the **SELF
TEST**. See page 15 for a reminder on how to access the
self-test for this section.

EXERCISE 8.5a:

Imaging—Lower Respiratory System

SELECT TOPIC	SELECT VIEW	
Lower Respiratory ➤	Bronchogram	GO

- *Click the* **TURN TAGS ON** *button, and you will see the
  following image:*

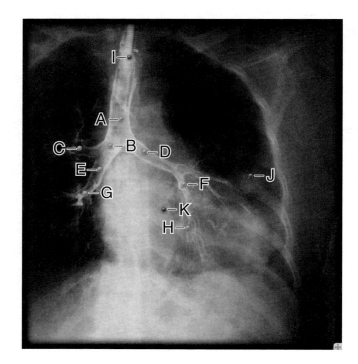

- *Mouse-over the green pins on the screen to find the information necessary to identify the following structures:*

A. _____

B. _____

C. _____

D. _____

E. _____

F. _____

G. _____

H. _____

Non-respiratory System Structures (blue pins)

I. _____

J. _____

K. _____

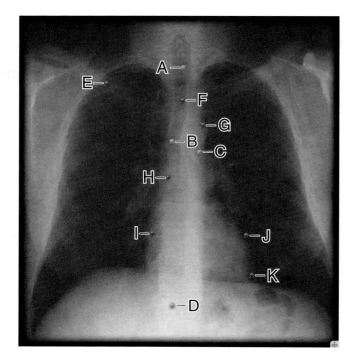

EXERCISE 8.5b:

Imaging—Lower Respiratory System

SELECT TOPIC	SELECT VIEW	
Lower Respiratory ➤	X-ray	GO

- *Click the* **TURN TAGS ON** *button, and you will see the
  following image:*

- *Mouse-over the green pins on the screen to find the information necessary to identify the following structures:*

A. _____

B. _____

C. _____

D. _____

Non-respiratory System Structures (blue pins)

E. _____

F. _____

G. _____

H. _____

I. _____

J. _____

K. _____

I N   R E V I E W

What Have I Learned?

The following questions cover the material that you have just learned, the lower respiratory system. Apply what you have learned to answer these questions on a separate piece of paper.

1. Name the serous membrane that covers the lungs.

2. Name the space formed between the visceral and parietal pleura.

3. What is the primary muscle of respiration?

4. What pressure changes occur when the primary muscle contracts?

5. What nerve is responsible for these contractions?

6. The oblique fissure of the left lung separates the _____ lobes, while in the right lung it separates the _____ lobes.

7. The horizontal fissure of the right lung separates the _____ lobes.

The Trachea

EXERCISE 8.6:

Respiratory System—Trachea, Histology (Low Magnification)

SELECT TOPIC	SELECT VIEW	
Trachea	▶ **Low Magnification**	**GO**

- *Click the* **TURN TAGS ON** *button, and you will see the following image:*

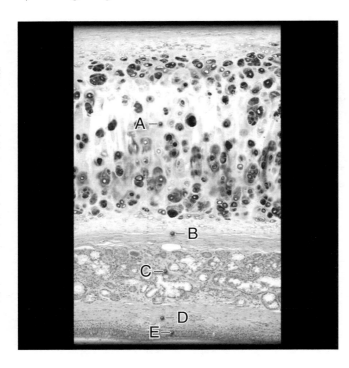

- *Mouse-over the pins on the screen to find the information necessary to identify the following structures:*

A. _____

B. _____

C. _____

D. _____

E. _____

CHECK POINT:

Trachea, Histology (Low Magnification)

1. What type of cartilage provides support for the majority of the respiratory tract?
2. Name the fibrocartilage layer that covers the hyaline cartilage.
3. Name the layer of the trachea that contains numerous seromucous glands and is rich in blood and lymph vessels.
4. Name the layer of the trachea that provides support for the epithelial cells. Name the lymphoid elements it may contain.
5. What tissue type makes up the epithelial layer? What is its function?

EXERCISE 8.7:

Respiratory System—Trachea, Histology (High Magnification)

SELECT TOPIC	SELECT VIEW	
Trachea	▶ **High Magnification**	**GO**

- *Click the* **TURN TAGS ON** *button, and you will see the following image:*

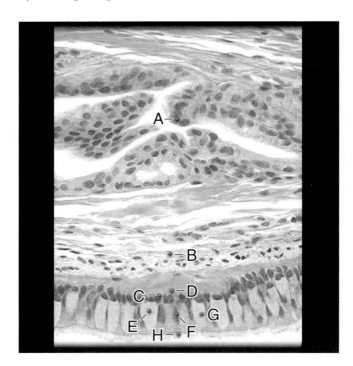

- *Mouse-over the pins on the screen to find the information necessary to identify the following structures:*

A. _____

B. _____

C. _____

D. _____

E. _____

F. _____

G. _____

H. _____

CHECK POINT:

Trachea, Histology (High Magnification)

1. Name the tissue layer that provides support for most epithelia in the body.
2. Name the stem cell that can produce new epithelial cells.
3. Name the mucous-producing epithelial cells. What is the purpose of the mucus?
4. What is the function of the cilia on the ciliated epithelial cells?

IN REVIEW

What Have I Learned?

The following questions cover the material that you have just learned, the trachea. Apply what you have learned to answer these questions on a separate piece of paper.

1. What type of cartilage provides support for the majority of the respiratory tract?

2. What tissue type makes up the epithelial layer of the trachea? What is its function?

3. Name the tissue layer that attaches the epithelium to the muscularis mucosa.

4. With your knowledge of the function of the cilia in the respiratory tract, explain why cigarette smoking, which paralyzes these cilia, causes the smoker's hacking cough.

The Alveolus and Alveolar Duct

EXERCISE 8.8:

Respiratory System—Alveolar Duct, Histology

| SELECT TOPIC Alveolar Duct | ➤ | SELECT VIEW High Magnification | GO |

- *Click the* **TURN TAGS ON** *button, and you will see the following image:*

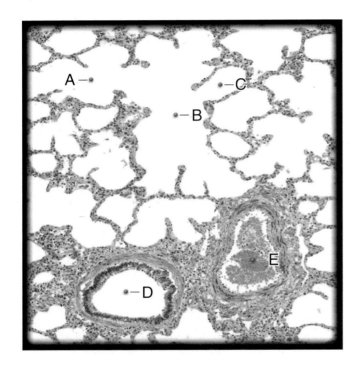

- *Mouse-over the pins on the screen to find the information necessary to identify the following structures:*

A. _____

B. _____

C. _____

D. _____

E. _____

CHECK POINT:

Alveolar Duct, Histology

1. Name the structure that provides the respiratory surface for gas exchange.
2. Name the structure that contains two or more alveoli.
3. Name the structure that conducts air to the alveoli.

EXERCISE 8.9:

Respiratory System—Alveolus, Histology

| SELECT TOPIC Alveolus | ➤ | SELECT VIEW High Magnification | GO |

- *Click the* **TURN TAGS ON** *button, and you will see the following image:*

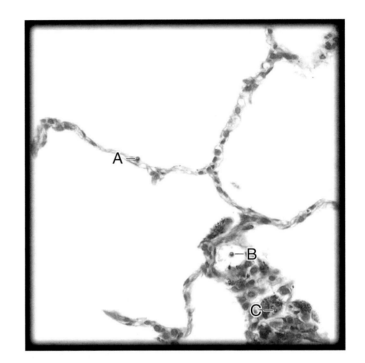

- *Mouse-over the pins on the screen to find the information necessary to identify the following structures:*

A. _____

B. _____

C. _____

CHECK POINT:

Alveolus, Histology

1. What is the function of the interalveolar wall? What cells line it?
2. How thick (or thin) is it?
3. What is the function of the segmental branch of the pulmonary artery?

Self Test

Take this opportunity to quiz yourself by taking the **SELF TEST.** See page 15 for a reminder on how to access the self-test for this section.

IN REVIEW

What Have I Learned?

The following questions cover the material that you have just learned, the alveolus and alveolar duct. Apply what you have learned to answer these questions on a separate piece of paper.

1. What are alveoli? Where are they located?

2. What is the diameter of an alveolus?

3. What are the two types of specialized cells in the wall of the alveoli?

4. What vessels transport deoxygenated blood toward the alveolar capillary plexus?

EXERCISE 8.10:

Respiratory System—Larynx, Anterior View

| SELECT TOPIC **Larynx** | ▶ | SELECT VIEW **Anterior** | GO |

• Click **LAYER 1** in the **LAYER CONTROLS** window, and you will see the following image:

• *Mouse-over the green pins on the screen to find the information necessary to identify the following structures:*

A. _____
B. _____
C. _____
D. _____
E. _____
F. _____

Non-respiratory System Structures (blue pins)

G. _____
H. _____
I. _____
J. _____

CHECK POINT:

Larynx, Anterior View

1. Name the cartilaginous structure that closes over the laryngeal inlet when swallowing.
2. Name a muscle that elevates the larynx and depresses the hyoid.
3. Name a muscle that produces waves of contractions during swallowing.

- *Click* **LAYER 2** *in the* **LAYER CONTROLS** *window, and you will see the following image:*

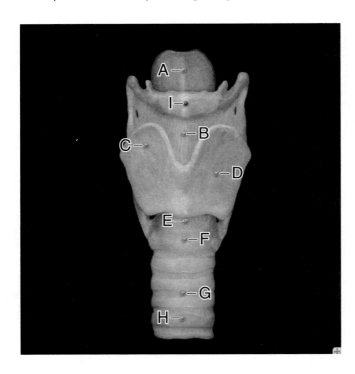

- *Mouse-over the green pins on the screen to find the information necessary to identify the following structures:*

A. _____

B. _____

C. _____

D. _____

E. _____

F. _____

G. _____

H. _____

Non-respiratory System Structure (blue pin)

I. _____

CHECK POINT:

Larynx, Anterior View, cont'd

4. Name the bone that does not articulate with any other bone.
5. Name the largest laryngeal cartilage.
6. Name the only cartilage of the respiratory tree that is a complete ring.

- *Click* **LAYER 3** *in the* **LAYER CONTROLS** *window, and you will see the following image:*

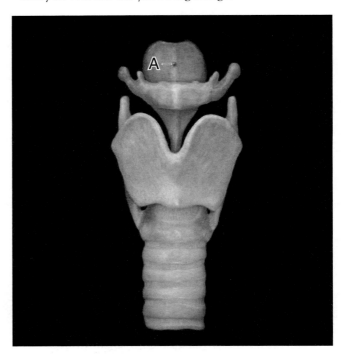

- *Mouse-over the pin on the screen to find the information necessary to identify the following structure:*

A. _____

EXERCISE 8.11:

Respiratory System—Larynx, Lateral View

SELECT TOPIC	SELECT VIEW	
Larynx ▶	**Lateral**	**GO**

- *Click* **LAYER 1** *in the* **LAYER CONTROLS** *window, and you will see the following image:*

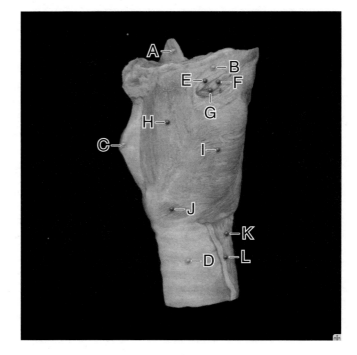

- *Mouse-over the green pins on the screen to find the information necessary to identify the following structures:*

A. _____

B. _____

C. _____

D. _____

Non-respiratory System Structures (blue pins)

E. _____

F. _____

G. _____

H. _____

I. _____

J. _____

K. _____

L. _____

C H E C K   P O I N T :

Larynx, Lateral View

1. Name the cranial nerve that contributes to the internal laryngeal nerve.
2. Name the sensory innervation of the internal laryngeal nerve.
3. Name the companion artery and vein to the internal laryngeal nerve.

- *Click* **LAYER 2** *in the* **LAYER CONTROLS** *window, and you will see the following image:*

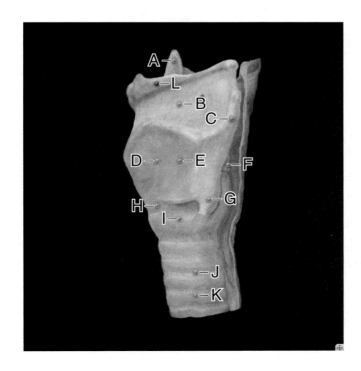

- *Mouse-over the green pins on the screen to find the information necessary to identify the following structures:*

A. _____

B. _____

C. _____

D. _____

E. _____

F. _____

G. _____

H. _____

I. _____

J. _____

K. _____

Non-respiratory System Structure (blue pin)

L. _____

C H E C K   P O I N T :

Larynx, Lateral View, cont'd

4. Name the superior projection of the posterior thyroid cartilage.
5. Name the inferior projection of the posterior thyroid cartilage.

- *Click* **LAYER 3** *in the* **LAYER CONTROLS** *window, and you will see the following image:*

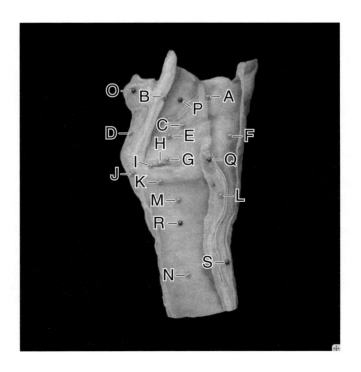

• *Mouse-over the green pins on the screen to find the information necessary to identify the following structures:*

A. _____

B. _____

C. _____

D. _____

E. _____

F. _____

G. _____

H. _____

I. _____

J. _____

K. _____

L. _____

M. _____

N. _____

Non-respiratory System Structures (blue pins)

O. _____

P. _____

Q. _____

R. _____

S. _____

CHECK POINT:

Larynx, Lateral View, cont'd

6. Name two structures with the major role in sound production.
7. Name the portion of the pharynx posterior to the larynx.
8. Name the sensory and motor innervations of the recurrent laryngeal nerve.

EXERCISE 8.12:

Respiratory System—Larynx, Posterior View

| SELECT TOPIC | SELECT VIEW | |
| Larynx | Posterior | GO |

• *Click **LAYER 1** in the **LAYER CONTROLS** window, and you will see the following image:*

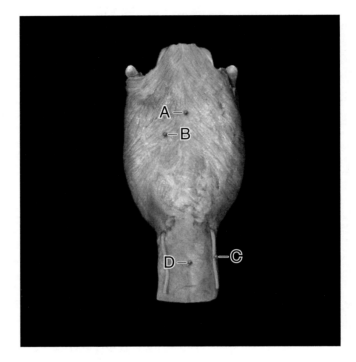

• *Mouse-over the blue pins on the screen to find the information necessary to identify the following non-respiratory system structures:*

A. _____

B. _____

C. _____

D. _____

CHECK POINT:

Larynx, Posterior View

1. What is a raphe?
2. Name the innervation for the inferior pharyngeal constrictor muscle.
3. Name the nerve responsible for the innervation of all but one intrinsic laryngeal muscle.

• *Click* **LAYER 2** *in the* **LAYER CONTROLS** *window, and you will see the following image:*

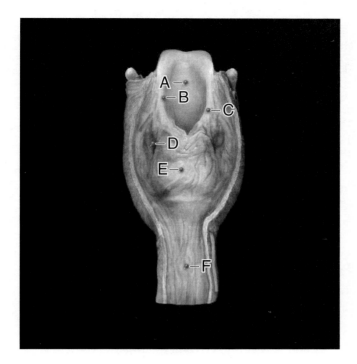

• *Mouse-over the green pins on the screen to find the information necessary to identify the following structures:*

A. _____

B. _____

C. _____

D. _____

E. _____

Non-respiratory System Structure (blue pin)

F. _____

CHECK POINT:

Larynx, Posterior View, cont'd

4. Name the superior opening of the larynx.
5. When food is "caught in the throat," where is it usually lodged?
6. Name the mucous membrane fold that surrounds the structure in question 4.

• *Click* **LAYER 3** *in the* **LAYER CONTROLS** *window, and you will see the following image:*

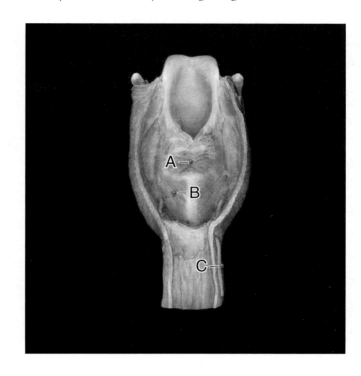

• *Mouse-over the blue pins on the screen to find the information necessary to identify the following non-respiratory system structures:*

A. _____

B. _____

C. _____

CHECK POINT:

Larynx, Posterior View, cont'd

7. Name two muscles that abduct the vocal folds.
8. Name a nerve of the larynx that is a branch of the vagus nerve (CN X).

- *Click* **LAYER 4** *in the* **LAYER CONTROLS** *window, and you will see the following image:*

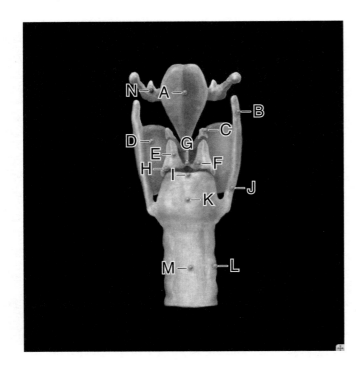

- *Mouse-over the green pins on the screen to find the information necessary to identify the following structures:*

A. _____

B. _____

C. _____

D. _____

E. _____

F. _____

G. _____

H. _____

I. _____

J. _____

K. _____

L. _____

M. _____

Non-respiratory System Structure (blue pin)

N. _____

C H E C K   P O I N T: _____

Larynx, Posterior View, cont'd

9. Name the process that is the posterior attachment site for the vocal ligament.
10. When is the vocal ligament called the vocal fold or vocal cord?
11. Name the cartilaginous structure that provides attachment for the lateral and posterior cricoarytenoid muscles.

I N   R E V I E W

What Have I Learned?

The following questions cover the material that you have just learned, the larynx. Apply what you have learned to answer these questions on a separate piece of paper.

1. Name three pharyngeal constrictors.

2. Name a muscle that tenses and elongates the vocal ligaments.

3. Describe the cricothyroid ligament.

4. Describe the thyrohyoid membrane.

5. Name the structure also known as the "Adam's apple."

6. Name the U-shaped structures that provide a rigid skeleton for the trachea.

7. What are the corniculate cartilages? Where are they embedded?

CHAPTER 9

The Digestive System

Overview: The Digestive System

Think of everything that you have eaten today, or this week for that matter. After you swallowed it, where did it go? Out of sight, out of mind. We don't even think about food again until it's meal time or our stomach growls during class (or worse yet, during a test). What happens to the food that you eat, between the time you swallow it and the time that the waste is eliminated? What structures are involved with this transformation?

The nutrients in your food are mostly unavailable to your body when they are swallowed. The primary nutrients—proteins, carbohydrates, and fats must be broken down into their basic building blocks before they can be absorbed into the bloodstream.

Only then can your body build proteins, carbohydrates, and fats from these available building blocks. It's like moving a large desk into a small office. The desk won't fit through the door, so you need to disassemble it first into smaller parts that *will* fit through the door. Once the parts have passed through the door and inside the office, they can be reassembled into a desk. Likewise, proteins for example, are too large to pass through the epithelium of the small intestine and into the capillaries located there.

Digestive enzymes must first break these proteins down into their smallest building blocks—amino acids. These amino acids can then pass through the epithelium of the small intestine and into the bloodstream. Then, when your body needs to make a protein molecule—say for example your growing hair, it puts together the required amino acids and constructs hair. This is where the office desk analogy breaks down. When you reassembled the desk in the small office, it was more or less the same desk that you had taken apart. The only difference would be any missing pieces lost in the move. In your body, the proteins that you create to build structures are unique and completely unlike the original protein in your food. But the point is the same, they must be broken down into smaller pieces to make it through the "door" of your digestive system.

To learn the structures of the digestive system, we will begin with an animated overview of the entire system. Then, we will look at the anatomical structures of the alimentary canal sequentially, from proximal to distal. Next we will look at the accessory digestive organs, and we will conclude by learning how we get energy out of the food we eat.

- *From the* **Home screen,** *click the drop-down box on the* **Select system** *menu.*

- *From the systems listed, click on* **Digestive,** *and you will see the image to the right:*

- *This is the opening screen for the* **Digestive System.**

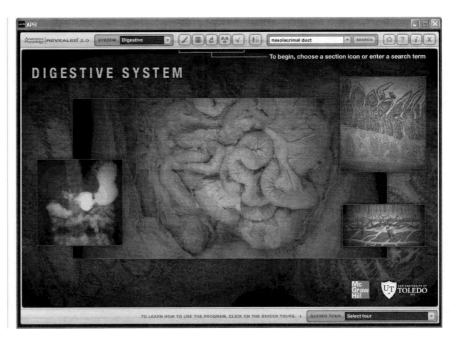

Animation: Digestive System Overview

SELECT ANIMATION
Digestive System Overview
PLAY

- *After viewing the animation, answer these questions:*

1. What are the four main functions of the digestive system?

2. Name the two types of digestion.

3. Where does digestion begin? How does this occur?

4. Another name for chewing is _____.

5. What is the function of the salivary glands? Of saliva?

6. What prevents food from entering the nasal cavity while swallowing?

7. What muscles push food particles into the pharynx?

8. Name the structure that prevents food from entering the respiratory system.

9. Name the structure that connects the pharynx to the stomach.

10. Once it has been swallowed, the food mass is called a _____.

11. What is the term for the involuntary wavelike contractions that propel the bolus to the stomach?

12. What are rugae? What are their functions?

13. The stomach cells secrete . . .

14. What effect do these secretions have on the bolus?

15. The bolus, mixed with stomach secretions, is now called _____.

16. _____ exits the stomach through the _____ and enters the _____.

17. Name the major site of nutrient absorption.

18. Name the three parts of the small intestine, from proximal to distal.

19. What digestive aids enter the duodenum? Where do they originate?

20. How are nutrients absorbed?

21. What is the destination of the chyme not absorbed in the small intestine?

22. List the sequence of structures the chyme passes through as it becomes feces. What has been absorbed from the chyme as it passes through the colon?

23. Where are the feces stored? What causes fecal elimination?

Self Test

Take this opportunity to quiz yourself by taking the **SELF TEST.** See page 15 for a reminder on how to access the self-test for this section.

The Oral Cavity

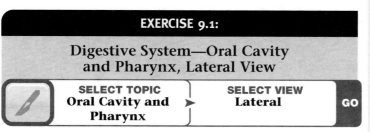

EXERCISE 9.1:

Digestive System—Oral Cavity and Pharynx, Lateral View

| SELECT TOPIC | SELECT VIEW | |
| Oral Cavity and Pharynx | Lateral | GO |

• *Click* **LAYER 1** *in the* **LAYER CONTROLS** *window, and you will see the following image:*

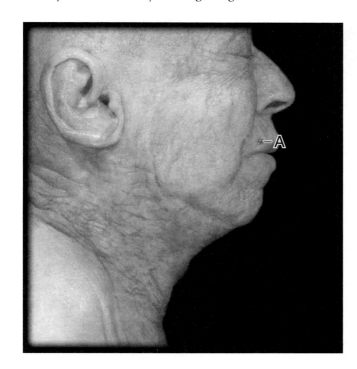

• *Mouse-over the pin on the screen to find the information necessary to identify the following structure:*

A. _____

CHECK POINT:

Oral Cavity and Pharynx, Lateral View

1. Name the fleshy folds surrounding the mouth.
2. Name the midline vertical groove of the upper lip.
3. Name the muscle contained in the lips.

• *Click* **LAYER 2** *in the* **LAYER CONTROLS** *window, and you will see the following image:*

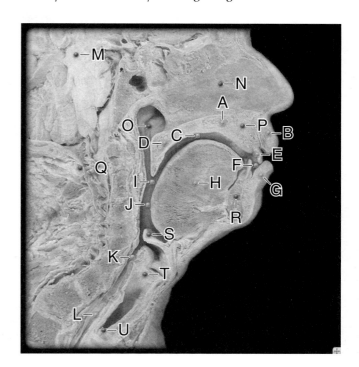

• *Mouse-over the green pins on the screen to find the information necessary to identify the following structures:*

A. _____

B. _____

C. _____

D. _____

E. _____

F. _____

G. _____

H. _____

I. _____

J. _____

K. _____

L. _____

Non-digestive System Structures (blue pins)

M. _____

N. _____

O. _____

P. _____

Q. _____

R. _____

S. _____

T. _____

U. _____

CHECK POINT:

Oral Cavity and Pharynx, Lateral View, cont'd

4. Name the narrow space between the dental arches, lips, and cheeks.
5. What is its function?
6. Name the space bounded by the teeth, tongue, and hard palate.

I N   R E V I E W

What Have I Learned?

The following questions cover the material that you have just learned, the oral cavity. Apply what you have learned to answer the following questions on a separate piece of paper.

1. Name the three functions of the lips.

2. What is the function of the oral cavity?

3. What is the pharynx? What are the three subdivisions?

4. Name the divisions that are part of the respiratory system, the digestive system, or both.

5. Describe the structure of the tongue.

6. What is the tongue's function?

7. Name the structure involved in oral breathing.

8. The Latin term pharynx means _____.

9. Name the structure that separates the oropharynx from the nasopharynx.

10. Name the structure that separates the oral cavity from the nasal cavity. Which bones form this structure?

11. Name the tube-shaped structure located between the oropharynx and the esophagus.

The Salivary Glands and Teeth

EXERCISE 9.2:

Digestive System—Salivary Glands, Lateral View

SELECT TOPIC	SELECT VIEW	
Salivary Glands ➤	**Lateral**	**GO**

- *Click* **LAYER 1** *in the* **LAYER CONTROLS** *window, and you will see the following image:*

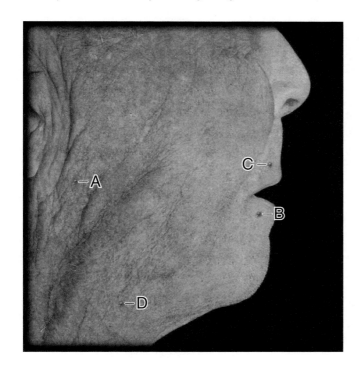

- *Mouse-over the blue pins on the screen to find the information necessary to identify the following non-digestive system structures:*

A. _____

B. _____

C. _____

D. _____

C H E C K   P O I N T :

Salivary Glands, Lateral View

1. The superficial part of which salivary gland is located anterior to the auricle?
2. Name the salivary gland located inferior and medial to the body of the mandible.
3. Name the three major paired salivary glands.

- *Click* **LAYER 2** *in the* **LAYER CONTROLS** *window, and you will see the following image:*

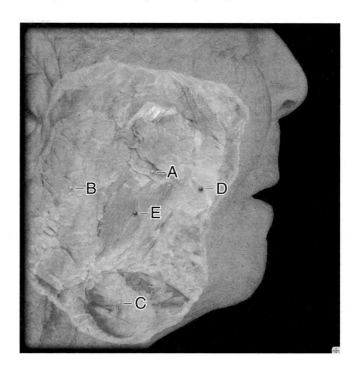

- *Mouse-over the green pins on the screen to find the information necessary to identify the following structures:*

A. _____

B. _____

C. _____

Non-digestive System Structures (blue pins)

D. _____

E. _____

C H E C K   P O I N T :

Salivary Glands, Lateral View, cont'd

4. Which salivary gland produces 25–30 percent of your saliva? Where is it located?
5. Which salivary gland produces 60–70 percent of your saliva? Where is it located?
6. Name the salivary duct that ends in the vestibule of the oral cavity opposite the second maxillary molar.

- *Click* **LAYER 3** *in the* **LAYER CONTROLS** *window, and you will see the following image:*

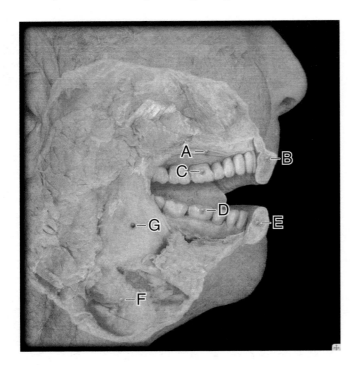

- *Mouse-over the green pins on the screen to find the information necessary to identify the following structures:*

A. _____

B. _____

C. _____

D. _____

E. _____

F. _____

Non-digestive System Structure (blue pin)

G. _____

CHECK POINT:

Salivary Glands, Lateral View, cont'd

7. Describe the gingivae. What is another name for them? What is gingivitis?
8. Describe the permanent mandibular teeth. What is their function?
9. Describe the permanent maxillary teeth. Are they identical in number and function as the mandibular teeth? What do you suppose is the reason for this?

- *Click* **LAYER 4** *in the* **LAYER CONTROLS** *window, and you will see the following image:*

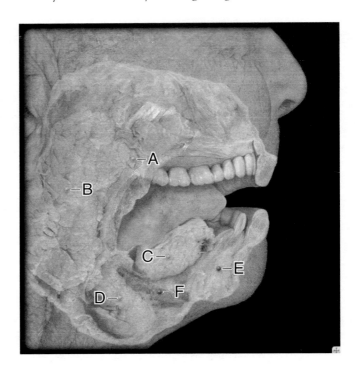

- *Mouse-over the green pins on the screen to find the information necessary to identify the following structures:*

A. _____

B. _____

C. _____

D. _____

Non-digestive System Structures (blue pins)

E. _____

F. _____

CHECK POINT:

Salivary Glands, Lateral View, cont'd

10. Name the salivary gland inferior to the tongue. What percentage of your saliva does it secrete?
11. What is the sublingual fossa?
12. Name a muscle responsible for the elevation of the floor of the mouth.

- *Click **LAYER 5** in the **LAYER CONTROLS** window, and you will see the following image:*

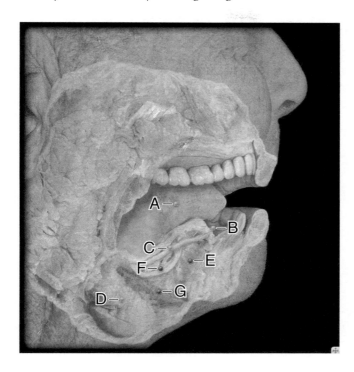

- *Mouse-over the green pins on the screen to find the information necessary to identify the following structures:*

A. _____

B. _____

C. _____

D. _____

Non-digestive System Structures (blue pins)

E. _____

F. _____

G. _____

CHECK POINT:

Salivary Glands, Lateral View, cont'd

13. Name the three functions of the tongue.
14. Where are the taste buds located?
15. What gives the dorsal surface of the tongue a "feltlike" appearance?

EXERCISE 9.3:

Digestive System—Teeth, Superior and Inferior Views

	SELECT TOPIC **Teeth**	SELECT VIEW **Superior-Inferior**	GO

- *Click **LAYER 1** in the **LAYER CONTROLS** window, and you will see the following image:*

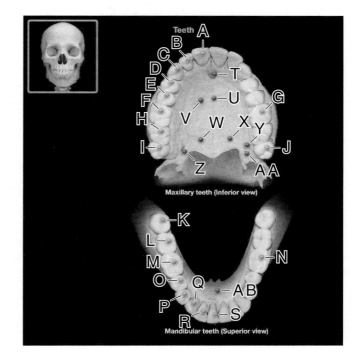

Maxillary teeth (inferior view)

Mandibular teeth (Superior view)

- *Mouse-over the green pins on the screen to find the information necessary to identify the following structures:*

A. _____

B. _____

C. _____

D. _____

E. _____

F. _____

G. _____

H. _____

I. _____

J. _____

K. _____

L. _____

M. _____

N. _____

O. _____

P. _____

Q. _____

R. _____

S. _____

Non-digestive System Structures (blue pins)

T. _____

U. _____

V. _____

W. _____

X. _____

Y. _____

Z. _____

AA. _____

AB. _____

CHECK POINT:

Teeth, Superior and Inferior Views

1. Describe the Universal/National System used for numbering adult teeth.
2. Name the teeth, on both bones, important for biting and cutting.
3. Which teeth are the longest? What is their function?

IN REVIEW

What Have I Learned?

The following questions cover the material that you have just learned, the salivary glands and teeth. Apply what you have learned to answer the following questions on a separate piece of paper.

1. Name the globular, encapsulated fat body prominent in the cheeks of infants.

2. What is its function?

3. Name a muscle involved with elevation of the mandible.

4. Name the secretions of each extrinsic salivary gland.

5. What is mastication?

6. What is deglutition?

7. What is the name for the "sockets" of the teeth?

8. Describe the submandibular duct. What is its function?

9. What is the general and special sensory innervation of the lingual nerve?

10. The lingual frenulum has been mentioned several times in this topic. What is the lingual frenulum?

11. Describe the Universal/National System used for numbering adult teeth.

12. Name the teeth, on both bones, important for biting and cutting.

13. Which teeth are the longest? What is their function?

14. Name all teeth important for grinding and crushing.

15. What foramina transmit the lesser palatine nerves and blood vessels?

16. Name the largest molar. Hint: There is a pair on each bone.

17. Name the pairs of teeth known as the wisdom teeth.

18. What nerves and blood vessels are transmitted through the incisive fossa?

The Esophagus

EXERCISE 9.4:
Digestive System—Esophagus, Anterior View

| | SELECT TOPIC
Esophagus | ▶ | SELECT VIEW
Anterior | GO |

- *Click **LAYER 1** in the **LAYER CONTROLS** window, and you will see the following image:*

- *Mouse-over the blue pins on the screen to find the information necessary to identify the following non-digestive system structures:*

A. _____

B. _____

C. _____

CHECK POINT:

Esophagus, Anterior View

1. Where is the esophagus located?
2. Describe the esophagus.
3. What is its function?

- *Click **LAYER 2** in the **LAYER CONTROLS** window, and you will see the following image:*

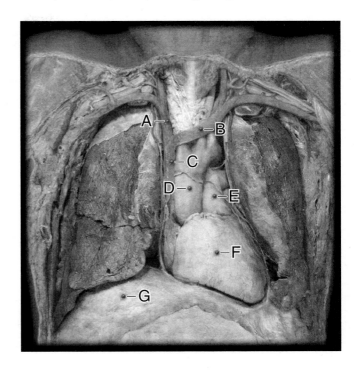

- *Mouse-over the blue pins on the screen to find the information necessary to identify the following non-digestive system structures:*

A. _____

B. _____

C. _____

D. _____

E. _____

F. _____

G. _____

• *Click **LAYER 3** in the **LAYER CONTROLS** window, and you will see the following image:*

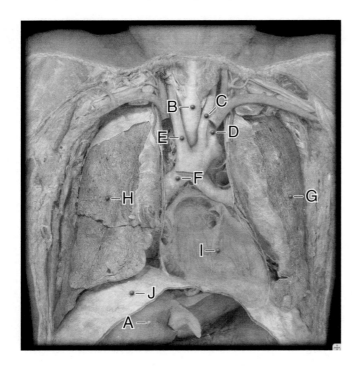

• *Mouse-over the green pin on the screen to find the information necessary to identify the following structure:*

A. _____

Non-digestive System Structures (blue pins)

B. _____

C. _____

D. _____

E. _____

F. _____

G. _____

H. _____

I. _____

J. _____

CHECK POINT:

Esophagus, Anterior View, cont'd

4. List the functions of the liver.
5. What are hepatocytes?
6. What can lead to the destruction of hepatocytes?

• *Click **LAYER 4** in the **LAYER CONTROLS** window, and you will see the following image:*

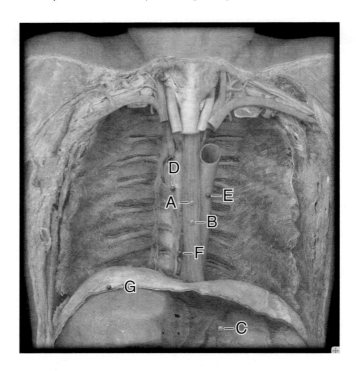

• *Mouse-over the green pins on the screen to find the information necessary to identify the following structures:*

A. _____

B. _____

C. _____

Non-digestive System Structures (blue pins)

D. _____

E. _____

F. _____

G. _____

CHECK POINT:

Esophagus, Anterior View, cont'd

7. The esophagus is located in what three body regions?
8. Describe the structure of the esophagus.
9. What is peristalsis?

- *Click* **LAYER 5** *in the* **LAYER CONTROLS** *window, and you will see the following image:*

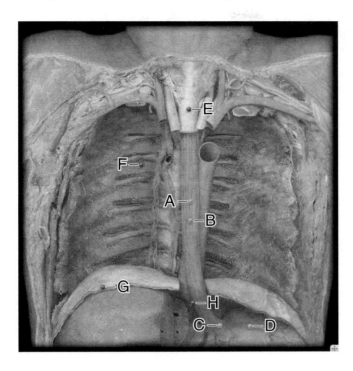

- *Mouse-over the green pins on the screen to find the information necessary to identify the following structures:*

A. _____

B. _____

C. _____

D. _____

Non-digestive System Structures (blue pins)

E. _____

F. _____

G. _____

H. _____

CHECK POINT:

Esophagus, Anterior View, cont'd

10. The esophagus conveys food to what digestive organ?
11. What is reflux esophagitis?
12. Name the "hole" in the diaphragm for the passage of the esophagus.

The Abdominal Cavity

> **EXERCISE 9.5:**
>
> **Digestive System—Abdominal Cavity, Anterior View**
>
> | SELECT TOPIC | SELECT VIEW | |
> | --- | --- | --- |
> | Abdominal Cavity ➤ | Anterior | GO |

- *Click* **LAYER 1** *in the* **LAYER CONTROLS** *window, and you will see the following image:*

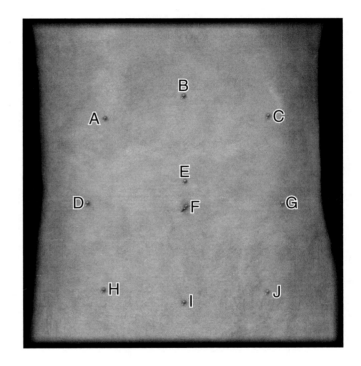

- *Mouse-over the blue pins on the screen to find the information necessary to identify the following non-digestive system structures:*

A. _____

B. _____

C. _____

D. _____

E. _____

F. _____

G. _____

H. _____

I. _____

J. _____

CHECK POINT:

Abdominal Cavity, Anterior View

1. What is the umbilicus? Which abdominal region contains the umbilicus?
2. Name the abdominal region located superior to that region.
3. Name the abdominal region located inferior to the region in question 1.

CHECK POINT:

Abdominal Cavity, Anterior View, cont'd

4. What is the greater omentum? What is its function?
5. What is the lesser omentum? Where is it located?
6. Describe the structure of the gallbladder. Where is it located?

- *Click **LAYER 2** in the **LAYER CONTROLS** window, and you will see the following image:*

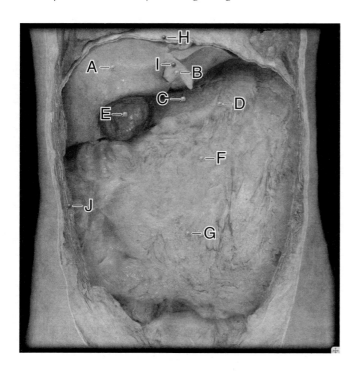

- *Click **LAYER 3** in the **LAYER CONTROLS** window, and you will see the following image:*

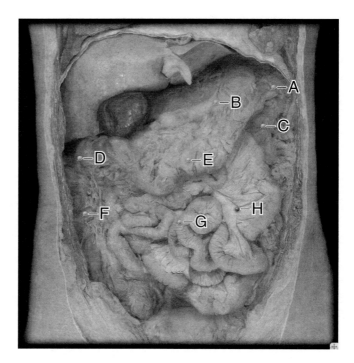

- *Mouse-over the green pins on the screen to find the information necessary to identify the following structures:*

A. _____

B. _____

C. _____

D. _____

E. _____

F. _____

G. _____

Non-digestive System Structures (blue pins)

H. _____

I. _____

J. _____

- *Mouse-over the green pins on the screen to find the information necessary to identify the following structures:*

A. _____

B. _____

C. _____

D. _____

E. _____

F. _____

G. _____

Non-digestive System Structure (blue pin)

H. _____

CHECK POINT:

Abdominal Cavity, Anterior View, cont'd

7. Name the four parts of the stomach from proximal to distal.
8. Name the location where the ascending colon joins the transverse colon. What is another name for this location? To what does this name refer?
9. Name the location where the transverse colon joins the descending colon. What is another name for this location? What does this name refer to?

• *Click* **LAYER 4** *in the* **LAYER CONTROLS** *window, and you will see the following image:*

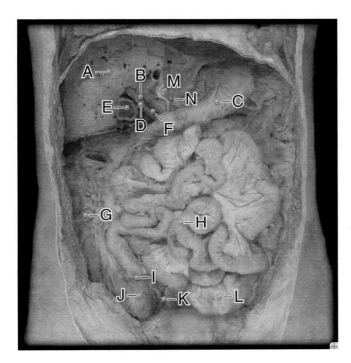

• *Mouse-over the green pins on the screen to find the information necessary to identify the following structures:*

A. _____

B. _____

C. _____

D. _____

E. _____

F. _____

G. _____

H. _____

I. _____

J. _____

K. _____

L. _____

Non-digestive System Structures (blue pins)

M. _____

N. _____

CHECK POINT:

Abdominal Cavity, Anterior View, cont'd

10. Name the structure at the junction of the small intestine and large intestine. What part of each intestine is joined here?
11. Name the pouch of the large intestine into which the small intestine empties. What muscle regulates the flow from one to the other?
12. Name the slender hollow appendage attached to this pouch. What is its function? What structures does it contain? What is it called when the lumen of this structure becomes obstructed?

• *Click* **LAYER 5** *in the* **LAYER CONTROLS** *window, and you will see the following image:*

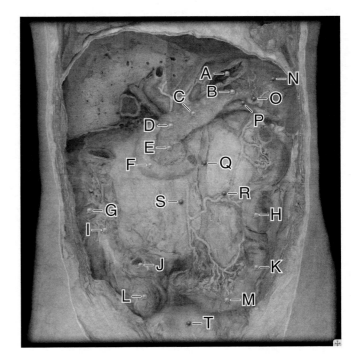

• *Mouse-over the green pins on the screen to find the information necessary to identify the following structures:*

A. _____

B. _____

C. _____

D. _____

E. _____

F. _____

G. _____

H. _____

I. _____

J. _____

K. _____

L. _____

M. _____

Non-digestive System Structures (blue pins)

N. _____

O. _____

P. _____

Q. _____

R. _____

S. _____

T. _____

CHECK POINT:

Abdominal Cavity, Anterior View, cont'd

13. What are the taeniae coli? What is their function?
14. What are haustra?
15. Name the peritoneal appendages filled with fat. What is their function?

- *Click* **LAYER 6** *in the* **LAYER CONTROLS** *window, and you will see the following image:*

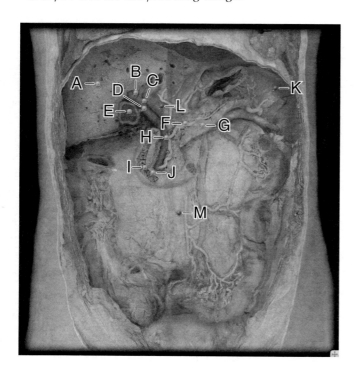

- *Mouse-over the green pins on the screen to find the information necessary to identify the following structures:*

A. _____

B. _____

C. _____

D. _____

E. _____

F. _____

G. _____

H. _____

I. _____

J. _____

Non-digestive System Structures (blue pins)

K. _____

L. _____

M. _____

CHECK POINT:

Abdominal Cavity, Anterior View, cont'd

16. Name and describe the largest lymphatic organ.
17. What is its function?
18. What protects this delicate structure?

IN REVIEW

What Have I Learned?

The following questions cover the material that you have just learned, the abdominal cavity. Apply what you have learned to answer the following questions on a separate piece of paper.

1. Using the following figure, label the nine abdominal regions.

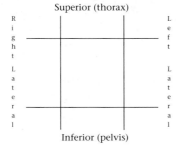

Superior (thorax)

Right Lateral

Left Lateral

Inferior (pelvis)

2. What is the function of the gallbladder?

3. Name the muscles of the abdominal wall from superficial to deep.

4. Name the four parts of the colon from proximal to distal.

5. Name the three parts of the small intestine, from proximal to distal.

6. Name the most mobile part of the large intestine.

7. Which parts of the colon absorb water? Which parts of the colon absorb electrolytes? Which parts of the colon store feces?

The Stomach

Animation: Stomach

 SELECT ANIMATION
Stomach   PLAY

- *After viewing the animation, answer these questions:*

1. Where is the stomach located? Between which two organs?

2. What is the function of the stomach? What two processes contribute to this function?

3. What structures of the stomach form the superior and inferior borders?

4. Describe the four stomach regions.

5. How is the distal part of the stomach subdivided?

6. What is the function of the pyloric sphincter?

7. What are gastric rugae? What is their function?

8. Describe the four layers of the stomach.

9. Name the layers of the stomach muscularis. How does it compare to the rest of the digestive tract?

10. What substance is secreted by the mucous cells of the epithelium?

11. List three functions of gastric mucus.

12. Describe the gastric pits.

13. What are the four different cells found in gastric glands? What are their functions?

Self Test

Take this opportunity to quiz yourself by taking the **SELF TEST.** See page 15 for a reminder on how to access the self-test for this section.

EXERCISE 9.6:

Digestive System—Stomach, Histology

SELECT TOPIC	SELECT VIEW	
Stomach	▶ **Low Magnification**	**GO**

• *Click the* **TURN TAGS ON** *button, and you will see the following image:*

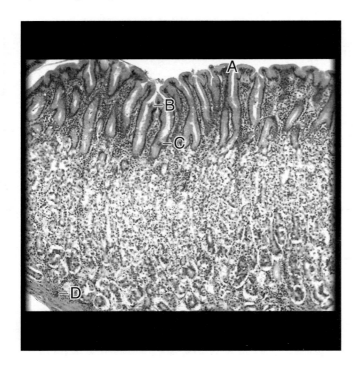

• *Mouse-over the pins on the screen to find the information necessary to identify the following structures:*

A. _____

B. _____

C. _____

D. _____

CHECK POINT:

Stomach, Histology

1. Describe the muscularis mucosae. What is its function?
2. Describe the gastric pits. What is their function?
3. Describe the gastric glands. What is their function?

EXERCISE 9.7:

Digestive System—Stomach and Duodenum, Anterior View

SELECT TOPIC	SELECT VIEW	
Stomach and Duodenum	▶ **Anterior**	**GO**

• *Click* **LAYER 1** *in the* **LAYER CONTROLS** *window, and you will see the following image:*

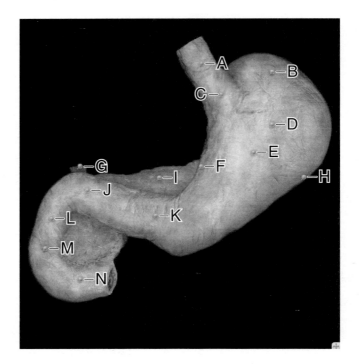

• *Mouse-over the pins on the screen to find the information necessary to identify the following structures:*

A. _____

B. _____

C. _____

D. _____

E. _____

F. _____

G. _____

H. _____

I. _____

J. _____

K. _____

L. _____

M. _____

N. _____

CHECK POINT:

Stomach and Duodenum, Anterior View

1. Name the part of the stomach located at the junction with the esophagus. What structure does it contain?
2. Name the dome-shaped superior part of the stomach. What is usually retained there?
3. Name the terminal part of the stomach. Name the ring of muscle located there. What is the function of this part of the stomach *and* the muscular ring?

• *Click* **LAYER 2** *in the* **LAYER CONTROLS** *window, and you will see the following image:*

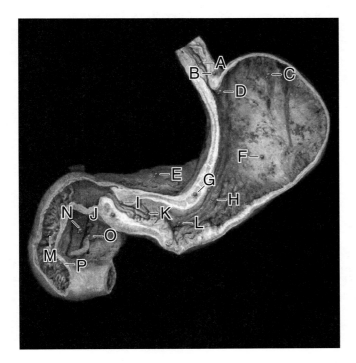

• *Mouse-over the pins on the screen to find the information necessary to identify the following structures:*

A. _____

B. _____

C. _____

D. _____

E. _____

F. _____

G. _____

H. _____

I. _____

J. _____

K. _____

L. _____

M. _____

N. _____

O. _____

P. _____

CHECK POINT:

Stomach and Duodenum, Anterior View, cont'd

4. Name the muscular structure that prevents reflux of stomach contents. Name two times that this muscle relaxes.
5. Name the structures that allow the stomach to expand as it fills.
6. What is the major duodenal papilla? What is its function?

Animation: Gastric Secretion

 SELECT ANIMATION
Gastric Secretion PLAY

• *After viewing the animation, answer these questions:*

1. List the three phases of gastric secretion.

2. What initiates the cephalic phase?

3. Describe the parasympathetic response involved in gastric secretion.

4. What role does gastrin play?

5. What initiates the gastric phase?

6. Describe the parasympathetic reflex and the direct stimulation that results in this phase.

7. What is the net result of these nervous stimuli?

8. What initiates the intestinal phase?

9. What two events inhibit gastric secretion?

10. Describe the three steps involved in this inhibition.

Self Test
Take this opportunity to quiz yourself by taking the **SELF TEST.** See page 15 for a reminder on how to access the self-test for this section.

Animation: HCl Production

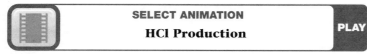

SELECT ANIMATION
HCl Production
PLAY

- *After viewing the animation, answer these questions:*

1. Which cells produce hydrochloric acid in the stomach? Where are these cells located?

2. What enzyme is involved in the formation of carbonic acid?

3. What two substrates are involved in this reaction?

4. Carbonic acid dissociates into . . .

5. Which ion is transported back to the bloodstream?

6. What exchange occurs at the ion exchange molecule in the plasma membrane?

7. How are the hydrogen ions transported into the duct of the gastric gland?

8. What movement occurs with the chloride ions?

9. What ions are transported into the parietal cell in exchange for the hydrogen ions?

10. What is the result of this movement?

Self Test
Take this opportunity to quiz yourself by taking the **SELF TEST.** See page 15 for a reminder on how to access the self-test for this section.

I N   R E V I E W

What Have I Learned?

The following questions cover the material that you have just learned, the stomach. Apply what you have learned to answer the following questions on a separate piece of paper.

1. Name each different type of gastric gland and the products that they produce.

2. Name the structure that conducts mucus and secretions from the gastric glands to the lumen of the stomach.

3. What is a lumen?

4. Name the elongated, nodular gland posterior to the stomach. Describe it.

5. What are the functions of this elongated, nodular gland?

6. Name the part of the small intestine to receive ingested material from the stomach. What is that ingested material called?

7. List the functions of the structure in question 6.

8. Name the four parts of this structure.

9. What is the function of the bile duct? Where is it located?

10. What is the function of the pancreatic duct? Where is it located?

The Small Intestine

EXERCISE 9.8:

Digestive System—Small Intestine (Duodenum), Histology

SELECT TOPIC
Small Intestine (Duodenum) ▶ SELECT VIEW
Low Magnification   GO

- *Click the* **TURN TAGS ON** *button, and you will see the following image:*

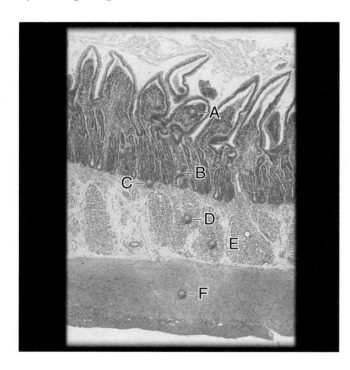

- *Mouse-over the pins on the screen to find the information necessary to identify the following structures:*

A. _____

B. _____

C. _____

D. _____

E. _____

F. _____

CHECK POINT:

Small Intestine (Duodenum), Histology

1. Name and describe the structures that increase the surface area of the small intestine. What is their function?
2. Describe the structures also known as the crypts of Lieberkühn. What are their functions?
3. What are duodenal glands? Where are they located? What is their function?

EXERCISE 9.9:

Digestive System—Small Intestine (Jejunum/Ileum), Histology

SELECT TOPIC
Small Intestine (Jejunum/Ileum) ▶ SELECT VIEW
Low Magnification   GO

- *Click the* **TURN TAGS ON** *button, and you will see the following image:*

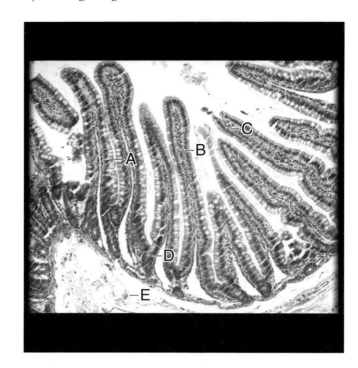

- *Mouse-over the pins on the screen to find the information necessary to identify the following structures:*

A. _____

B. _____

C. _____

D. _____

E. _____

CHECK POINT:

Small Intestine (Jejunum/Ileum), Histology

1. Name the cells that produce and release mucin into the intestinal lumen.
2. Where are these cells located?
3. What tissue-type covers the villi of the small intestine?

EXERCISE 9.10:

Imaging—Stomach and Small Intestine

SELECT TOPIC
Stomach and Small Intestine   ➤   SELECT VIEW   **GO**

• *Click the* **TURN TAGS ON** *button, and you will see the following image:*

• *Mouse-over the pins on the screen to find the information necessary to identify the following structures:*

A. _____

B. _____

C. _____

D. _____

E. _____

F. _____

G. _____

I N   R E V I E W

What Have I Learned?

The following questions cover the material that you have just learned, the small intestine. Apply what you have learned to answer these questions on a separate piece of paper.

1. What are villi? What is their function?

2. Which layer regulates the length of the villi?

3. The contraction of which layer compresses the glands of the mucosa to expel their contents?

4. Name and describe the layer responsible for peristalsis.

5. Which layer contains blood and lymph vessels, lymphoid nodules, and nerve plexus?

6. What is the function of the submucosal glands?

7. What cell types are found on the epithelium of the small intestine?

8. What are the functions of these cells?

9. Name two phenomena that change the shape of the villi.

10. What layer of the small intestine contains Peyer's patches? In what part of the small intestine are they found? What is their function?

The Colon

EXERCISE 9.11:

Digestive System—Colon, Histology

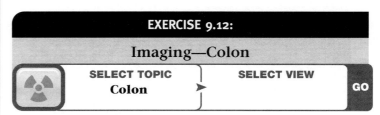

SELECT TOPIC	SELECT VIEW	
Colon	▶ Low Magnification	GO

- *Click the **TURN TAGS ON** button, and you will see the following image:*

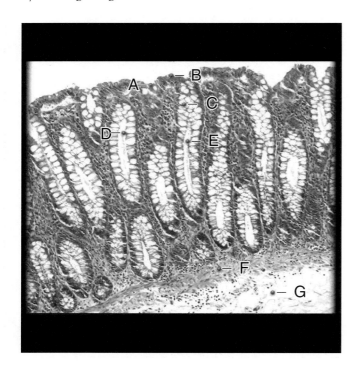

- *Mouse-over the pins on the screen to find the information necessary to identify the following structures:*

A. _____

B. _____

C. _____

D. _____

E. _____

F. _____

G. _____

C H E C K   P O I N T :

Colon, Histology

1. Name and describe the tubular glands located in the colon.
2. What is their function?
3. What is the dominant cell-type in these glands?
4. Name the tissue-type that lines the colon. What is its function?
5. The contraction of which muscles compresses the glands of the mucosa and expels their contents?
6. Name the layer of the colon containing blood and lymph vessels.

EXERCISE 9.12:

Imaging—Colon

SELECT TOPIC	SELECT VIEW	
Colon	▶	GO

- *Click the **TURN TAGS ON** button, and you will see the following image:*

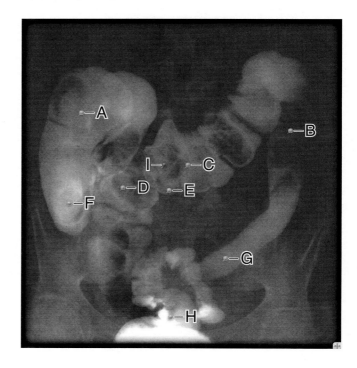

• *Mouse-over the green pins on the screen to find the information necessary to identify the following structures:*

A. _____

B. _____

C. _____

D. _____

E. _____

F. _____

G. _____

H. _____

Non-digestive System Structure (blue pin)

I. _____

IN   REVIEW

What Have I Learned?

The following questions cover the material that you have just learned, the histology of the colon. Apply what you have learned to answer the following questions on a separate piece of paper.

1. Name and describe the tubular glands located in the colon.

2. What is their function?

3. What is the dominant cell-type in these glands?

4. Name the tissue-type that lines the colon. What is its function?

5. The contraction of which muscles compresses the glands of the mucosa and expels their contents?

6. Name the layer of the colon containing blood and lymph vessels.

The Peritoneum

EXERCISE 9.13:

Digestive System—Peritoneum, Mid-sagittal View

| SELECT TOPIC **Peritoneum** | ▶ | SELECT VIEW **Mid-sagittal** | GO |

- *Click* **LAYER 1** *in the* **LAYER CONTROLS** *window, and you will see the following image:*

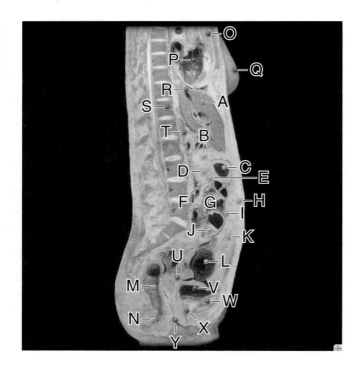

- *Mouse-over the green pins on the screen to find the information necessary to identify the following structures:*

A. _____

B. _____

C. _____

D. _____

E. _____

F. _____

G. _____

H. _____

I. _____

J. _____

K. _____

L. _____

M. _____

N. _____

Non-digestive System Structures (blue pins)

O. _____

P. _____

Q. _____

R. _____

S. _____

T. _____

U. _____

V. _____

W. _____

X. _____

Y. _____

CHECK POINT:

Peritoneum, Mid-sagittal View

1. Name the single layer of serous membrane that coats the outer surface of many abdominal organs.
2. What is its function?
3. Name the single layer of serous membrane that lines the wall of the abdomen.

Self Test

Take this opportunity to quiz yourself by taking the **SELF TEST.** See page 15 for a reminder on how to access the self-test for this section.

I N   R E V I E W

What Have I Learned?

The following questions cover the material you have just learned, the peritoneum. Apply what you have learned to answer these questions on a separate piece of paper.

1. Name the single layer of serous membrane that coats the outer surface of many abdominal organs.

2. What is its function?

3. Name the single layer of serous membrane that lines the wall of the abdomen.

4. Describe the transverse mesocolon.

5. What is its function?

6. What is the peritoneal cavity? What does this term mean?

7. Name the double-layered fold of peritoneum capable of storing large amounts of fat.

8. How is it associated with the immune system?

The Liver

Animation: Liver

 | SELECT ANIMATION **Liver** | PLAY

- *After viewing the animation, answer these questions:*

1. The liver is the _____ internal organ of the body.

2. Where is it located?

3. Name the four lobes of the liver. Which two are part of another lobe?

4. Name and describe the structure that separates the two anterior lobes.

5. What are the two fetal remnants found on the liver?

6. What structures are located in the porta hepatis?

7. Histologically, the liver is composed of functional units called _____.

8. The hub of these functional units is . . .

9. Describe the portal triad.

10. Describe the sinusoids. How are they arranged?

11. How are the hepatocytes arranged?

12. What are the two basic functions of the liver?

13. Name the three categories of this task.

14. Describe these three categories of tasks.

Self Test

Take this opportunity to quiz yourself by taking the **SELF TEST.** See page 15 for a reminder on how to access the self-test for this section.

EXERCISE 9.14:

Digestive System—Liver, Anterior and Postero-inferior Views

| SELECT TOPIC | | SELECT VIEW | |
| Liver | ▶ | Anterior-Postero-inferior | GO |

- *Click **LAYER 1** in the **LAYER CONTROLS** window, and you will see the following image:*

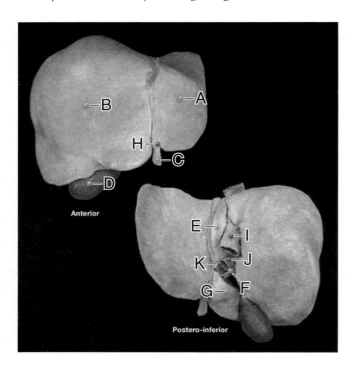

- *Mouse-over the green pins on the screen to find the information necessary to identify the following structures:*

A. _____

B. _____

C. _____

D. _____

E. _____

F. _____

G. _____

Non-digestive System Structures (blue pins)

H. _____

I. _____

J. _____

K. _____

CHECK POINT:

Liver, Anterior and Postero-inferior Views

1. Name the largest lobe of the liver. What two lobes make up this lobe?
2. Name the structures that separate the left and right lobes of the liver.
3. Name the functions of the liver.

EXERCISE 9.15:

Digestive System—Liver, Histology

| SELECT TOPIC | | SELECT VIEW | |
| Liver | ▶ | High Magnification | GO |

- *Click the **TURN TAGS ON** button, and you will see the following image:*

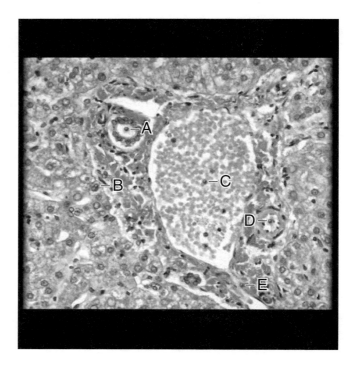

- *Mouse-over the pins on the screen to find the information necessary to identify the following structures:*

A. _____

B. _____

C. _____

D. _____

E. _____

CHECK POINT:

Liver, Histology

1. Name the liver cells arranged in cords within a liver lobule. What are their functions?
2. What are hepatic sinusoids? What is their function?
3. Name the structure that brings venous blood to the liver sinusoids.

EXERCISE 9.16:

Digestive System—Biliary Ducts, Anterior View

SELECT TOPIC	SELECT VIEW	
Biliary Ducts ▶	**Anterior**	GO

• Click **LAYER 1** in the **LAYER CONTROLS** window, and you will see the following image:

• Mouse-over the blue pins on the screen to find the information necessary to identify the following non-digestive system structures:

A. _____

B. _____

C. _____

D. _____

E. _____

F. _____

• Click **LAYER 2** in the **LAYER CONTROLS** window, and you will see the following image:

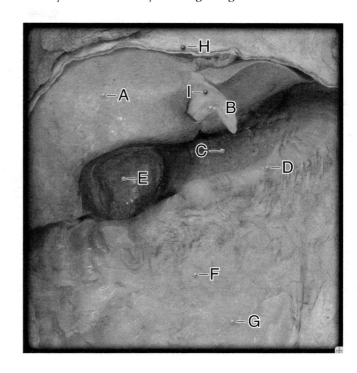

• Mouse-over the green pins on the screen to find the information necessary to identify the following structures:

A. _____

B. _____

C. _____

D. _____

E. _____

F. _____

G. _____

Non-digestive System Structures (blue pins)

H. _____

I. _____

- *Click* **LAYER 3** *in the* **LAYER CONTROLS** *window, and you will see the following image:*

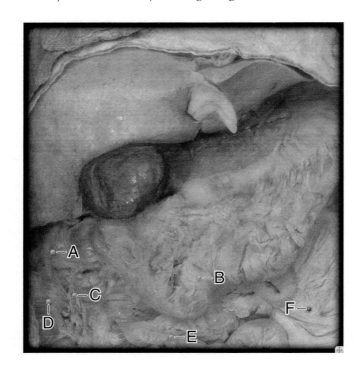

- *Mouse-over the green pins on the screen to find the information necessary to identify the following structures:*

A. _____

B. _____

C. _____

D. _____

E. _____

Non-digestive System Structure (blue pin)

F. _____

- *Click* **LAYER 4** *in the* **LAYER CONTROLS** *window, and you will see the following image:*

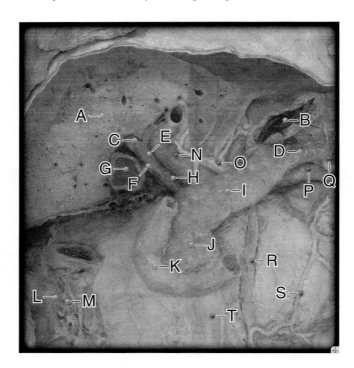

- *Mouse-over the green pins on the screen to find the information necessary to identify the following structures:*

A. _____

B. _____

C. _____

D. _____

E. _____

F. _____

G. _____

H. _____

I. _____

J. _____

K. _____

L. _____

M. _____

Non-digestive System Structures (blue pins)

N. _____

O. _____

P. _____

Q. _____

R. _____

S. _____

T. _____

CHECK POINT:

Biliary Ducts, Anterior View

1. Name the two structures that receive bile from the liver. What structure do they form when they merge?
2. Bile passes from the _____ to the _____ for storage and from the _____ to the _____ for _____ of _____.
3. What two ducts unite to form the bile duct?

• Click **LAYER 5** in the **LAYER CONTROLS** window, and you will see the following image:

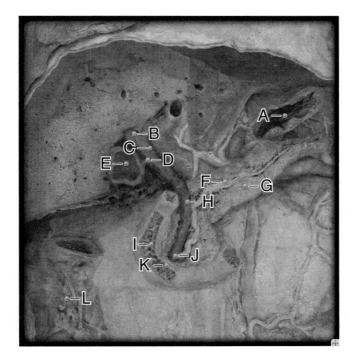

• *Mouse-over the pins on the screen to find the information necessary to identify the following structures:*

A. _____

B. _____

C. _____

D. _____

E. _____

F. _____

G. _____

H. _____

I. _____

J. _____

K. _____

L. _____

CHECK POINT:

Biliary Ducts, Anterior View, cont'd

4. Name the structure that transmits pancreatic secretions.
5. Name the structure formed by the joining of this structure and the bile duct.
6. Into which section of the small intestine does the structure in question 5 empty?

IN REVIEW

What Have I Learned?

The following questions cover the material that you have just learned, the liver. Apply what you have learned to answer the following questions on a separate piece of paper.

1. Name the structure that is the remnant of the umbilical vein of the fetus.

2. Name the blood vessel that carries absorbed products of digestion to the liver.

3. Name the organ that stores, concentrates, and releases bile.

4. Name the structure that brings arterial blood to the liver sinusoids.

5. Name the structure in the liver lobule that collects and transports bile toward the bile duct.

6. What is the hepatic portal triad?

7. Name the elevation on the wall of the duodenum that contains the hepatopancreatic ampulla.

8. What is the exocrine function of the pancreas?

9. What is the endocrine function of the pancreas?

10. What hormones are produced by the pancreas?

The Pancreas

EXERCISE 9.17:

Digestive System—Pancreas (Exocrine), Histology

SELECT TOPIC
Pancreas (Exocrine) ▶ SELECT VIEW
High Magnification **GO**

- *Click the **TURN TAGS ON** button, and you will see the following image:*

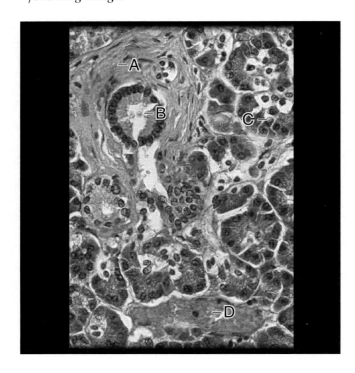

- *Mouse-over the pins on the screen to find the information necessary to identify the following structures:*

A. _____

B. _____

C. _____

D. _____

CHECK POINT:

Pancreas (Exocrine), Histology

1. Name the oval collection of secretory (acinar) cells in the exocrine pancreas.
2. What do these cells secrete? How are these secretions regulated?
3. Name the series of ducts that transport digestive enzymes through the exocrine pancreas. Where is the destination of these enzymes?
4. Name the structure that drains blood from, and contains hormones from, the endocrine pancreas.
5. Name the structure that separates the pancreatic lobules.

I N   R E V I E W

What Have I Learned?

The following questions cover the material that you have just learned, the histology of the pancreas. Apply what you have learned to answer the following questions on a separate piece of paper.

1. Name the oval collection of secretory (acinar) cells in the exocrine pancreas.

2. What do these cells secrete? How are these secretions regulated?

3. Name the series of ducts that transport digestive enzymes through the exocrine pancreas. Where is the destination of these enzymes?

4. Name the structure that drains blood from, and contains hormones from, the endocrine pancreas.

5. Name the structure that separates the pancreatic lobules.

ATP Synthesis

Animation: Hydrolysis of Sucrose

We have been discovering the structures involved in digestion—the process of breaking down our food into smaller, simple molecules that can be absorbed into the bloodstream. But that is not the end of the story. Amino acids will be recombined by your body to build proteins, fatty acids, and glycerol will be recombined into energy storing fat molecules, and glucose will either be stored as glycogen molecules in your liver and muscle tissues or used to fuel the production of energy in the form of ATP. Any of these molecules can serve to supply the fuel to feed the ATP production process, but they must be converted into glucose first. The following animations will walk you through the process of harvesting energy from the foods you eat. As you view these animations, fill in the blanks in the chart to keep track of the products produced in this process. Then, answer the questions that follow each animation to check your comprehension of the information presented.

Produced	Glycolysis	Krebs Cycle	Electron Transport Chain	Total
ATP				
NADH				
FADH$_2$				

SELECT ANIMATION
Hydrolysis of Sucrose   PLAY

- *After viewing the animation, answer these questions:*

1. The enzyme sucrase breaks the disaccharide _____ into two monosaccharides: _____, or _____ sugar, and _____, or _____ sugar.

2. Where does this reaction occur?

3. For hydrolysis to occur, the sucrose must bind to what part of the sucrase enzyme?

4. What happens to the enzyme when this occurs? What happens to the sucrose?

5. What molecule breaks the bond? Do you see why the process is called hydrolysis? hydro = water and lysis = to break. The bond is broken by water.

6. After the bond is broken and the two monosaccharides are released, what happens to the enzyme?

7. How many times can this process be repeated?

8. What three events can occur to end this process?

Self Test
Take this opportunity to quiz yourself by taking the **SELF TEST.** See page 15 for a reminder on how to access the self-test for this section.

Animation: NADH/Oxidation-Reduction Reactions

 SELECT ANIMATION
NADH/Oxidation- Reduction Reactions **PLAY**

- *After viewing the animation, answer these questions:*

1. Cells obtain energy through the process of . . .

2. How is this energy obtained? What is the energy molecule that results from this process?

3. What is oxidation?

4. A hydrogen atom consists of . . .

5. Whenever a molecule is oxidized, another molecule must be _____, which means . . .

6. What is the enzyme's role in the oxidation-reduction reaction in the cell?

7. What structures does the enzyme have that facilitate this reaction?

8. In this reaction, the substrate is _____ (loses a hydrogen) and the _____ is reduced to _____.

9. What two things occur after the reaction is complete?

10. What is NADH? What is this molecule available to do now?

Self Test
Take this opportunity to quiz yourself by taking the **SELF TEST.** See page 15 for a reminder on how to access the self-test for this section.

Animation: Glycolysis

 SELECT ANIMATION
Glycolysis **PLAY**

- *After viewing the animation, answer these questions:*

1. Cells derive energy from the _____ of nutrients, such as _____.

2. The oxidation of _____ to _____ occurs through a series of steps called _____.

3. How many carbons are in a molecule of glucose?

4. The energy released during these _____ reactions is used to form _____ (_____), the _____ of the cell.

5. Name the two initial steps in glycolysis.

6. What are the three molecules that result?

7. What then occurs to the 6-carbon molecule?

8. These 3-carbon molecules are converted to _____.

9. What happens to the electrons in this reaction? What two molecules are formed?

10. What happens to the pyruvate under aerobic conditions?

11. What happens to the pyruvate under anaerobic conditions?

Self Test
Take this opportunity to quiz yourself by taking the **SELF TEST.** See page 15 for a reminder on how to access the self-test for this section.

Animation: Krebs Cycle

 SELECT ANIMATION
Krebs Cycle
PLAY

- *After viewing the animation, answer these questions:*
1. Name the product of glycolysis that enters the Krebs cycle. How many of these molecules are produced per glucose molecule?

2. Name the two-carbon fragment that enters the Krebs cycle.

3. What is another name for the Krebs cycle? Where in the cell does it occur? Where in the cell did glycolysis occur?

4. What two products are formed during the conversion of pyruvate to acetyl-CoA?

5. What reaction occurs to release the CoA carrier molecule?

6. What two products are formed as the 6-carbon molecule is converted to a 5-carbon molecule?

7. What three products are produced during the second oxidation and decarboxylation? This process forms a _____-carbon molecule.

8. Finally, the _____-carbon molecule is further oxidized, and the _____ removed are used to form _____ and _____. What is regenerated by these reactions?

9. Each glucose molecule is broken down into _____ _____ molecules during glycolysis. Then each _____ molecule is converted to _____ and enters the Krebs cycle.

10. For each glucose molecule, how many circuits of the Krebs cycle are completed to completely break down the two pyruvate molecules?

11. For each circuit of the Krebs cycle, how many ATP are produced? Therefore, how many ATP are produced in the Krebs cycle for every glucose molecule? (Remember, the products from each glucose molecule require two circuits of the Kreb's cycle.)

Self Test
Take this opportunity to quiz yourself by taking the **SELF TEST.** See page 15 for a reminder on how to access the self-test for this section.

Animation: Electron Transport and ATP Synthesis

 SELECT ANIMATION
Electron Transport and ATP Synthesis **PLAY**

- *After viewing the animation, answer these questions:*

1. In the mitochondrion, the energy stored in NADH is _____.

2. This energy of the _____ is used to form _____.

3. What changes occur to NAD^+ and FAD during glycolysis and the Krebs cycle?

4. Where are the electrons from NADH transferred? How? What happens to the protons?

5. Where are the electrons carried? By what molecule? What else occurs as this is happening?

6. Where are the electrons from $FADH_2$ transferred? And the protons?

7. Where do the electrons travel next? And the protons?

8. What is the terminal electron acceptor? What molecule forms when the electrons are transferred (accepted)?

9. What force is generated by the transfer of protons to the intermembrane space? Where is this force generated?

10. Name the special proton channel proteins that allow the protons to re-enter the matrix of the mitochondrion.

11. What energy is used to synthesize the ATP molecules? What products are used to make the ATP?

12. What name is given to this method of ATP production?

Self Test

Take this opportunity to quiz yourself by taking the **SELF TEST.** See page 15 for a reminder on how to access the self-test for this section.

EXERCISE 9.18:

Coloring Exercise

Look up the organs of the digestive system. Then, with colored pens or pencils, color in these organs.

Note: Many organs are shown separated from one another. In life, many digestive organs overlap one another when viewed from this perspective.

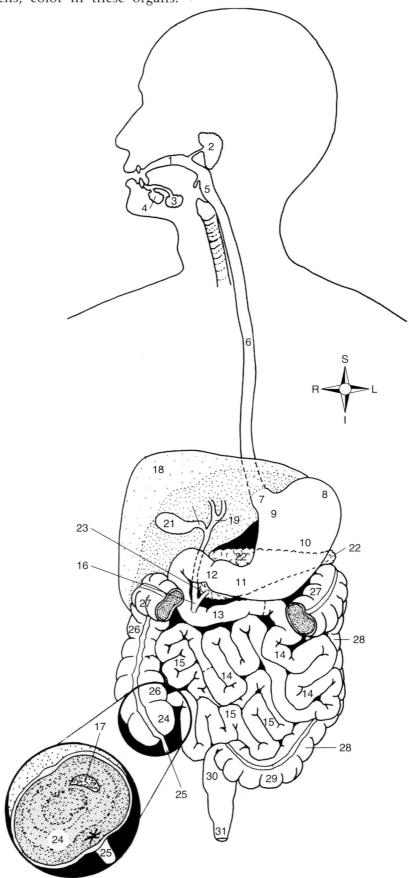

☐ 1. _____

☐ 2. _____

☐ 3. _____

☐ 4. _____

☐ 5. _____

☐ 6. _____

☐ 7. _____

☐ 8. _____

☐ 9. _____

☐10. _____

☐11. _____

☐12. _____

☐13. _____

☐14. _____

☐15. _____

☐16. _____

☐17. _____

☐18. _____

☐19. _____

☐20. _____

☐21. _____

☐22. _____

☐23. _____

☐24. _____

☐25. _____

☐26. _____

☐27. _____

☐28. _____

☐29. _____

☐30. _____

☐31. _____

CHAPTER **10**

The Urinary System

Overview: The Urinary System

Metabolic wastes, toxins, drugs, hormones, salts, hydrogen ions, and water are all excreted through the urinary system of your body. To do this, it must filter your blood. All of your blood plasma is filtered 60 times per day. That's once every 24 minutes. The average plasma filtration rate is 125 mL per minute for adults with two kidneys. This produces 180 liters or 45 gallons of filtrate every day. Aren't you glad that you don't have to urinate all of that fluid? Fortunately, your body reabsorbs 99 percent of this fluid, so that you only excrete 1.8 liters or 0.45 gallons of urine daily.

The urinary system also regulates blood volume and pressure; regulates the amounts of dissolved solutes in body fluids—called osmolarity; helps to control blood pressure, red blood cell count and the blood's ability to carry oxygen, the pH of body fluids, calcium homeostasis, and, free radical detoxification; and it has the ability to convert amino acids to glucose during times of starvation. Talk about multitasking!

We will begin our study with an overview of the structures of the urinary system. We will then take a sequential look at the organs in great detail, from the kidneys through the urethra. Finally, we will look at how molecules move through the body and pass through the cell membrane.

- *From the* **Home screen,** *click the drop-down box on the* **Select system** *menu.*
- *From the systems listed, click on* **Urinary** *and you will see the image to the right:*
- *This is the opening screen for the* **Urinary System.**

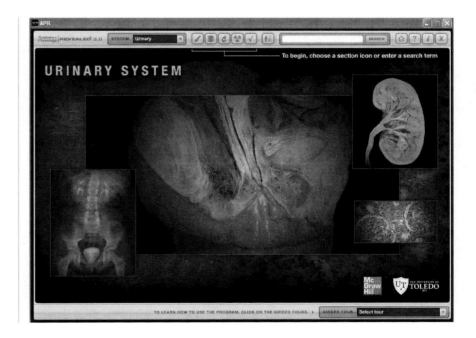

Animation: Urinary System Overview

SELECT ANIMATION
Urinary System Overview

PLAY

- *After viewing the animation, answer these questions:*

1. What are the major functions of the urinary system?

2. What are the organs of the urinary system?

3. Where are the kidneys located?

4. What is the renal hilum?

5. Blood to be filtered is transported to the kidney by the _____.

6. Filtered blood leaves the kidney via the _____.

7. Name the structure that covers the outer surface of the kidney.

8. Describe the structure of the interior of a kidney.

9. Fluids from the _____ ultimately are funneled into the _____ of the _____.

10. What is the function of the ureters?

11. How is urine propelled through the ureters?

12. What is the urinary bladder? Where is it located?

13. Where do the ureters drain into the urinary bladder?

14. Name the muscle of the urinary bladder wall.

15. Another name for urination is _____.

16. How is urine expelled from the urinary bladder?

17. Compare the functions of the male and female urethras.

18. What is the function of the internal urethral sphincter muscle? Is it under voluntary or involuntary control?

19. What is the function of the external urethral sphincter muscle? Is it under voluntary or involuntary control?

20. How are both of these sphincters involved with urination?

Self Test

Take this opportunity to quiz yourself by taking the **SELF TEST.** See page 15 for a reminder on how to access the self-test for this section.

The Upper Urinary System

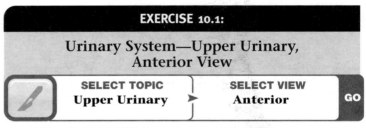

EXERCISE 10.1:

Urinary System—Upper Urinary, Anterior View

SELECT TOPIC   SELECT VIEW
Upper Urinary ▶ Anterior   GO

- *After clicking* **LAYERS 1** *through* **3** *to orient yourself to our location, click* **LAYER 4** *in the* **LAYER CON- TROLS** *window, and you will see the following image:*

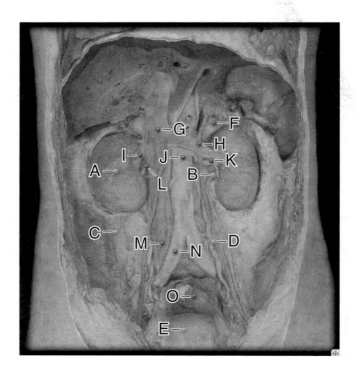

- *Mouse-over the green pins on the screen to find the information necessary to identify the following structures:*

A. _____

B. _____

C. _____

D. _____

E. _____

Non-urinary System Structures (blue pins)

F. _____

G. _____

H. _____

I. _____

J. _____

K. _____

L. _____

M. _____

N. _____

O. _____

CHECK POINT:

Upper Urinary, Anterior View

1. Describe the structure of the kidneys.
2. What is their function?
3. Name the structure that conveys urine from the renal pelvis.

- *Click **LAYER 5** in the **LAYER CONTROLS** window, and you will see the following image:*

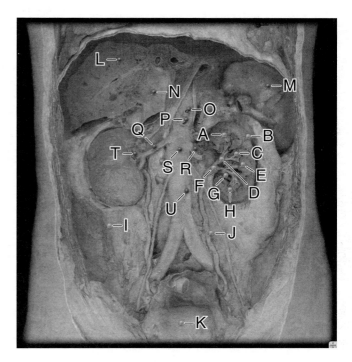

- *Mouse-over the pins on the screen to find the information necessary to identify the following structures:*

A. _____

B. _____

C. _____

D. _____

E. _____

F. _____

G. _____

H. _____

I. _____

J. _____

K. _____

Non-urinary System Structures (blue pins)

L. _____

M. _____

N. _____

O. _____

P. _____

Q. _____

R. _____

S. _____

T. _____

U. _____

CHECK POINT:

Upper Urinary, Anterior View, cont'd

4. What is the function of the renal pyramids? How many are located in each kidney?
5. Name the tip of the renal pyramids. Where does it project?
6. What is a minor calyx? What is its function?

- *Click* **LAYER 6** *in the* **LAYER CONTROLS** *window, and you will see the following image:*

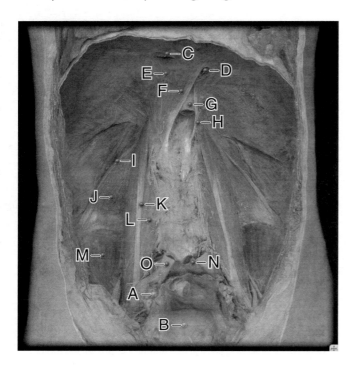

- *Mouse-over the green pins on the screen to find the information necessary to identify the following structures:*

A. _____

B. _____

Non-urinary System Structures (blue pins)

C. _____

D. _____

E. _____

F. _____

G. _____

H. _____

I. _____

J. _____

K. _____

L. _____

M. _____

N. _____

O. _____

CHECK POINT:

Upper Urinary, Anterior View, cont'd

7. What differences exist in the pathway of the ureter for males and females?
8. How long is the ureter?
9. List the pathway of urine from the kidney to the urethra.

Self Test

Take this opportunity to quiz yourself by taking the **SELF TEST.** See page 15 for a reminder on how to access the self-test for this section.

I N R E V I E W

What Have I Learned?

The following questions cover the material that you have just learned, the upper urinary system. Apply what you have learned to answer these questions on a separate piece of paper.

1. Name the arteries that supply blood to the kidneys.

2. Name the veins that drain blood from the kidneys.

3. Which is anterior to the other?

4. Name the funnel-shaped structure that drains urine from the kidneys to the ureter.

5. Name the structure that conducts urine from the minor calyx to the renal pelvis.

6. How many of each calyx are located in each kidney?

7. Name the outer layer of the kidney. What are its extensions into the middle part of the kidney called? What structures are contained in this outer layer?

8. Name the inner layer of the kidney.

9. Name the part of the kidney where blood vessels and the renal pelvis enter and exit the kidney.

10. Name the part of the kidney between the structure in question 9 and the renal papillae.

11. Name the muscle in the walls of the urinary bladder.

12. What is the trigone? Name the openings that define the trigone.

13. The size and position of the urinary bladder varies with . . .

14. The volume of the urine also effects the position of . . .

The Kidney

Animation: Kidney—Gross Anatomy

SELECT ANIMATION
Kidney—Gross Anatomy
PLAY

- *After viewing the animation, answer these questions:*

1. Where are the kidneys located?

2. What is the renal hilum? Name three structures that pass through the hilum.

3. The hilum is continuous with the _____, which is . . .

4. How is the kidney tissue divided? Name the structures of the one tissue division that project into the other.

5. The base of each pyramid is located at the junction of these two tissues, the . . .

6. Name the structure at the apex of the renal pyramids. Where does it project?

7. Several _____ _____ merge to form the larger _____ _____, which merge to form a single, funnel-shaped _____.

8. How many lobes are in each kidney? Each lobe consists of . . .

9. How is the blood carried to the kidneys for filtration?

10. List the series of branches that form from this artery up to and including the glomerulus. Give the locations where each artery is found.

11. Name the initial filtering component of the kidney. Name the functional filtration unit of the kidney. What are its components?

12. Name the arteriole that leaves each glomerulus. Upon leaving the glomerulus, this vessel enters . . .

13. For nephrons in the renal cortex, a _____ forms around the . . .

14. List the series of vessels that carry blood as it drains from these capillary networks.

15. What is the destination of the efferent arterioles associated with nephrons at the corticomedullary junction?

16. These capillaries are known as the . . .

17. Name the sequence of vessels for the blood draining these capillaries.

Self Test

Take this opportunity to quiz yourself by taking the **SELF TEST**. See page 15 for a reminder on how to access the self-test for this section.

EXERCISE 10.2:

Urinary System—Kidney, Anterior View

| SELECT TOPIC | SELECT VIEW | |
| Kidney | Anterior | GO |

- *Click **LAYER 1** in the **LAYER CONTROLS** window, and you will see the following image:*

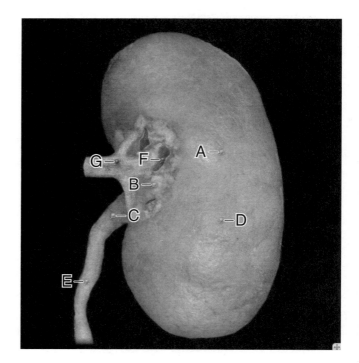

- *Mouse-over the green pins on the screen to find the information necessary to identify the following structures:*

A. _____

B. _____

C. _____

D. _____

E. _____

Non-urinary System Structures (blue pins)

F. _____

G. _____

C H E C K   P O I N T :

Kidney, Anterior View

1. Name the connective tissue coat of the kidney. What is its function?
2. Describe the structure of the renal hilum. What is its function?
3. What is the function of the renal pelvis?

- *Click* **LAYER 2** *in the* **LAYER CONTROLS** *window, and you will see the following image:*

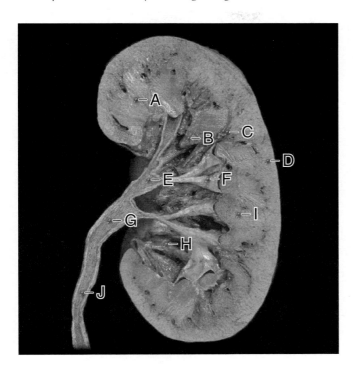

- *Mouse-over the pins on the screen to find the information necessary to identify the following structures:*

A. _____

B. _____

C. _____

D. _____

E. _____

F. _____

G. _____

H. _____

I. _____

J. _____

CHECK POINT:

Kidney, Anterior View, cont'd

4. Describe the structure of the renal pyramids.
5. Name the structures that separate the renal pyramids. These structures are an extension of what part of the kidney?
6. Which nephron parts are located in the renal cortex? Which are located in the renal medulla?

EXERCISE 10.3:

Imaging—Kidney

SELECT TOPIC	SELECT VIEW	
Kidney		GO

- *Click the* **TURN TAGS ON** *button, and you will see the following image:*

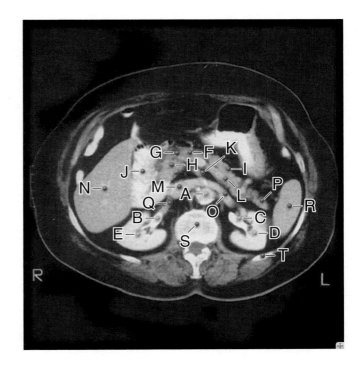

- *Mouse-over the green pins on the screen to find the information necessary to identify the following structures:*

A. _____

B. _____

C. _____

D. _____

E. _____

Non-urinary System Structures (blue pins)

F. _____

G. _____

H. _____

I. _____

J. _____

K. _____

L. _____

M. _____

N. _____

O. _____

P. _____

Q. _____

R. _____

S. _____

T. _____

Self Test
Take this opportunity to quiz yourself by taking the **SELF TEST.** See page 15 for a reminder on how to access the self-test for this section.

I N   R E V I E W

What Have I Learned?

The following questions cover the material that you have just learned, the kidney. Apply what you have learned to answer these questions on a separate piece of paper.

1. What is the distribution of the renal artery?

2. Describe its branching before it enters the kidney.

3. List the structures drained by the renal vein.

4. Describe the ureter. What is its function?

5. Trace the pathway of urine as it passes through and then out of the kidney through the urethra.

The Renal Corpuscle

EXERCISE 10.4:
Urinary System—Renal Corpuscle, Histology

SELECT TOPIC	SELECT VIEW	
Renal Corpuscle ▶		GO

*   Click the **TURN TAGS ON** button, and you will see the following image:

*   *Mouse-over the pins on the screen to find the information necessary to identify the following structures:*

A. _____

B. _____

C. _____

D. _____

E. _____

F. _____

C H E C K   P O I N T :

Renal Corpuscle, Histology

1. Name the two structures that make up the renal corpuscle.
2. What is the function of the renal corpuscle?
3. Name all structures that constitute a nephron.

Movement Through the Cell Membrane

Animation: Diffusion

SELECT ANIMATION
Diffusion

PLAY

- *After viewing the animation, answer these questions:*

1. What causes the constant random motion of dissolved molecules?

2. What is one result of this random motion?

3. This tendency of molecules to spread out is an example of _____.

4. Even in a solid state, molecules are _____.

5. What two motions characterize the movement of the dissolving sugar molecules?

6. As they dissolve, the sugar molecules move from the area where they are _____ to the area where they are _____.

7. This type of motion, from an area of higher concentration to an area of lower concentration, is called _____.

8. Diffusion continues until . . .

9. What three things affect the rate of diffusion?

Self Test
Take this opportunity to quiz yourself by taking the **SELF TEST.** See page 15 for a reminder on how to access the self-test for this section.

Animation: Osmosis

SELECT ANIMATION
Osmosis

PLAY

- *After viewing the animation, answer these questions:*

1. What is diffusion?

2. What does this process allow?

3. Do most polar molecules freely cross the lipid cell membrane? Name two groups of polar molecules.

4. What is the name for the special case of diffusion that involves the movement of water molecules across a membrane?

5. Why is a molecule of urea unable to diffuse across the membrane?

6. How does a urea molecule interact with water molecules? Why?

7. Why is there now a net movement of water molecules? Which direction do they move?

8. What happens to the water level on the side of the beaker where the water molecules are moving into?

9. Define isotonic, hypertonic, and hypotonic.

Self Test
Take this opportunity to quiz yourself by taking the **SELF TEST.** See page 15 for a reminder on how to access the self-test for this section.

Animation: Facilitated Diffusion

SELECT ANIMATION **Facilitated Diffusion**		PLAY

- *After viewing the animation, answer these questions:*

1. What occurs in the process of facilitated diffusion?

2. What is unique about the carrier molecules and the molecules to which they bind?

3. Once the molecule binds to the carrier protein, the protein will facilitate the diffusion process by . . .

4. Facilitated diffusion and simple diffusion are similar in that both . . .

5. How is facilitated diffusion different from simple diffusion?

6. What determines which direction facilitated diffusion occurs?

Self Test
Take this opportunity to quiz yourself by taking the **SELF TEST.** See page 15 for a reminder on how to access the self-test for this section.

Animation: Cotransport

SELECT ANIMATION **Cotransport**		PLAY

- *After viewing the animation, answer these questions:*

1. Which direction can small molecules, such as sugars and amino acids, be transported?

2. How does the sugar move? How does the concentration of sugar compare inside and outside the cell?

3. How is this transport of sugar driven through a coupled transport protein? Are these counterions moving from a higher to a lower concentration or from a lower to a higher concentration?

4. What is a symport? What occurs there?

5. How is a low concentration of sodium maintained inside the cell? How is it powered?

6. What is counter-transport?

7. What is an antiport? What occurs there? How is this different than what occurs at a symport?

8. How does the sodium-potassium pump come into play in this process?

Self Test
Take this opportunity to quiz yourself by taking the **SELF TEST.** See page 15 for a reminder on how to access the self-test for this section.

Animation: Kidney—Microscopic Anatomy

SELECT ANIMATION **Kidney—Microscopic Anatomy**		PLAY

- *After viewing the animation, answer these questions:*

1. Name the functional filtration unit of the kidney. Approximately how many of these are located in each kidney?

2. Name the two parts to a nephron.

3. Describe the structure of the renal corpuscle. What are its two poles? What structures are located at each pole?

4. Describe the visceral layer of the glomerular capsule. What are podocytes and pedicels?

5. Describe the parietal layer of the glomerular capsule. What is the capsular space?

6. Name and describe the three components of the filtration membrane of the glomerulus.

7. Name and describe the three parts of the renal tubule.

8. What are the structural differences between the thick and thin segments of the nephron loop?

9. Several distal convoluted tubules drain into _____ that pass through the _____. These merge to form _____ that drain into a _____.

10. What events occur as the fluid passes through the renal tubule?

11. Describe the two types of nephrons. What percent of the total number of nephrons consist of each type?

12. Describe the blood pathway to the vascular pole of the glomerulus. Name the apparatus located at this point.

13. What are the three parts of this apparatus? What is its function?

14. The efferent arteriole from each glomerulus enters a _____.

15. For cortical nephrons, where is the peritubular capillary network located? Where does it drain?

16. The efferent arterioles associated with juxtaglomerular nephrons are known as _____ that surround _____. Where does this blood drain?

Self Test
Take this opportunity to quiz yourself by taking the **SELF TEST**. See page 15 for a reminder on how to access the self-test for this section.

Animation: Urine Formation

- *After viewing the animation, answer these questions:*

1. What are the primary functions of the kidneys?

2. In the process of these functions, they regulate. . . . How?

3. Where is urine formed? What are the three processes in its formation?

4. Where does glomerular filtration occur? How does this occur?

5. What is tubular fluid? What occurs to it as it flows through the renal tubule?

6. What occurs to the fluid during tubular reabsorption?

7. How do the solutes move across the tubule wall into the interstitial fluid? Where do they go from there?

8. How does water move through the tubule wall? What percent of water is reabsorbed from each portion of the renal tubule?

9. What percent of the water in the glomerular filtrate returns to the bloodstream? What happens to the remaining water?

10. What event occurs during the process of tubular secretion?

11. Give some examples of waste products involved in this process.

12. What happens to the waste products to prepare them for excretion?

13. Tubular fluid that enters the collecting ducts is called _____.

14. What determines whether the urine produced is dilute or concentrated?

15. Describe the urine produced when water intake is high and also when water intake is limited.

16. What are the two key factors that determine the kidneys' ability to concentrate urine?

Self Test
Take this opportunity to quiz yourself by taking the **SELF TEST.** See page 15 for a reminder on how to access the self-test for this section.

IN REVIEW

What Have I Learned?

The following questions cover the material that you have just learned, the renal corpuscle. Apply what you have learned to answer these questions on a separate piece of paper.

1. Name the three parts of the renal tubule.

2. Name the specialized cells of the distal convoluted tubule that are part of the juxtaglomerular device.

3. What is the function of the juxtaglomerular device?

4. Name the structure that regulates the final volume and electrolyte content of the urine. What hormone influences this regulation?

5. What is the function of the capsular space?

The Ureter

SELECT TOPIC	SELECT VIEW	
Ureter	Low Magnification	GO

- *Click the* **TURN TAGS ON** *button, and you will see the following image:*

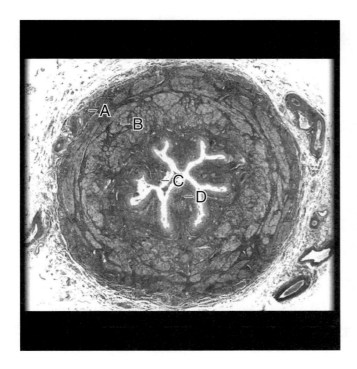

- *Mouse-over the pins on the screen to find the information necessary to identify the following structures:*

A. _____

B. _____

C. _____

D. _____

CHECK POINT:

Ureter, Histology

1. Name the tissue that loosely attaches the ureter to adjacent structures.
2. What structures are found in this tissue?
3. How does the muscularis change in composition along the length of the ureter? What is its function?
4. What is the mucosa? What is its function?
5. Describe the transitional epithelium. What is unique about its function?

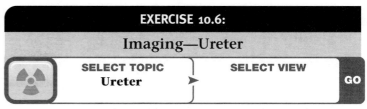

SELECT TOPIC	SELECT VIEW	
Ureter		GO

- *Click the* **TURN TAGS ON** *button, and you will see the following image:*

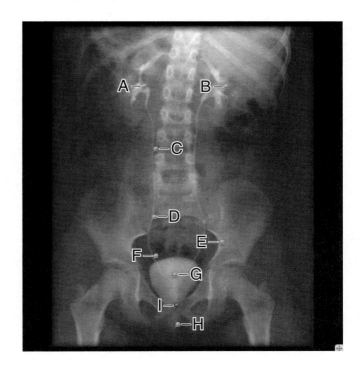

- *Mouse-over the green pins on the screen to find the information necessary to identify the following structures:*

A. _____

B. _____

C. _____

D. _____

E. _____

F. _____

G. _____

H. _____

Non-urinary System Structure (blue pin)

I. _____

The Female Lower Urinary System

EXERCISE 10.7:

Urinary System—Lower Urinary—Female, Sagittal View

SELECT TOPIC	SELECT VIEW	
Lower Urinary— Female ▶	**Sagittal**	**GO**

• *Click* **LAYER 1** *in the* **LAYER CONTROLS** *window, and you will see the following image:*

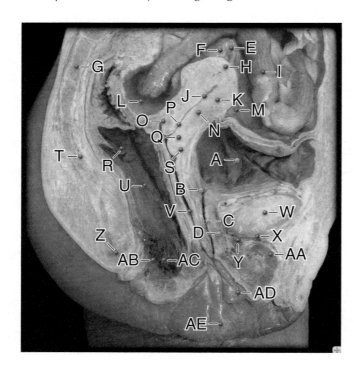

• *Mouse-over the green pins on the screen to find the information necessary to identify the following structures:*

A. _____

B. _____

C. _____

D. _____

Non-urinary System Structures (blue pins)

E. _____

F. _____

G. _____

H. _____

I. _____

J. _____

K. _____

L. _____

M. _____

N. _____

O. _____

P. _____

Q. _____

R. _____

S. _____

T. _____

U. _____

V. _____

W. _____

X. _____

Y. _____

Z. _____

AA. _____

AB. _____

AC. _____

AD. _____

AE. _____

CHECK POINT:

Lower Urinary—Female, Sagittal View

1. Describe the structure of the urinary bladder.
2. What are the two functions of this organ?
3. What is the function of the female urethra?
4. Give two reasons why urinary tract infections are more common in females.

Self Test

Take this opportunity to quiz yourself by taking the **SELF TEST.** See page 15 for a reminder on how to access the self-test for this section.

The Male Lower Urinary System

EXERCISE 10.8:

**Urinary System—Lower Urinary—Male,
Sagittal View**

SELECT TOPIC	SELECT VIEW	
Lower Urinary—
Male	▶ Sagittal	GO

- *Click* **LAYER 1** *in the* **LAYER CONTROLS** *window,
and you will see the following image:*

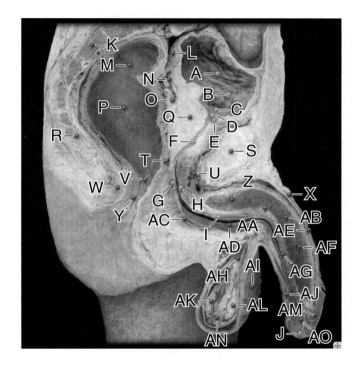

- *Mouse-over the green pins on the screen to find the information necessary to identify the following structures:*

A. _____

B. _____

C. _____

D. _____

E. _____

F. _____

G. _____

H. _____

I. _____

J. _____

Non-urinary System Structures (blue pins)

K. _____

L. _____

M. _____

N. _____

O. _____

P. _____

Q. _____

R. _____

S. _____

T. _____

U. _____

V. _____

W. _____

X. _____

Y. _____

Z. _____

AA. _____

AB. _____

AC. _____

AD. _____

AE. _____

AF. _____

AG. _____

AH. _____

AI. _____

AJ. _____

AK. _____

AL. _____

AM. _____

AN. _____

AO. _____

CHECK POINT:

Lower Urinary—Male, Sagittal View

1. Name the structures that form the trigone of the
   urinary bladder.
2. Name the first part of the male urethra.
3. Where are sperm formed?
4. Name the four different parts of the male urethra.
   What is the function of the male urethra?
5. Name the external orifice of the urinary tract.

Self Test
Take this opportunity to quiz yourself by taking the **SELF
TEST.** See page 15 for a reminder on how to access the
self-test for this section.

IN REVIEW

What Have I Learned?

The following questions cover the material that you have just learned, the lower urinary tract. Apply what you have learned to answer these questions on a separate piece of paper.

1. Describe the structure of the urinary bladder.

2. What are the two functions of this organ?

3. What is the function of the female urethra?

4. Give two reasons why urinary tract infections are more common in females.

5. Name the structures that form the trigone of the urinary bladder.

6. Name the first part of the male urethra.

7. Where are sperm formed?

8. Name the four different parts of the male urethra. What is the function of the male urethra?

9. Name the external orifice of the urinary tract.

The Urinary Bladder

EXERCISE 10.9:
Urinary System—Urinary Bladder, Histology (Low Magnification)

SELECT TOPIC	SELECT VIEW	
Urinary Bladder ▶	**Low Magnification**	**GO**

- *Click the* **TURN TAGS ON** *button, and you will see the following image:*

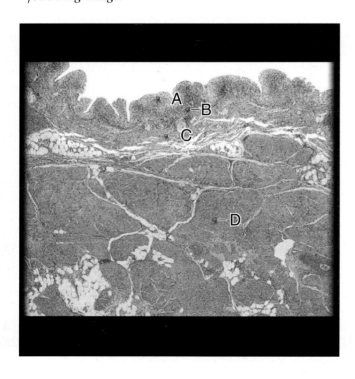

- *Mouse-over the pins on the screen to find the information necessary to identify the following structures:*

A. _____

B. _____

C. _____

D. _____

CHECK POINT:

Urinary Bladder, Histology (Low Magnification)

1. Name the connective tissue layer of the urinary bladder that contains blood, lymph vessels, and nerves.
2. Name the three-layered muscle of the urinary bladder.
3. What is this muscle's function?

EXERCISE 10.10:

Urinary System—Urinary Bladder, Histology (High Magnification)

SELECT TOPIC
Urinary Bladder ➤ **SELECT VIEW**
Low Magnification   **GO**

- *Click the* **TURN TAGS ON** *button, and you will see the following image:*

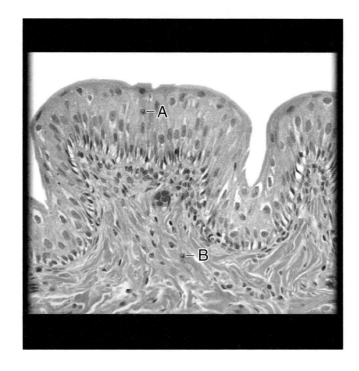

- *Mouse-over the pins on the screen to find the information necessary to identify the following structures:*

A. _____

B. _____

CHECK POINT:

Urinary Bladder, Histology (High Magnification)

1. Describe the transition epithelium of the urinary bladder.
2. What is the function of this epithelium?
3. Name the connective tissue layer that attaches the epithelium to the muscularis mucosa.

Animation: Micturition Reflex

SELECT ANIMATION
Micturition Reflex   **PLAY**

- *After viewing the animation, answer these questions:*

1. What type of reflex is the micturition reflex? What pathway do the impulses take?

2. How is the reflex coordinated? What signals can influence this reflex?

3. What causes the increase in the frequency of action potentials along this pathway?

4. What is the parasympathetic response to these action potentials? How does the urinary bladder respond to these stimuli?

5. What causes a conscious desire to urinate?

6. What happens if urination is not appropriate?

7. What causes the external urethral sphincter to remain contracted? What does this prevent?

8. When urination is desired, what stimulates the micturition reflex?

9. How does the brain then initiate urination?

Self Test
Take this opportunity to quiz yourself by taking the **SELF TEST.** See page 15 for a reminder on how to access the self-test for this section.

I N   R E V I E W

What Have I Learned?

The following questions cover the material that you have just learned, the urinary bladder. Apply what you have learned to answer these questions on a separate piece of paper.

1. Name the connective tissue layer of the urinary bladder that contains blood, lymph vessels, and nerves.

2. Name the three-layered muscle of the urinary bladder.

3. What is this muscle's function?

4. Describe the transition epithelium of the urinary bladder.

5. What is the function of this epithelium?

6. Name the connective tissue layer that attaches the epithelium to the muscularis mucosa.

Self Test

Take this opportunity to quiz yourself by taking the **SELF TEST.** See page 15 for a reminder on how to access the self-test for this section.

EXERCISE 10.11:

Coloring Exercise

Look up the features of the urinary system. Then color them in using colored pens or pencils.

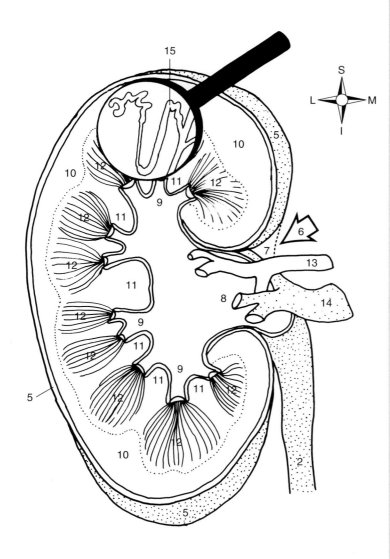

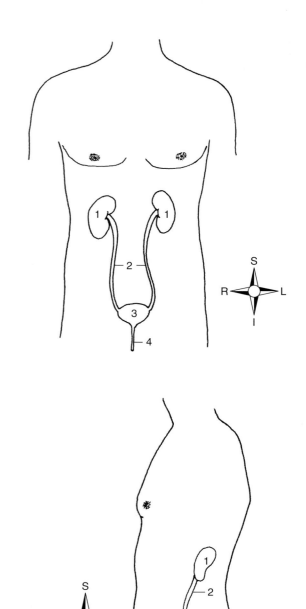

☐ 1. _____

☐ 2. _____

☐ 3. _____

☐ 4. _____

☐ 5. _____

☐ 6. _____

☐ 7. _____

☐ 8. _____

☐ 9. _____

☐10. _____

☐11. _____

☐12. _____

☐13. _____

☐14. _____

☐15. _____

CHAPTER 11

The Reproductive System

Overview: The Reproductive System

Without the reproductive system, you wouldn't be here! This is the one organ system that doesn't demand much attention until those wonderful years of puberty—and then it takes over like a raging fire. Hormone levels fluctuate and structures change as the body transitions from childhood into adulthood. This is a time when life takes on a whole different perspective.

Our perspective with *Anatomy & Physiology | Revealed*® will be to look at the reproductive system from a macroscopic as well as a microscopic level. We will begin with the female reproductive system, focusing on the structures unique to the human female. We will then explore how the body orchestrates her monthly reproductive cycles. We will conclude with the male reproductive system, which, although not as complex as the female's, is every bit the product of the same precision engineering.

- *From the* **Home screen**, *click the drop-down box on the* **Select system** *menu.*

- *From the systems listed, click on* **Reproductive,** *and you will see the image to the right:*

- *This is the opening screen for the* **Reproductive System.**

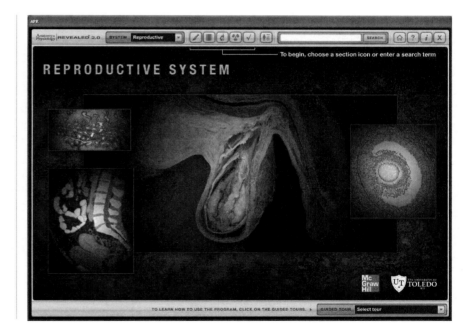

Animation: Female Reproductive System Overview

SELECT ANIMATION
Female Reproductive System Overview   PLAY

- *After viewing the animation, answer these questions:*

1. Name the internal organs of the female reproductive system.

2. Name the other organs of the female reproductive system.

3. Describe the structure of the ovaries.

4. Name the ligaments of the ovaries?

5. Name the three structures housed in each suspensory ligament.

6. What is the mesovarium? What is its function?

7. Name the capsule that surrounds each ovary. What tissue type is it?

8. Name the two regions of the ovary. Which region contains the ovarian follicles?

9. What is located within each follicle?

10. What event triggers development of some follicles?

11. What events occur during ovulation?

12. The remnant of the ruptured follicle is the _____, which later degenerates into a _____.

13. Where are the uterine tubes located? What is the mesosalpinx?

14. Into where do the uterine tubes directly open? Why?

15. Name the four segments of each uterine tube.

16. What are the fimbriae? What is their function?

17. Where does fertilization normally occur?

18. What is the shape and function of the uterus?

19. Name the three regions of the uterus.

20. The uterus' hollow lumen connects to the vagina via . . .

21. Describe the three layers of the uterus.

22. Describe the two layers of the lamina propria. Which layer is shed during menstruation? How is it regenerated?

23. Name the three functions of the uterus.

24. Describe the three layers of the vaginal wall.

25. How do bacteria help inhibit the growth of pathogens in the vagina?

The Female Breast

EXERCISE 11.1:

Reproductive System—Breast—Female, Anterior View

SELECT TOPIC	SELECT VIEW	
Breast—Female ▶	**Anterior**	**GO**

- *Click* **LAYER 1** *in the* **LAYER CONTROLS** *window, and you will see the following image:*

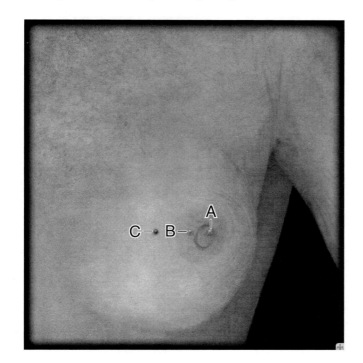

- *Mouse-over the green pins on the screen to find the information necessary to identify the following structures:*

A. _____

B. _____

Non-reproductive System Structure (blue pin):

C. _____

C H E C K   P O I N T :

Breast—Female, Anterior View

1. Name the elevation of skin containing lactiferous ducts for transferring milk during nursing.
2. Name the structure that maintains a permanent pigment increase during the first pregnancy.
3. What is the function of this structure?

- *Click* **LAYER 2** *in the* **LAYER CONTROLS** *window, and you will see the following image:*

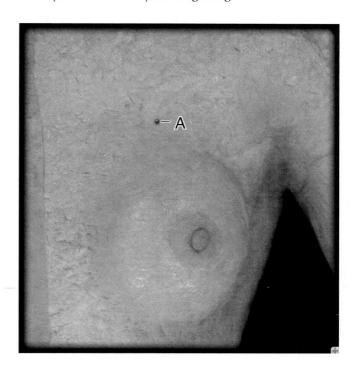

- *Mouse-over the blue pin on the screen to find the information necessary to identify the following non-reproductive system structure:*

A. _____

C H E C K   P O I N T :

Breast—Female, Anterior View, cont'd

4. What tissue-types make up the pectoral superficial fascia?
5. What structures do they contain?
6. Name four functions of this structure.

- *Click **LAYER 3** in the **LAYER CONTROLS** window, and you will see the following image:*

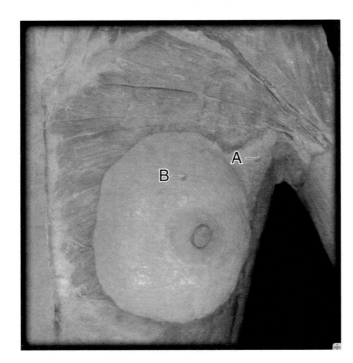

- *Mouse-over the pins on the screen to find the information necessary to identify the following structures:*

A. _____

B. _____

CHECK POINT:

Breast—Female, Anterior View, cont'd

7. Where is the base of the breast located? What is located at its apex?
8. Where is the axillary process of the breast located?
9. What three variables affect the size and shape of the breast?

- *Click **LAYER 4** in the **LAYER CONTROLS** window, and you will see the following image:*

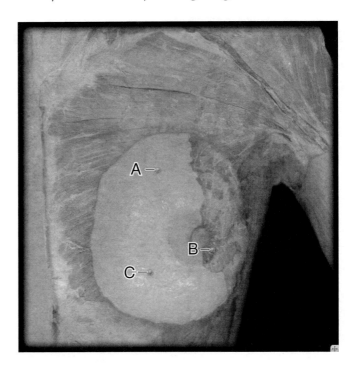

- *Mouse-over the pins on the screen to find the information necessary to identify the following structures:*

A. _____

B. _____

C. _____

CHECK POINT:

Breast—Female, Anterior View, cont'd

10. Name the structures that support the breast tissue.
11. What affect does age have on these structures?
12. Describe the lactiferous ducts.

• *Click* **LAYER 5** *in the* **LAYER CONTROLS** *window, and you will see the following image:*

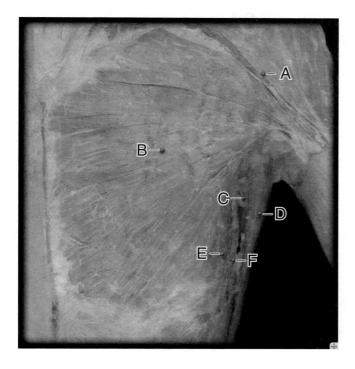

• *Mouse-over the blue pins on the screen to find the information necessary to identify the following non-reproductive system structures:*

A. _____

B. _____

C. _____

D. _____

E. _____

F. _____

I N   R E V I E W

What Have I Learned?

The following questions cover the material that you have just learned, the breast. Apply what you have learned to answer these questions on a separate piece of paper.

1. What is the composition of the breast?

2. What muscle is deep to the base of the breast?

3. The "tail" of the breast extends into the _____.

4. Into what structures do the suspensory ligaments divide the breast?

5. What tissue partially replaces the breast fat tissue during pregnancy and lactation?

6. The _____ allows movement of the breast independent of the thoracic wall muscles.

7. What are the functions of the suspensory ligaments?

8. What is the function of lactiferous ducts?

9. Where does each duct open?

The Female Pelvis

EXERCISE 11.2:	

Reproductive System—Pelvis—Female, Sagittal View

SELECT TOPIC	SELECT VIEW	
Pelvis—Female ▶	**Sagittal**	**GO**

- *Click* **LAYER 1** *in the* **LAYER CONTROLS** *window, and you will see the following image:*

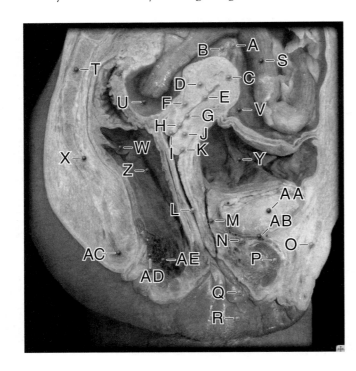

- *Mouse-over the green pins on the screen to find the information necessary to identify the following structures:*

A. _____

B. _____

C. _____

D. _____

E. _____

F. _____

G. _____

H. _____

I. _____

J. _____

K. _____

L. _____

M. _____

N. _____

O. _____

P. _____

Q. _____

R. _____

Non-reproductive System Structures (blue pins)

S. _____

T. _____

U. _____

V. _____

W. _____

X. _____

Y. _____

Z. _____

AA. _____

AB. _____

AC. _____

AD. _____

AE. _____

CHECK POINT:

Pelvis—Female, Sagittal View

1. Name the primary organ of female sexual response. Describe its structures.
2. Describe the structure of the crus of the clitoris. What is its function?
3. Describe the structure of the labium minus. What is its function?

EXERCISE 11.3:

Reproductive System—Pelvis—Female, Superior View

SELECT TOPIC	SELECT VIEW	
Pelvis—Female	**Superior**	**GO**

• *Click on* **LAYER 1** *in the* **LAYER CONTROLS** *window, and you will see the following image:*

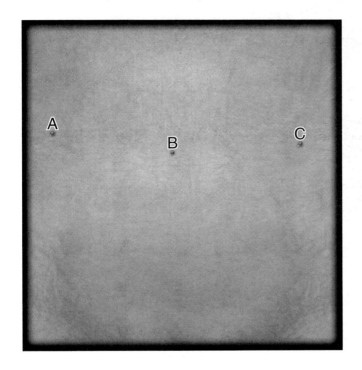

• *Mouse-over the blue pins on the screen to find the information necessary to identify the following non-reproductive system structures:*

A. _____

B. _____

C. _____

CHECK POINT:

Pelvis—Female, Superior View

1. Name the lower medial abdominal region.
2. What is another name for this region?
3. Name the two regions that flank this region on either side.

• *Click* **LAYER 2** *in the* **LAYER CONTROLS** *window, and you will see the following image:*

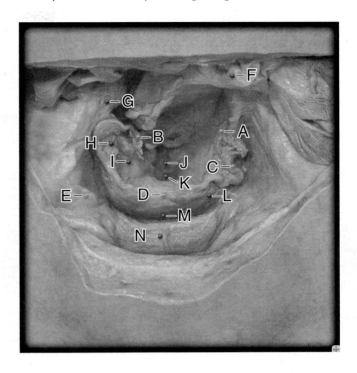

• *Mouse-over the green pins on the screen to find the information necessary to identify the following structures:*

A. _____

B. _____

C. _____

D. _____

E. _____

Non-reproductive System Structures (blue pins)

F. _____

G. _____

H. _____

I. _____

J. _____

K. _____

L. _____

M. _____

N. _____

CHECK POINT:

Pelvis—Female, Superior View, cont'd

4. What is a hysterectomy?
5. Name the blind recess that lies between the urinary bladder and the vagina.
6. Name the structure that is the embryonic remnant of the gubernaculum of the ovary. What was its embryonic function?

EXERCISE 11.4:

Imaging—Uterus and Vagina

SELECT TOPIC	SELECT VIEW	
Uterus and Vagina ▶	**Sagittal**	**GO**

- Click the **TURN TAGS ON** button, and you will see the following image:

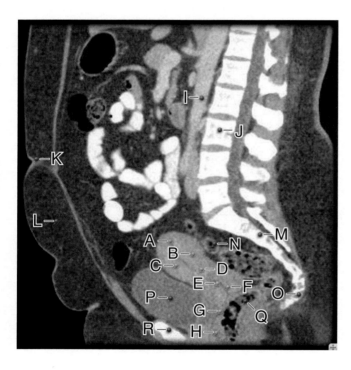

- Mouse-over the green pins on the screen to find the information necessary to identify the following structures:

A. _____

B. _____

C. _____

D. _____

E. _____

F. _____

G. _____

H. _____

Non-reproductive System Structures (blue pins)

I. _____

J. _____

K. _____

L. _____

M. _____

N. _____

O. _____

P. _____

Q. _____

R. _____

I N R E V I E W

What Have I Learned?

The following questions cover the material that you have just learned, the female pelvis. Apply what you have learned to answer these questions on a separate piece of paper.

1. Name the region between the labia minora.

2. What is the urogenital triangle?

3. Describe the structure of the labium minus. What is its function?

4. Describe the structure of the labium majus. What is its function?

5. Describe the structure of the vagina. What is its function?

6. Define parturition. What is the standard measure for timing parturition?

7. Name the structure that varies its position with the fullness of the urinary bladder and rectum. What is the function of that structure?

8. The junction of the cervical canal and what structure forms the internal os? What is the function of that structure?

9. Name the female gonads. What is their function?

10. Name the location for a tubal ligation. What is this procedure?

11. Name the location where the cartilage softens in late pregnancy to allow a slight separation of the pubic bones.

12. Name the structure that conducts the ovarian vessels and nerves.

The Uterus

EXERCISE 11.5:

Reproductive System—Uterus (Proliferative Phase), Histology

SELECT TOPIC
Uterus (Proliferative Phase) ▸

SELECT VIEW
Low Magnification

GO

- *Click the* **TURN TAGS ON** *button, and you will see the following image:*

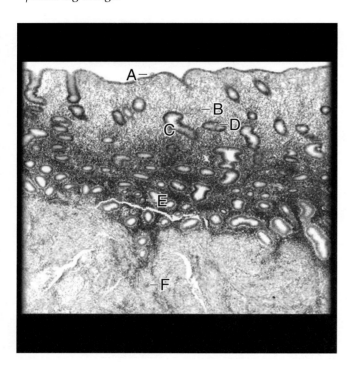

- *Mouse-over the pins on the screen to find the information necessary to identify the following structures:*

A. _____

B. _____

C. _____

D. _____

E. _____

F. _____

CHECK POINT:

Uterus (Proliferative Phase), Histology

1. What is unique for the cervical endometrium compared to the endometrium of the body and fundus of the uterus?
2. Describe the endometrial glands. What is their function?
3. The endometrium is formed from what two layers?
4. Name the thick, smooth muscle layer of the uterus.
5. Name the superficial part of the endometrium sloughed during menstruation.

The Ovarian Follicle

Animation: Comparison of Meiosis and Mitosis

SELECT ANIMATION
Comparison of Meiosis and Mitosis

PLAY

- *After viewing the animation, answer these questions:*

1. What is the normal human chromosome number? This number is referred to as _____.

2. There are _____ pairs of chromosomes, known as _____ pairs.

3. What process occurs before both meiosis and mitosis? What happens to the chromosomes afterward?

4. How many successive divisions occur in meiosis?

5. List the events of the first division.

6. List the events of the second division.

7. Define haploid.

8. How many cell divisions occur during mitosis? What is the result of mitosis?

Animation: Meiosis

| | SELECT ANIMATION | |
| | **Meiosis** | PLAY |

- *After viewing the animation, answer these questions:*

1. Sperm and egg cells are formed through the process of _____. Where are these cells found?

2. Are the germ-cell lines diploid or haploid? Through the process of meiosis, the cells produced will be _____, having _____ set(s) of chromosomes.

3. During fertilization, these _____ cells _____ to form a _____ offspring.

4. Name the phase of meiosis when the DNA replicates and each chromosome becomes doubled.

5. These doubled chromosomes consist of . . .

6. The first division of meiosis, meiosis I, . . .

7. In the second division, meiosis II, . . .

8. What is the result of meiosis?

9. What changes have occurred to the chromosomes during prophase I?

10. Name the structure that joins identical sister chromatids.

11. What is crossing-over? What events occur just prior to crossing-over?

12. As a result of crossing-over, how do the two sister chromatids of one chromosome compare?

13. List the events of metaphase I.

14. What is independent assortment?

15. List the events of anaphase I. Describe the chromosomes during this phase.

16. How many members of the homologous chromosome pair will be in each new cell at the end of anaphase I? Are these new cells haploid or diploid?

17. List the events in telophase I.

18. Define cytokinesis.

19. What are the results of meiosis I?

20. How does meiosis II compare to mitosis?

21. List the events of prophase II.

22. What action occurs among the chromosomes during metaphase II? How does metaphase II compare with metaphase I?

23. What events occur during anaphase II? During telophase II?

24. How many cells result from meiosis? How do they compare genetically?

Animation: Unique Features of Meiosis

SELECT ANIMATION	
Unique Features of Meiosis	PLAY

- *After viewing the animation, answer these questions:*

1. Name the three unique features of meiosis.

2. Describe the events of synapsis.

3. Describe the events of crossing-over.

4. What is another name for crossing-over?

5. The name for the first nuclear division is _____.

6. What is the result of reduction division?

7. What effect does crossing-over have on the sister chromatids of each daughter cell?

EXERCISE 11.6:

Reproductive System—Primordial Follicle, Histology

SELECT TOPIC **Primordial Follicle**	▶	SELECT VIEW **High Magnification**	GO

- *Click the* **TURN TAGS ON** *button, and you will see the following image:*

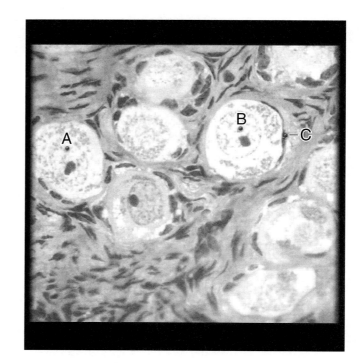

- *Mouse-over the pins on the screen to find the information necessary to identify the following structures:*

A. _____

B. _____

C. _____

CHECK POINT:

Primordial Follicle, Histology

1. A primordial follicle is one that has not responded to _____ to become a _____.
2. Where are these follicles located?
3. What structure surrounds the oocyte?

Animation: Female Reproductive Cycles

| SELECT ANIMATION
Female Reproductive Cycles | PLAY |

- *After viewing the animation, answer these questions:*

1. Female reproductive cycles are initiated usually between the ages of _____ and _____, a period known as _____.

2. How long is the average female reproductive cycle? Where do the changes occur during the cycle?

3. During each cycle, the hypothalamus releases _____, which stimulates the _____ to release two _____ hormones, _____ and _____.

4. Name the target organ for these hormones.

5. Name the two components of the female reproductive cycle.

6. Ovulation occurs on day _____ of the _____-day ovarian cycle.

7. What are the 14 days prior to ovulation called in the ovarian cycle? The 14 days following ovulation?

8. What regulates the ovarian cycle?

9. Ovulation occurs on day _____ of the _____-day uterine cycle.

10. Describe the two subdivisions of the uterine cycle that occur before ovulation. And the 14 days following ovulation?

11. What controls the uterine cycle?

12. Describe the first five days of the follicular stage.

13. What occurs during this time in the uterine cycle? This is the _____ phase, commonly referred to as _____.

14. Describe the ovarian events during days 6–13 of the follicular phase.

15. Describe the ovarian events two days before ovulation. What hormones influence these events?

16. What causes the final maturation of the follicle? How soon before ovulation does this occur? What is another name for the mature follicle?

17. What event occurs in the primary oocyte just prior to ovulation? What does this form?

18. What event occurs in the uterus during the proliferative stage? Name the hormone responsible for this event and where that hormone is produced.

19. Describe the events of ovulation. What hormone peaks at this time?

20. During the luteal stage, what structure is formed from the remaining luteal cells? What is the function of this structure and the hormones that it produces?

21. After the ovum is fertilized and implants in the uterine wall, what hormone is produced by the cells of the implantation site? What is the function of this hormone?

22. How long does the corpus luteum continue hormone production? What transition takes place after this time? What structures produce hormones in the place of the corpus luteum?

23. If fertilization does not occur, when does the corpus luteum become the corpus albicans? What effect does this have on hormone levels? What events do these changes initiate?

CHECK POINT:

Primary Follicle, Histology

1. Name the cells responsible for estrogen secretion. Where are they located?
2. Name the glycoprotein produced by the oocyte.
3. Is the primary oocyte diploid or haploid? What do these terms mean?
4. Name and describe the structure that surrounds the primary oocyte.

EXERCISE 11.7:

Reproductive System—Primary Follicle, Histology

SELECT TOPIC   SELECT VIEW
Primary Follicle ➤ **High Magnification**   GO

- *Click the* **TURN TAGS ON** *button, and you will see the following image:*

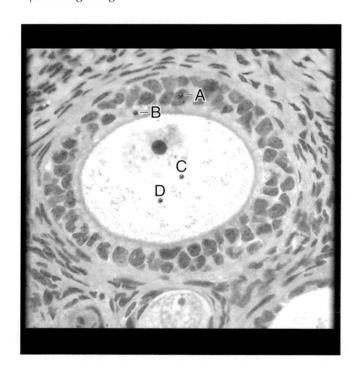

- *Mouse-over the pins on the screen to find the information necessary to identify the following structures:*

A. _____

B. _____

C. _____

D. _____

EXERCISE 11.8:

Reproductive System—Secondary Follicle, Histology

SELECT TOPIC   SELECT VIEW
Secondary Follicle ➤ **High Magnification**   GO

- *Click the* **TURN TAGS ON** *button, and you will see the following image:*

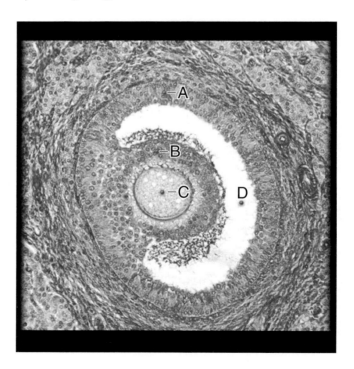

- *Mouse-over the pins on the screen to find the information necessary to identify the following structures:*

A. _____

B. _____

C. _____

D. _____

CHECK POINT:

Secondary Follicle, Histology

1. Is the secondary oocyte haploid or diploid?
2. Name the fluid-filled cavity rich in estrogen. Where is it located?
3. Name the innermost layer of granulosa cells. These cells are in contact with which layer?

Animation: Ovulation Through Implantation

 SELECT ANIMATION
Ovulation Through Implantation PLAY

• *After viewing the animation, answer these questions:*

1. What are oocytes? Where are they produced? What is the process of their production?

2. How many primary oocytes are produced before birth? Are they diploid or haploid?

3. How many primary oocytes survive until puberty? Of these, how many will become secondary oocytes?

4. What meiotic stage do the primary oocytes remain in until just before ovulation?

5. Describe the structure of the follicle.

6. What events occur in the primary oocyte just before ovulation? What do these events form?

7. How do the ovaries alternate between ovulations?

8. What events occur during ovulation? Name the structures that receive the ovulated secondary oocyte.

9. What processes move the secondary oocyte toward the uterus?

10. Where does fertilization occur?

11. How many sperm are deposited in the vagina during ejaculation? How many reach the ampulla? How soon do they arrive?

12. How many sperm penetrate the secondary oocyte? What does this event trigger?

13. What events occur on a cellular level during fertilization? What cell do they form? How many chromosomes does this cell contain? Is it diploid or haploid?

14. What is the term for a fertilized oocyte (ovum)? What term describes the mitotic divisions of this fertilized secondary oocyte? Where is the fertilized ovum located while these mitotic divisions occur?

15. What is a morula? What is the term for the hollow sphere of cells into which the morula transforms? Where is the morula located when this transition occurs?

16. What are a trophoblast and embryoblast? What structures are formed from each?

17. What role do the hormones estrogen and progesterone play in the implantation of the blastocyst?

18. What is implantation? When does it occur?

19. What changes occur in the trophoblast as implantation occurs? What are the functions of these new tissues?

20. When does the blastocyst become embedded superficially in the endometrium?

EXERCISE 11.9:

Reproductive System—Corpus Albicans, Histology

SELECT TOPIC	SELECT VIEW	
Corpus Albicans ➤	**High Magnification**	**GO**

- *Click the* **TURN TAGS ON** *button, and you will see the following image:*

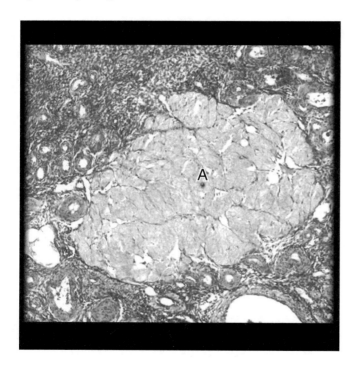

- *Mouse-over the pin on the screen to find the information necessary to identify the following structure:*

A. _____

C H E C K   P O I N T :

Corpus Albicans, Histology

1. The corpus albicans is a connective tissue _____ at the surface of the _____ .
2. The corpus albicans is a remnant of _____ .
3. What does it identify?

Did you know that the name corpus albicans means "white body" and the name corpus luteum means "yellow body"? These are descriptive terms for their visual appearance.

The Uterine Tube

EXERCISE 11.10:

Reproductive System—Uterine Tube, Histology (Low Magnification)

SELECT TOPIC	SELECT VIEW	
Uterine Tube ➤	**Low Magnification**	**GO**

- *Click the* **TURN TAGS ON** *button, and you will see the following image:*

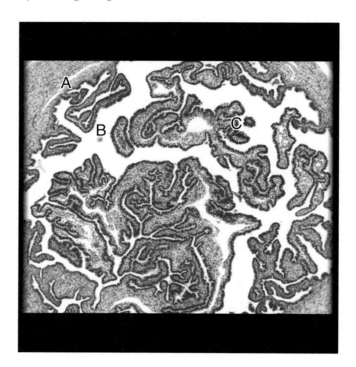

- *Mouse-over the pins on the screen to find the information necessary to identify the following structures:*

A. _____

B. _____

C. _____

C H E C K   P O I N T :

Uterine Tube, Histology (Low Magnification)

1. What is the central tubular cavity of the uterine tube? What is its function?
2. What is the lining of the uterine tube? It consists of which tissue type?
3. Name the smooth muscle layers in the wall of the uterine tube. What is their function?

EXERCISE 11.11:

Reproductive System—Uterine Tube, Histology (High Magnification)

| SELECT TOPIC **Uterine Tube** | ▶ | SELECT VIEW **High Magnification** | GO |

- *Click the* **TURN TAGS ON** *button, and you will see the following image:*

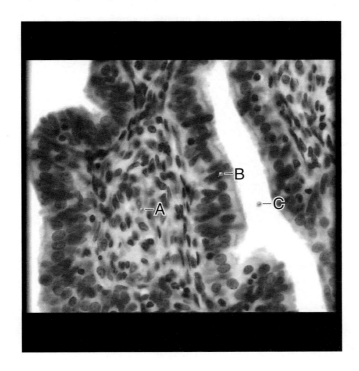

- *Mouse-over the pins on the screen to find the information necessary to identify the following structures:*

A. _____

B. _____

C. _____

CHECK POINT:

Uterine Tube, Histology (High Magnification)

1. Name the tissue type lining the uterine tubes.
2. What is the function of this tissue?
3. Name the layer that attaches the epithelium to the muscularis of the uterine tube. Name two structural layers not present in the uterine tube.

The Female Perineum

EXERCISE 11.12:

Reproductive System—Perineum—Female, Inferior View

| SELECT TOPIC **Perineum—Female** | ▶ | SELECT VIEW **Inferior** | GO |

- *Click on* **LAYER 1** *in the* **LAYER CONTROLS** *window, and you will see the following image:*

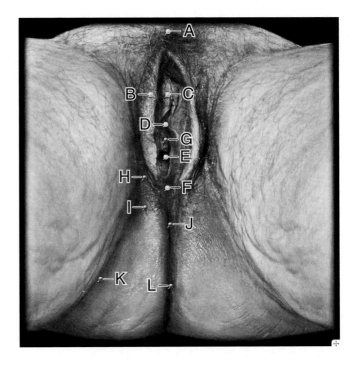

- *Mouse-over the green pins on the screen to find the information necessary to identify the following structures:*

A. _____

B. _____

C. _____

D. _____

E. _____

F. _____

Non-reproductive System Structures (blue pins)

G. _____

H. _____

I. _____

J. _____

K. _____

L. _____

C H E C K P O I N T :

Perineum—Female, Inferior View

1. Name the subcutaneous fat pad anterior to the pubic symphysis and pubic bones. When does the fat layer increase and decrease?
2. Name the limiting structures of the urogenital triangle. What is contained in this area?
3. What is the hymen? What is its function?

- *Click on* **LAYER 2** *in the* **LAYER CONTROLS** *window, and you will see the following image:*

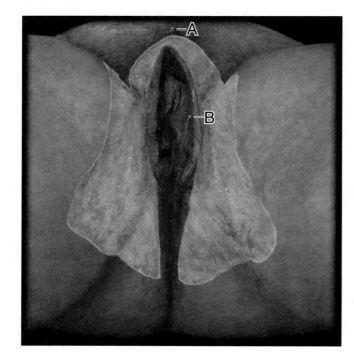

- *Mouse-over the pins on the screen to find the information necessary to identify the following structures:*

A. _____

B. _____

- *Click on* **LAYER 3** *in the* **LAYER CONTROLS** *window, and you will see the following image:*

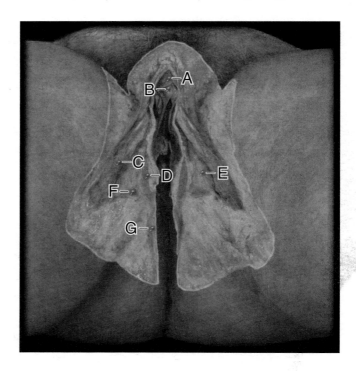

- *Mouse-over the green pins on the screen to find the information necessary to identify the following structures:*

A. _____

B. _____

C. _____

Non-reproductive System Structures (blue pins)

D. _____

E. _____

F. _____

G. _____

C H E C K P O I N T :

Perineum—Female, Inferior View, cont'd

4. Name the hood of thin skin over the glans clitoris. What are its functions?
5. Describe the structure of the glans clitoris. What is its function?
6. What is the perineal membrane? What is its function?

- *Click* **LAYER 4** *in the* **LAYER CONTROLS** *window, and you will see the following image:*

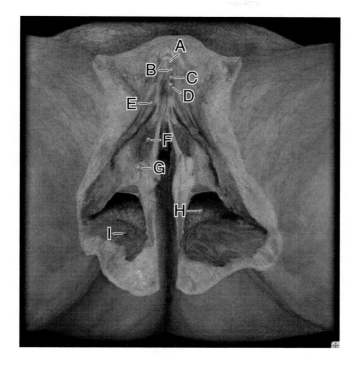

- *Mouse-over the green pins on the screen to find the information necessary to identify the following structures:*

A. _____

B. _____

C. _____

D. _____

E. _____

F. _____

G. _____

H. _____

Non-reproductive System Structure (blue pin)

I. _____

CHECK POINT:

Perineum—Female, Inferior View, cont'd

7. What is the function of the clitoris? What occurs within the clitoris during sexual arousal?
8. Name the three sections of the clitoris.
9. What is the function of the bulb of the vestibule?

IN REVIEW

What Have I Learned?

The following questions cover the material that you have just learned, the female perineum. Apply what you have learned to answer these questions on a separate piece of paper.

1. What is an episiotomy? Why is this procedure done?

2. What is the anal triangle?

3. What is the intergluteal cleft?

4. What is the gluteal fold?

5. Name the potential space between the labia minora. Name the contents of this space from anterior to posterior.

6. Name the inferior opening of the vagina.

7. Name the structure that protects the external genitalia.

8. What are the functions of the labia minus?

9. What are the actions of the ischiocavernosus muscles?

10. What are the actions of the bulbospongiosus muscles?

11. Name the paired mucus-producing glands lateral to the vaginal opening. What is their function?

12. Describe the structures of the clitoris. What is the function of each?

The Male Pelvis

Animation: Male Reproductive
System Overview

```
SELECT ANIMATION
Male Reproductive System Overview        PLAY
```

- *After viewing the animation, answer these questions:*

1. The primary male sex organs, or _____, are the _____.

2. List the pathway of sperm from the primary sex organs through the penis.

3. Name the accessory sex glands. What is their function?

4. Describe the location of the testes. How does the temperature there compare to core body temperature? Why?

5. How is the temperature of the testes regulated? What muscles are involved? What are the actions of these muscles when the temperature is too warm *and* too cold?

6. Describe the counter-current heat exchanger mechanism. What effect does this mechanism have on the arterial blood temperature as it reaches the testes?

7. The testes are the site of what events?

8. Name the two connective tissue coats of the testes?

9. Internally, the testes are divided into _____ wedge-shaped _____. Each _____ contains up to _____ ducts called _____. What occurs in these ducts?

10. What is equivalent to the combined length of these ducts in both testes?

11. How many sperm are in the average ejaculate?

12. Describe the epithelium that lines the seminiferous tubule.

13. Name the cells responsible for testosterone production. Where are they located?

14. Describe the pathway of the sperm as they leave the seminiferous tubule, mature, and are stored.

15. How many sperm form each day?

16. What structure is continuous with the tail of the epididymis? Where is this structure located?

17. The _____ joins the duct of the _____ to form the _____, which passes through the _____ to empty into the _____ part of the _____.

18. Name the accessory gland that produces 60 percent of the semen volume. Describe the secretion produced.

19. Name the accessory gland that produces 30 percent of the semen volume. Describe the secretion produced.

20. Name the accessory gland that produces mucus that neutralizes the acidic urethra. What percent of the total semen volume consists of this fluid?

21. What are the dual functions of the penis?

22. What are the divisions of the penis?

23. Describe the erectile bodies of the penis.

24. Which structure encloses the spongy urethra? What is the terminal end of this structure?

25. How is the size and shape of the penis determined?

26. What events occur in the process of an erection?

• *Mouse-over the green pins on the screen to find the information necessary to identify the following structures:*

A. _____

B. _____

C. _____

D. _____

E. _____

F. _____

G. _____

H. _____

I. _____

J. _____

K. _____

L. _____

M. _____

N. _____

O. _____

P. _____

Q. _____

R. _____

S. _____

T. _____

U. _____

V. _____

Non-reproductive System Structures (blue pins)

W. _____

X. _____

Y. _____

Z. _____

AA. _____

AB. _____

AC. _____

AD. _____

AE. _____

AF. _____

AG. _____

AH. _____

AI. _____

AJ. _____

EXERCISE 11.13:

Reproductive System—Pelvis—Male, Sagittal View

| SELECT TOPIC **Pelvis—Male** | ▶ | SELECT VIEW **Sagittal** | GO |

• *Click* **LAYER 1** *in the* **LAYER CONTROLS** *window, and you will see the following image:*

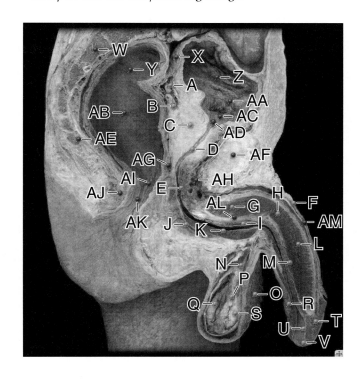

AK. _____

AL. _____

AM. _____

CHECK POINT:

Pelvis—Male, Sagittal View

1. Name the paired and unpaired erectile tissues of the penis. What is their function?
2. What structure is responsible for most of the increased size and rigidity of the penis during an erection?
3. What is circumcision? How does the structure of the prepuce compare between circumcised and uncircumcised males?

IN REVIEW

What Have I Learned?

The following questions cover the material that you have just learned, the male pelvis. Apply what you have learned to answer these questions on a separate piece of paper.

1. Describe the structure of the crus of the penis. What is its function?

2. Describe the structure of the tunica albuginea of the penis. What is its function?

3. Describe the structure of the bulb of the penis. What is its function?

4. What are the actions of the bulbospongiosus muscle in the male? What are the results of these actions?

5. Describe the parts of the male urethra from proximal to distal. What is the function of the male urethra?

6. Name the accessory reproductive organ that contributes 30 percent of the semen volume. What is BPH?

7. Name the accessory reproductive organ that contributes 60 percent of the semen volume. What is the pH of this contribution?

8. Name the network of veins that drain the erectile tissues of the penis. What other structures are drained by these veins?

The Prostate

EXERCISE 11.14:

Reproductive System—Prostate, Histology

SELECT TOPIC
Prostate Gland ▶

SELECT VIEW
High Magnification

GO

- *Click the* **TURN TAGS ON** *button, and you will see the following image:*

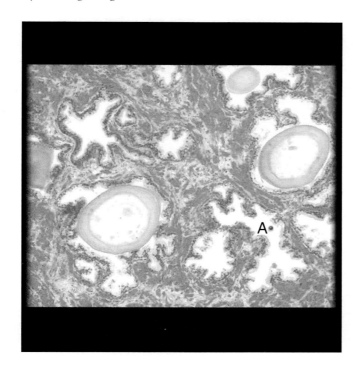

- *Mouse-over the pin on the screen to find the information necessary to identify the following structure:*

A. _____

C H E C K   P O I N T :

Prostate, Histology

1. Describe the structure of the tubuloalveolar glands of the prostate.
2. What is the function of prostatic secretions?

The Male Perineum

EXERCISE 11.15:

Reproductive System—Perineum—Male, Inferior View

SELECT TOPIC
Perineum—Male ▶

SELECT VIEW
Inferior

GO

- *Click* **LAYER 1** *in the* **LAYER CONTROLS** *window, and you will see the following image:*

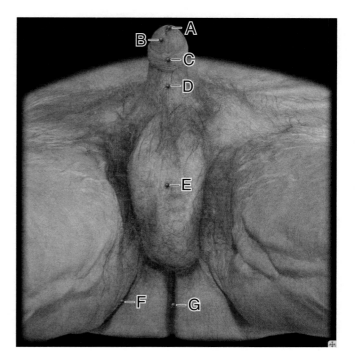

- *Mouse-over the blue pins on the screen to find the information necessary to identify the following non-reproductive system structures:*

A. _____

B. _____

C. _____

D. _____

E. _____

F. _____

G. _____

C H E C K   P O I N T :

Perineum—Male, Inferior View

1. Name the pouch of skin that contains the testes. What muscle is in the wall of the pouch?
2. What changes occur when this muscle contracts? Why?
3. The size of this pouch varies with _____.

- *Click **LAYER 2** in the **LAYER CONTROLS** window, and you will see the following image:*

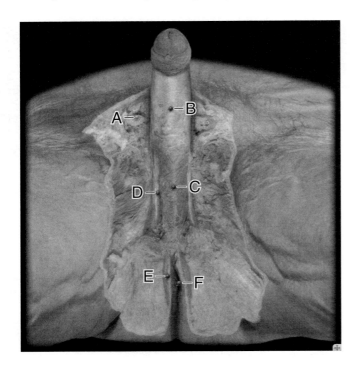

- *Mouse-over the green pin on the screen to find the information necessary to identify the following structure:*

A. _____

Non-reproductive System Structures (blue pins)

B. _____

C. _____

D. _____

E. _____

F. _____

CHECK POINT:

Perineum—Male, Inferior View, cont'd

4. Describe the actions of the ischiocavernosus muscle in the male.
5. Describe the actions of the bulbospongiosus muscle in the male. What are the results of these actions?
6. What is the function of the deep fascia of the penis?

- *Click **LAYER 3** in the **LAYER CONTROLS** window, and you will see the following image:*

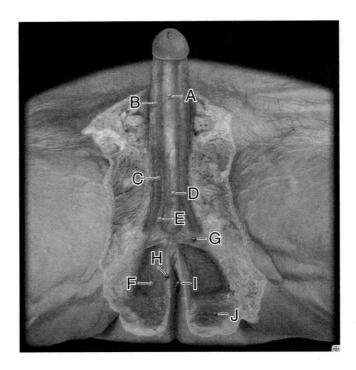

- *Mouse-over the green pins on the screen to find the information necessary to identify the following structures:*

A. _____

B. _____

C. _____

D. _____

E. _____

F. _____

Non-reproductive System Structures (blue pins)

G. _____

H. _____

I. _____

J. _____

CHECK POINT:

Perineum—Male, Inferior View, cont'd

7. Name the single erectile body on the ventral aspect of the penis. This is a continuation of what structure?
8. Name the paired erectile body on the dorsal aspect of the penis. This is a continuation of what structure?
9. What is the function of these erectile tissues?

I N   R E V I E W

What Have I Learned?

The following questions cover the material that you have just learned, the male perineum. Apply what you have learned to answer these questions on a separate piece of paper.

1. How does the scrotal temperature compare with body temperature? Why?

2. How is the scrotum divided?

The Penis and Scrotum

EXERCISE 11.16:

Reproductive System—Penis and Scrotum, Anterior View

| SELECT TOPIC **Penis and Scrotum** | ▶ | SELECT VIEW **Anterior** | **GO** |

- *Click* **LAYER 1** *in the* **LAYER CONTROLS** *window, and you will see the following image:*

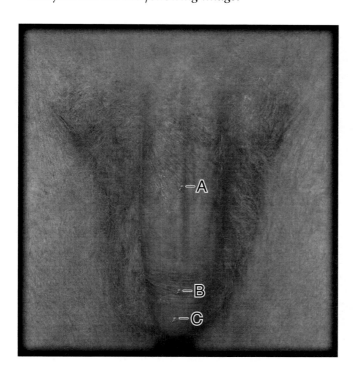

- *Mouse-over the pins on the screen to find the information necessary to identify the following structures:*

A. _____

B. _____

C. _____

C H E C K   P O I N T :

Penis and Scrotum, Anterior View

1. Name the unattached portion of the penis.
2. Name the distal expansion of the penis.
3. Define flaccid. How does the dorsal surface compare when the penis is flaccid and when it is erect?

- *Click on* **LAYER 2** *in the* **LAYER CONTROLS** *window, and you will see the following image:*

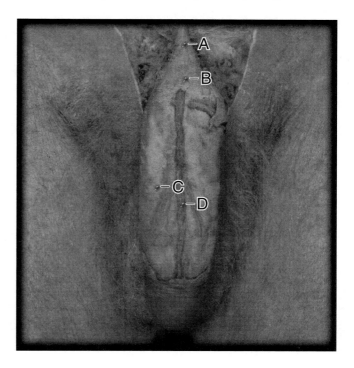

- *Mouse-over the pins on the screen to find the information necessary to identify the following structures:*

A. _____

B. _____

C. _____

D. _____

CHECK POINT:

Penis and Scrotum, Anterior View, cont'd

4. Name the large vein on the dorsal surface of the penis that is *not* involved in an erection.
5. Describe the deep fascia of the penis. What is its function?
6. Name the sling of deep fascia that suspends the body of the penis.

- *Click on* **LAYER 3** *in the* **LAYER CONTROLS** *window, and you will see the following image:*

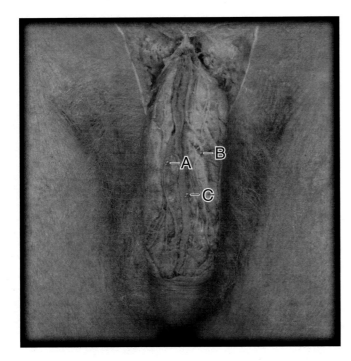

- *Mouse-over the pins on the screen to find the information necessary to identify the following structures:*

A. _____

B. _____

C. _____

CHECK POINT:

Penis and Scrotum, Anterior View, cont'd

7. What prevents blood from escaping the erectile tissue?
8. Name the paired arteries on the dorsal side of the penis that distribute to the skin and glans of the penis.
9. Name the nerve that supplies innervation to the body of the penis.

- *Click* **LAYER 4** *in the* **LAYER CONTROLS** *window, and you will see the following image:*

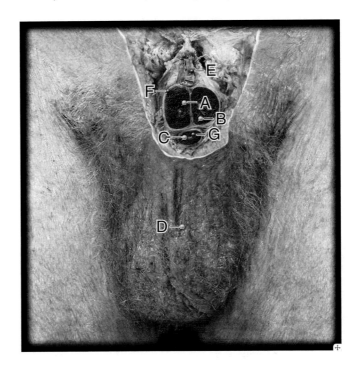

- *Mouse-over the green pins on the screen to find the information necessary to identify the following structures:*

A. _____

B. _____

C. _____

D. _____

Non-reproductive System Structures (blue pins)

E. _____

F. _____

G. _____

CHECK POINT:

Penis and Scrotum, Anterior View, cont'd

10. Name the distal end cap of the corpora cavernosa.
11. Name the distal end of the corpus spongiosum.
12. Name the vein that drains the erectile tissues of the body of the penis.

- Click **LAYER 5** in the **LAYER CONTROLS** window, and you will see the following image:

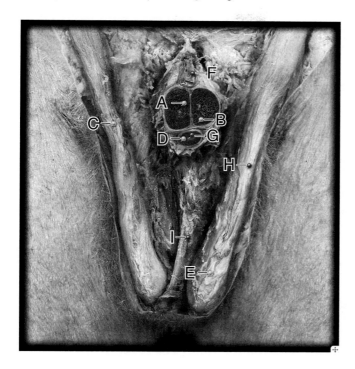

- Mouse-over the green pins on the screen to find the information necessary to identify the following structures:

A. _____

B. _____

C. _____

D. _____

E. _____

Non-reproductive System Structures (blue pins)

F. _____

G. _____

H. _____

I. _____

CHECK POINT:

Penis and Scrotum, Anterior View, cont'd

13. Name the structures located in the spermatic cord. What muscle is located in the middle fascial layer? What is the function of this muscle?

14. Name the structure that separates the right and left testes.

15. Name the structures that enclose the spermatic cord. What are the three layers that make up these structures?

- Click **LAYER 6** in the **LAYER CONTROLS** window, and you will see the following image:

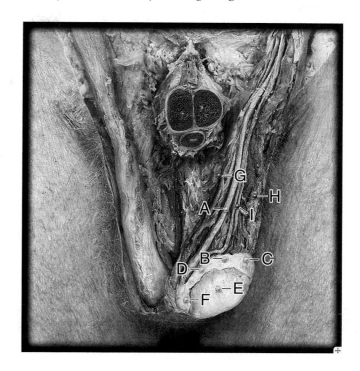

- Mouse-over the green pins on the screen to find the information necessary to identify the following structures:

A. _____

B. _____

C. _____

D. _____

E. _____

F. _____

Non-reproductive System Structures (blue pins)

G. _____

H. _____

I. _____

CHECK POINT:

Penis and Scrotum, Anterior View, cont'd

16. Name the paired male gonads. What are their functions?

17. Name the structure for maturation and storage of spermatozoa. How long can spermatozoa be stored?

18. What are the three parts of the structure in question 17?

I N   R E V I E W

What Have I Learned?

The following questions cover the material that you have just learned, the penis and scrotum. Apply what you have learned to answer these questions on a separate piece of paper.

1. What is emission? Name the structure that conducts sperm and testicular fluid during emission.

2. Name the arteries located within the spermatic cord. What is their distribution?

3. Name the network of veins that ascend within the spermatic cord. Where do these drain? (Hint: right is different from left)

The Seminiferous Tubule

EXERCISE 11.17:
Reproductive System—Seminiferous Tubule, Histology

SELECT TOPIC
Seminiferous Tubule

▶ **SELECT VIEW**
High Magnification **GO**

- *Click the* **TURN TAGS ON** *button, and you will see the following image:*

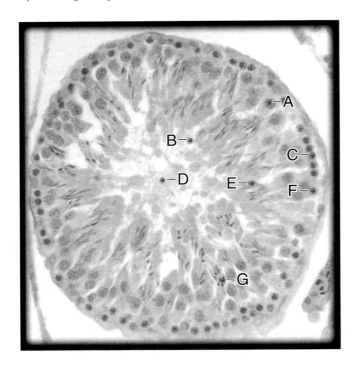

- *Mouse-over the pins on the screen to find the information necessary to identify the following structures:*

A. _____

B. _____

C. _____

D. _____

E. _____

F. _____

G. _____

C H E C K   P O I N T :

Seminiferous Tubule, Histology

1. Name the site of spermatogenesis. Where is this site located? How many are in each location?
2. Name the central tubular cavity of the seminiferous tubules. What is usually contained in this cavity?
3. What are the functions of the sustentacular cells? What other names are they known by?

Animation: Spermatogenesis

| SELECT ANIMATION **Spermatogenesis** | PLAY |

- *After viewing the animation, answer these questions:*

1. Where are sperm formed?

2. Describe the structure and lining of the seminiferous tubules. What specialized cell types are located there?

3. Name the germ cells from which sperm cells arise.

4. What differences occur in the daughter cells of the spermatogonia? Are they haploid or diploid? How many chromosomes do they each contain?

5. The primary spermatocytes divide by _____ to form _____ _____ spermatocytes, each containing _____ chromosomes. The _____ divide again to form _____.

6. What is spermiogenesis? Where does it occur? What changes occur during this process?

7. Sperm cells are _____ when they leave the seminiferous tubules and testis and mature _____.

I N   R E V I E W

What Have I Learned?

The following questions cover the material that you have just learned, the seminiferous tubule. Apply what you have learned to answer these questions on a separate piece of paper.

1. Name the most primitive cell in the male germ line. Diploid or haploid?

2. Name the diploid cells derived from the cells in question 1. These cells give rise to_____, which are (haploid/diploid). These cells in turn give rise to _____, which are (haploid/diploid).

3. Name and describe the structure of the male germ cells. What is their function?

Self Test

Take this opportunity to quiz yourself by taking the **SELF TEST.** See page 15 for a reminder on how to access the self-test for this section.

EXERCISE 11.18:

Coloring Exercise

Look up the features of the female reproductive system. Then color in the features with colored pens or pencils.

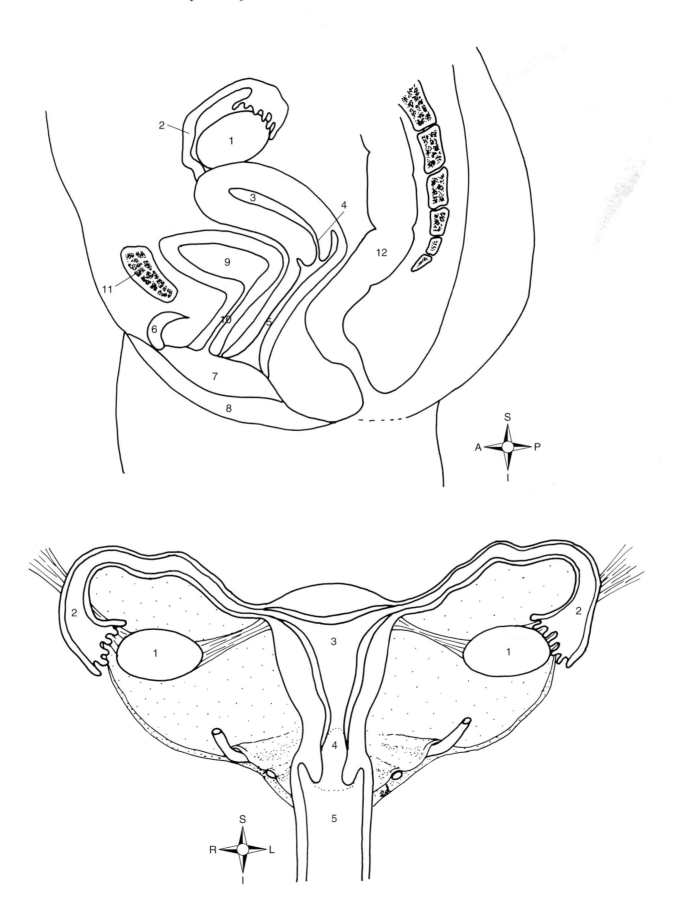

☐ 1. _____   ☐ 7. _____

☐ 2. _____   ☐ 8. _____

☐ 3. _____   ☐ 9. _____

☐ 4. _____   ☐10. _____

☐ 5. _____   ☐11. _____

☐ 6. _____   ☐12. _____

EXERCISE 11.19:

Coloring Exercise

Look up the features of the male reproductive system. Then color in the features with colored pens or pencils.

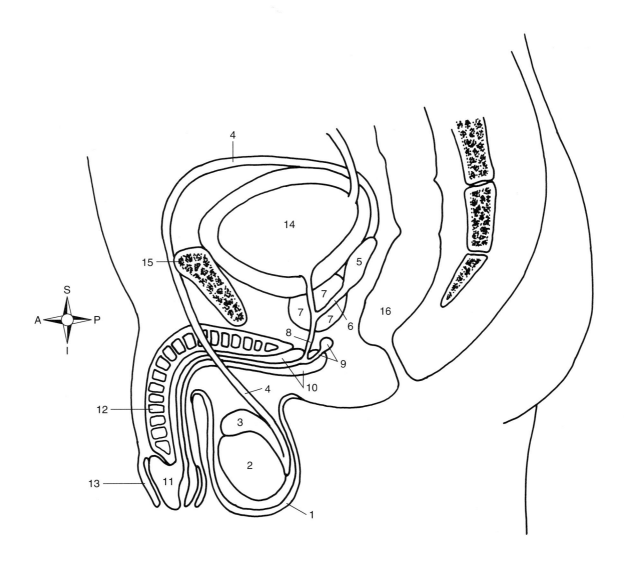

☐ 1. _____

☐ 2. _____

☐ 3. _____

☐ 4. _____

☐ 5. _____

☐ 6. _____

☐ 7. _____

☐ 8. _____

☐ 9. _____

☐10. _____

☐11. _____

☐12. _____

☐13. _____

☐14. _____

☐15. _____

☐16. _____

CHAPTER 12

The Endocrine System

Overview: The Endocrine System

The endocrine system is one of two organ systems involved with internal communication. It often works hand-in-hand with the other—the nervous system. An example of this complement between the two systems occurs between the autonomic nervous system and the suprarenal (adrenal) gland of the endocrine system during the fight-or-flight response.

We touched on the function of the endocrine system when we discussed the reproductive system in the last chapter. The body orchestrates the dynamic events of puberty and reproduction with intercellular messengers called hormones. But hormones regulate many other bodily functions, including but not limited to digestion, blood glucose levels, mood, and growth. Your textbook has an excellent discussion of this topic.

We will begin with a discussion of what has been termed the master gland because of its regulation of other endocrine glands and conclude with the endocrine system's involvement with reproduction.

- *From the **Home screen,** click the drop-down box on the **Select system** menu.*

- *From the systems listed, click on **Endocrine,** and you will see the image to the right:*

- *This is the opening screen for the **Endocrine System.***

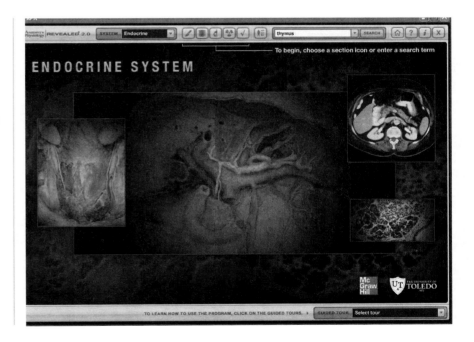

Animation: Hypothalamus and Pituitary Gland

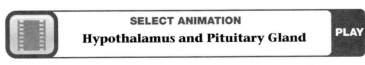

SELECT ANIMATION
Hypothalamus and Pituitary Gland   PLAY

- *After viewing the animation, answer these questions:*

1. The hypothalamus is sometimes referred to as the _____. Why?

2. Where in the brain is the hypothalamus located?

3. Describe the structure of the hypothalamus.

4. What is the infundibulum? What is its function?

5. Where is the pituitary gland located? How is it divided?

6. What is another name for the anterior pituitary? How is it connected to the hypothalamus?

7. What travels along this pathway? What is their function?

8. What is another name for the posterior pituitary? How is it connected to the hypothalamus?

9. What travels along this pathway? How are they transported? What is their destination?

10. Name the two classes of hypothalamic hormones that regulate the anterior pituitary. How do they reach the anterior pituitary? What is their function?

11. How do anterior pituitary hormones arrive at their target tissues?

12. Describe an example of these hormones and their function.

13. Name the hormones produced by the posterior pituitary. What is the source of posterior pituitary hormones?

14. Name two posterior pituitary hormones. How do they arrive at the posterior pituitary?

15. Name the structures that store the posterior pituitary hormones. What causes their release? Where are they released?

16. Name the functions of each posterior pituitary hormone.

Self Test
Take this opportunity to quiz yourself by taking the **SELF TEST.** See page 15 for a reminder on how to access the self-test for this section.

Animation: Hormonal Communication

 SELECT ANIMATION
Hormonal Communication PLAY

- *After viewing the animation, answer these questions:*

1. In general, how does hormonal communication begin? What reaction then occurs?

2. How are hormones transported to target cells?

3. What occurs when the hormones arrive at their target cells?

4. What then triggers changes in the target cells?

Self Test
Take this opportunity to quiz yourself by taking the **SELF TEST.** See page 15 for a reminder on how to access the self-test for this section.

Animation: Intracellular Receptor Model

 SELECT ANIMATION
Intracellular Receptor Model PLAY

- *After viewing the animation, answer these questions:*

1. Describe aldosterone, the hormone used in the animation.

2. What does aldosterone bind with in the cytoplasm of the cell?

3. Where does the aldosterone-receptor complex go, and where does it bind?

4. This binding stimulates the synthesis of what molecule? What is the function of this molecule?

5. Where does this mRNA molecule go, and what does it do?

6. What is directed by this binding? What response is produced?

Self Test
Take this opportunity to quiz yourself by taking the **SELF TEST.** See page 15 for a reminder on how to access the self-test for this section.

Animation: Receptors and G proteins

SELECT ANIMATION
Receptors and G Proteins

PLAY

• *After viewing the animation, answer these questions:*

1. What is located on the membrane-bound receptor on the outside of the cell?

2. What is a ligand?

3. To what does the portion of the membrane-bound receptor on the inside of the cell bind?

4. What are the three subunits of this protein? What is attached to the alpha subunit?

5. What changes occur in the G protein when the ligand binds to the receptor site? What changes occur to the alpha subunit?

6. What now occurs with the activated alpha subunit? How long can this step be repeated?

7. What occurs when the ligand separates from the receptor site?

8. How is the alpha subunit inactivated?

9. What occurs with the G protein subunits after this inactivation?

Self Test
Take this opportunity to quiz yourself by taking the **SELF TEST.** See page 15 for a reminder on how to access the self-test for this section.

The Hypothalamus, Pituitary, and Pineal Glands

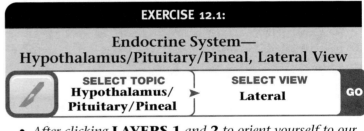

EXERCISE 12.1:

**Endocrine System—
Hypothalamus/Pituitary/Pineal, Lateral View**

SELECT TOPIC
**Hypothalamus/
Pituitary/Pineal** ▶ SELECT VIEW
Lateral   GO

• *After clicking* **LAYERS 1** *and* **2** *to orient yourself to our location, click* **LAYER 3** *in the* **LAYER CONTROLS** *window, and you will see the following image:*

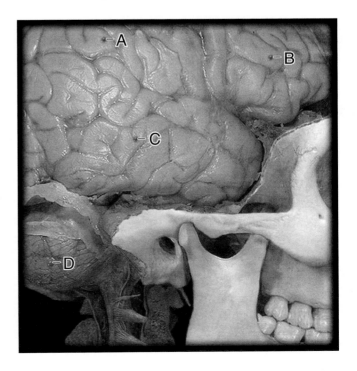

- *Mouse-over the blue pins on the screen to find the information necessary to identify the following non-endocrine system structures:*

A. _____

B. _____

C. _____

D. _____

- *Click **LAYER 4** in the **LAYER CONTROLS** window, and you will see the following image:*

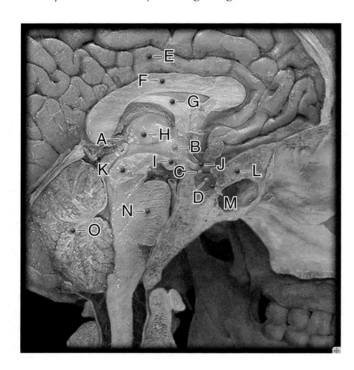

- *Mouse-over the green pins on the screen to find the information necessary to identify the following structures:*

A. _____

B. _____

C. _____

D. _____

Non-endocrine System Structures (blue pins)

E. _____

F. _____

G. _____

H. _____

I. _____

J. _____

K. _____

L. _____

M. _____

N. _____

O. _____

CHECK POINT:

Hypothalamus/Pituitary/Pineal Glands, Lateral View

1. Describe the structure of the hypothalamus.
2. Name the functions of the hypothalamus.
3. Describe the size and location of the pineal gland. What are its known functions?
4. Describe the location and structure of the pituitary gland.
5. List the functions of the hypothalamus.

EXERCISE 12.2:

Imaging—Hypothalamus and Pituitary Gland

SELECT TOPIC	SELECT VIEW	
Hypothalamus and ▶ Pituitary Gland	Sagittal	GO

- *Click the **TURN TAGS ON** button, and you will see the following image:*

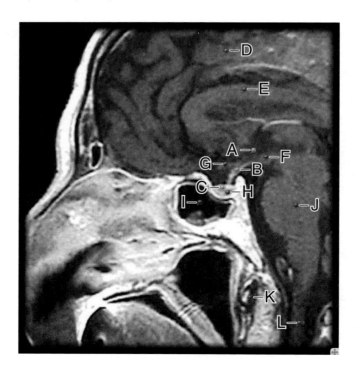

- *Mouse-over the green pins on the screen to find the information necessary to identify the following structures:*

A. _____

B. _____

C. _____

Non-endocrine System Structures (blue pins)

D. _____

E. _____

F. _____

G. _____

H. _____

I. _____

J. _____

K. _____

L. _____

The Pituitary Gland

EXERCISE 12.3:

Endocrine System—Anterior Pituitary, Histology

SELECT TOPIC	SELECT VIEW	
Anterior Pituitary	► High Magnification	GO

- *Click the **TURN TAGS ON** button, and you will see the following image:*

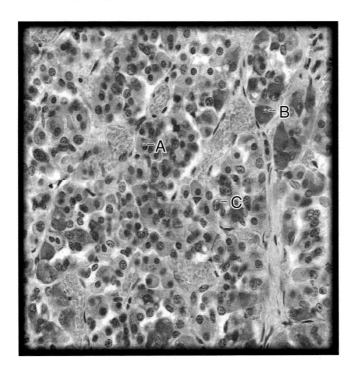

- *Mouse-over the pins on the screen to find the information necessary to identify the following structures:*

A. _____

B. _____

C. _____

C H E C K   P O I N T :

Anterior Pituitary, Histology

1. What are chromophils? Name the two subtypes of chromophils.
2. Which chromophils secrete somatotropin and prolactin? What is somatotropin?
3. Name the functions of the basophils.

EXERCISE 12.4:

Endocrine System—Posterior Pituitary, Histology

SELECT TOPIC	SELECT VIEW	
Posterior Pituitary	► High Magnification	GO

- *Click the **TURN TAGS ON** button, and you will see the following image:*

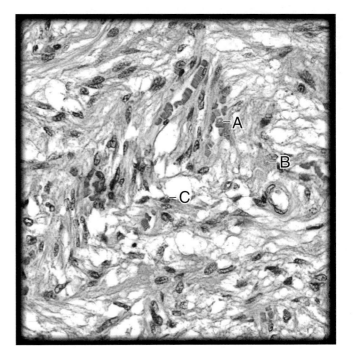

- *Mouse-over the pins on the screen to find the information necessary to identify the following structures:*

A. _____

B. _____

C. _____

CHECK POINT:

Posterior Pituitary, Histology

1. Name and describe the cells that occupy 25 percent of the volume of the posterior pituitary. What is their function?
2. What are herring bodies? What hormones are associated with them?
3. What is the function of the small blood vessels of the posterior pituitary?

The Thyroid Gland

Animation: Thyroid Gland

 SELECT ANIMATION
Thyroid Gland
PLAY

- *After viewing the animation, answer these questions:*

1. Name the largest endocrine gland. Describe its location and structure.

2. What causes the thyroid's reddish color? Name the blood vessels that distribute to and drain the thyroid gland. What other function do the veins have?

3. The thyroid gland is composed of spherical structures called _____. Describe the wall of these structures. What do they surround?

4. Describe what is located in the lumen of the follicles.

5. What are located in the loose connective tissue between the follicles? Which cells secrete hormones?

6. Name that hormone and its function. What is its antagonist?

7. Name the two different hormones referred to as thyroid hormone.

8. Name the hormone that maintains TH synthesis and secretion. Where is this hormone secreted?

9. What is contained in thyroglobulin?

10. Name the molecules that cross the follicular cell from the blood in the first phase of TH production.

11. What is the destination of these molecules? What are they converted to?

12. What two molecules combine in the lumen of the follicles? What is formed by this combination? How long can it be stored in the follicle lumen?

13. In the second phase of TH production, what becomes of the combined molecules formed in phase one?

14. Name the primary effect of TH. Describe its importance in children.

15. What is hyperthyroidism? What can result from this condition?

16. What is hypothyroidism? What can result from this condition?

Self Test
Take this opportunity to quiz yourself by taking the **SELF TEST.** See page 15 for a reminder on how to access the self-test for this section.

EXERCISE 12.5:

Endocrine System—Thyroid Gland, Anterior View

SELECT TOPIC	SELECT VIEW	
Thyroid Gland ▶	**Anterior**	**GO**

- *Click* **LAYER 1** *in the* **LAYER CONTROLS** *window, and you will see the following image:*

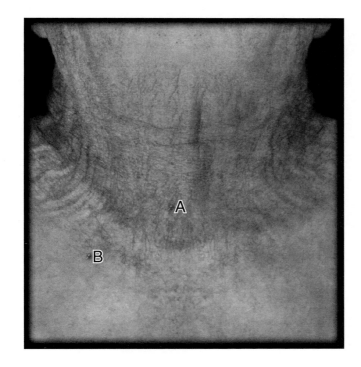

- *Mouse-over the blue pins on the screen to find the information necessary to identify the following non-endocrine system structures:*

A. _____

B. _____

- *Click* **LAYER 2** *in the* **LAYER CONTROLS** *window, and you will see the following image:*

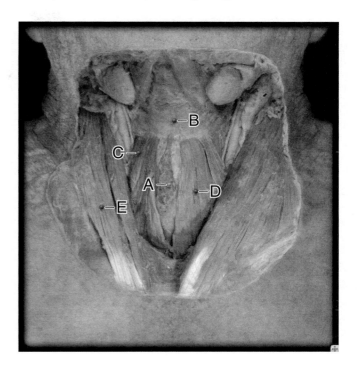

- *Mouse-over the green pin on the screen to find the information necessary to identify the following structure:*

A. _____

Non-endocrine System Structures (blue pins)

B. _____

C. _____

D. _____

E. _____

• Click **LAYER 3** in the **LAYER CONTROLS** window, and you will see the following image:

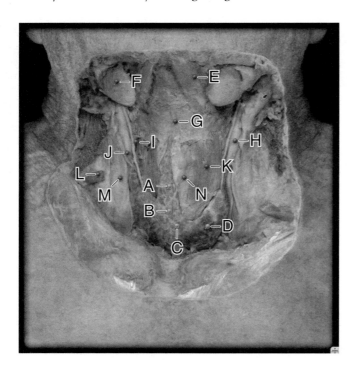

• Mouse-over the green pins on the screen to find the information necessary to identify the following structures:

A. _____

B. _____

C. _____

D. _____

Non-endocrine System Structures (blue pins)

E. _____

F. _____

G. _____

H. _____

I. _____

J. _____

K. _____

L. _____

M. _____

N. _____

CHECK POINT:

Thyroid Gland, Anterior View

1. Describe the structure of the thyroid gland.
2. What is the function of the thyroid gland?
3. What is the name for an enlarged thyroid gland? What is the cause of this condition?

EXERCISE 12.6:

Endocrine System—Thyroid Gland, Histology

SELECT TOPIC	SELECT VIEW	
Thyroid Gland	➤ High Magnification	GO

• Click the **TURN TAGS ON** button, and you will see the following image:

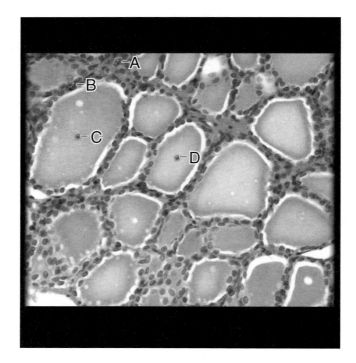

• Mouse-over the pins on the screen to find the information necessary to identify the following structures:

A. _____

B. _____

C. _____

D. _____

CHECK POINT:

Thyroid Gland, Histology

1. Name the thyroid cells that produce calcitonin. What is the function of calcitonin?
2. Describe the structure and function of the thyroid follicle.
3. What is follicular colloid? How long of a supply can be stored?

The Parathyroid Glands

Animation: Parathyroid Glands

SELECT ANIMATION
Parathyroid Glands   PLAY

- *After viewing the animation, answer these questions:*

1. Describe the location of the parathyroid glands. How many are there? What arteries supply these glands?

2. Name the two types of parathyroid gland cells. What are their functions?

3. What causes the release of PTH?

4. How does PTH raise blood calcium levels?

5. What are the symptoms that may result from hyperparathyroidism?

6. What is the most common cause of hypoparathyroidism?

7. What are the symptoms of hypoparathyroidism?

Self Test

Take this opportunity to quiz yourself by taking the **SELF TEST.** See page 15 for a reminder on how to access the self-test for this section.

Endocrine System—Parathyroid Gland, Histology

SELECT TOPIC | **SELECT VIEW**
Parathyroid Gland ➤ High Magnification   GO

- *Click the **TURN TAGS ON** button, and you will see the following image:*

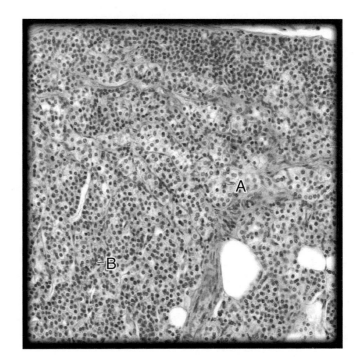

- *Mouse-over the pins on the screen to find the information necessary to identify the following structures:*

A. _____

B. _____

C H E C K P O I N T :

Parathyroid Gland, Histology

1. Name the most numerous functional cell type in the parathyroid gland. What hormone do they secrete?
2. What is the function of this hormone? What hormone is an antagonist to this one?
3. Name the parathyroid cells that increase in number with age.

The Pancreas

Animation: Pancreas

SELECT ANIMATION
Pancreas   PLAY

- *After viewing the animation, answer these questions:*

1. The organs of the endocrine system secrete _____ directly into the _____ to _____.

2. Describe the location of the pancreas. Is it endocrine, exocrine, or both?

3. The primary cells of the pancreas are _____. What do these cells secrete, and what is the function of these secretions?

4. Where are the endocrine cells of the pancreas located? What are they called?

5. Name the four types of cells located in the pancreatic islets. What type of substances do these cells release?

6. Which cells are activated by declining blood glucose levels? What hormone do they release? What action does this hormone stimulate?

7. Which cells are activated by increasing blood glucose levels? What hormone do they release? What action does this hormone stimulate?

8. Which cells release somatostatin? What actions does this hormone initiate?

9. Which cells secrete pancreatic polypeptide? What is the function of this secretion?

10. What do these pancreatic hormones together provide for?

Self Test

Take this opportunity to quiz yourself by taking the **SELF TEST**. See page 15 for a reminder on how to access the self-test for this section.

EXERCISE 12.8:

Endocrine System—Pancreas, Anterior View

SELECT TOPIC	SELECT VIEW	
Pancreas ▶	Anterior	GO

- *After clicking **LAYERS 1** through **3** to orient yourself to our location, click **LAYER 4** in the **LAYER CONTROLS** window and you will see the following image:*

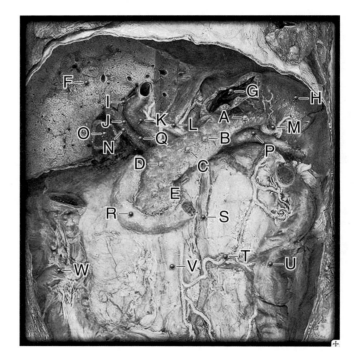

- *Mouse-over the green pins on the screen to find the information necessary to identify the following structures:*

A. _____

B. _____

C. _____

D. _____

E. _____

Non-endocrine System Structures (blue pins)

F. _____

G. _____

H. _____

I. _____

J. _____

K. _____

L. _____

M. _____

N. _____

O. _____

P. _____

Q. _____

R. _____

S. _____

T. _____

U. _____

V. _____

W. _____

C H E C K   P O I N T :

Pancreas, Anterior View

1. What is the exocrine function of the pancreas?
2. What is the endocrine function of the pancreas?
3. Name two events that can result in diabetes mellitus.

EXERCISE 12.9:

Endocrine System—Pancreas (Endocrine), Histology

| SELECT TOPIC | SELECT VIEW | |
| Pancreas (Endocrine) | ▶ High Magnification | GO |

- *Click the* **TURN TAGS ON** *button, and you will see the following image:*

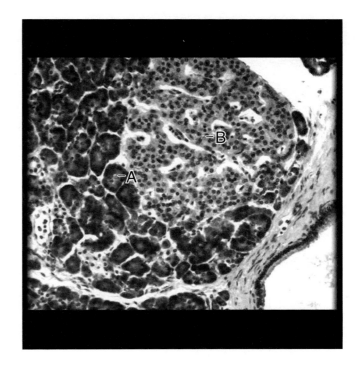

- *Mouse-over the pins on the screen to find the information necessary to identify the following structures:*

A. _____

B. _____

C H E C K   P O I N T :

Pancreas (Endocrine), Histology

1. Name the clusters of cells that form the endocrine portion of the pancreas.
2. Name the hormones produced by these cells.
3. Where are these hormones secreted?

EXERCISE 12.10:

Imaging—Pancreas

| SELECT TOPIC | SELECT VIEW | |
| Pancreas > | Axial | GO |

- *Click the* **TURN TAGS ON** *button, and you will see the following image:*

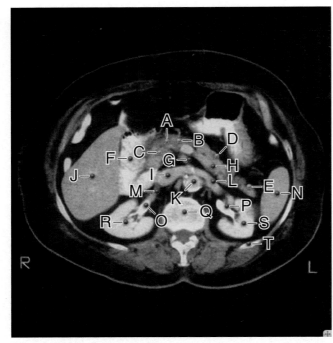

- *Mouse-over the green pins on the screen to find the information necessary to identify the following structures:*

A. _____

B. _____

C. _____

D. _____

E. _____

Non-endocrine System Structures (blue pins)

F. _____

G. _____

H. _____

I. _____

J. _____

K. _____

L. _____

M. _____

N. _____

O. _____

P. _____

Q. _____

R. _____

S. _____

T. _____

The Suprarenal (Adrenal) Gland

Animation: Suprarenal (Adrenal) Gland

| SELECT ANIMATION | |
| Suprarenal (Adrenal) Gland | PLAY |

- *After viewing the animation, answer these questions:*

1. The organs of the endocrine system secrete _____ directly into the _____ to _____.

2. Describe the location of the suprarenal glands. Where do they receive their blood supply? Name the single vein that drains them.

3. The suprarenal glands are composed of what two layers?

4. What are corticosteroids? Where are they synthesized?

5. What are mineralocorticoids? Where are they synthesized?

6. What is the principal mineralocorticoid? What is its function?

7. What are glucocorticoids? Where are they synthesized?

8. What are the two most common glucocorticoids?

9. What are the adrenal sex hormones? Where are they synthesized?

10. Name the inner core of the suprarenal glands. What cells are located in large numbers there?

11. What hormones are produced in this inner core area? What stimulates their release? What is their function?

Self Test

Take this opportunity to quiz yourself by taking the **SELF TEST**. See page 15 for a reminder on how to access the self-test for this section.

EXERCISE 12.11:

Endocrine System—Suprarenal (Adrenal) Gland, Anterior View

SELECT TOPIC	SELECT VIEW	
Suprarenal Gland ➤	**Anterior**	**GO**

- *Click* **LAYER 1** *in the* **LAYER CONTROLS** *window, and you will see the following image:*

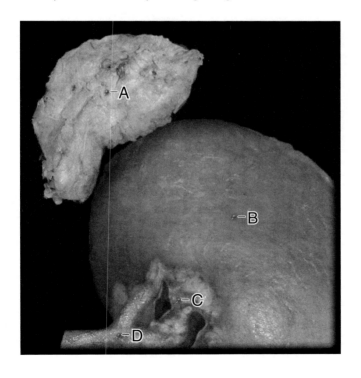

- *Mouse-over the green pin on the screen to find the information necessary to identify the following structure:*

A. _____

Non-endocrine System Structures (blue pins)

B. _____

C. _____

D. _____

CHECK POINT:

Suprarenal (Adrenal) Gland, Anterior View

1. Name the endocrine gland located superior to the kidney.
2. Name the two parts of this gland.
3. What are the functions of these two parts?

- *Click* **LAYER 2** *in the* **LAYER CONTROLS** *window, and you will see the following image:*

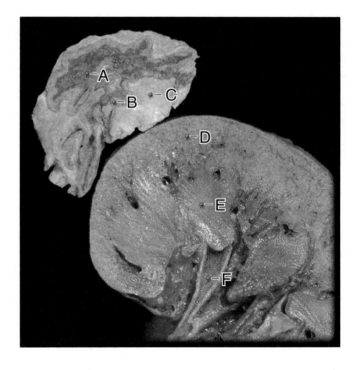

- *Mouse-over the green pins on the screen to find the information necessary to identify the following structures:*

A. _____

B. _____

C. _____

Non-endocrine System Structures (blue pins)

D. _____

E. _____

F. _____

CHECK POINT:

Suprarenal (Adrenal) Gland, Anterior View, cont'd

4. Name the connective tissue that surrounds the suprarenal gland.
5. Name the lipid-rich outer part of the suprarenal gland. What is the function of each of the three regions of this part?
6. Name the reddish-brown core of the suprarenal gland. What is its function?

CHECK POINT:

Suprarenal (Adrenal) Gland, Histology

1. Name the five layers of the suprarenal gland from superficial to deep.
2. Name the thick fibroblastic outer coat of the suprarenal gland. What is its function?
3. Name the layer of the suprarenal gland responsible for sex hormone production.

The Ovary

EXERCISE 12.12:

Endocrine System—Suprarenal Gland, Histology

SELECT TOPIC **Suprarenal Gland** ▶ SELECT VIEW **Low Magnification** GO

- Click the **TURN TAGS ON** button, and you will see the following image:

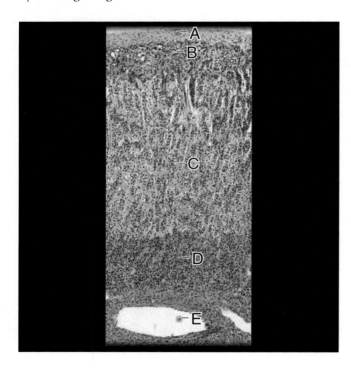

- Mouse-over the pins on the screen to find the information necessary to identify the following structures:

A. _____

B. _____

C. _____

D. _____

E. _____

EXERCISE 12.13:

Endocrine System—Ovary, Superior View

SELECT TOPIC **Ovary** ▶ SELECT VIEW **Superior** GO

- After clicking **LAYER 1** to orient yourself to our location, click **LAYER 2** in the **LAYER CONTROLS** window and you will see the following image:

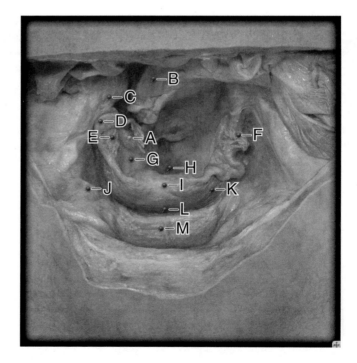

- Mouse-over the green pin on the screen to find the information necessary to identify the following structure:

A. _____

Non-endocrine System Structures (blue pins)

B. _____

C. _____

D. _____

E. _____

F. _____

G. _____

H. _____

I. _____

J. _____

K. _____

L. _____

M. _____

CHECK POINT:

Ovary, Superior View

1. Describe the structure of the female gonads.
2. Name the hormones produced by the female gonads.
3. What change occurs in these structures after menopause?

The Testis

EXERCISE 12.14:

Endocrine System—Testis and Spermatic Cord, Anterior View

| SELECT TOPIC
Testis and Spermatic Cord | ▶ | SELECT VIEW
Anterior | GO |

- *After clicking* **LAYERS 1** *through* **3** *to orient yourself to our location, click* **LAYER 4** *in the* **LAYER CONTROLS** *window and you will see the following image:*

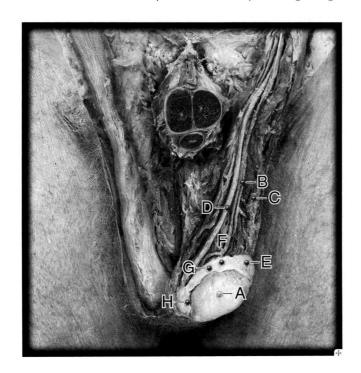

- *Mouse-over the green pin on the screen to find the information necessary to identify the following structure:*

A. _____

Non-endocrine System Structures (blue pins)

B. _____

C. _____

D. _____

E. _____

F. _____

G. _____

H. _____

CHECK POINT:

Testis and Spermatic Cord, Anterior View

1. Describe the structure of the male gonads.
2. Name the hormones produced by the male gonads.
3. What is the function of testosterone?

- *Click* **LAYER 5** *in the* **LAYER CONTROLS** *window, and you will see the following image:*

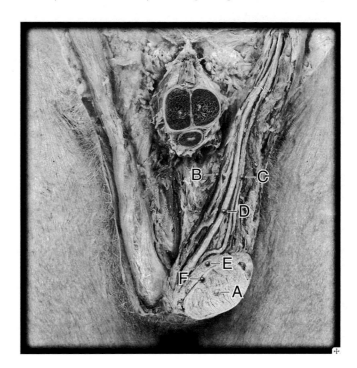

- *Mouse-over the green pin on the screen to find the information necessary to identify the following structure:*

A. _____

Non-endocrine System Structures (blue pins)

B. _____

C. _____

D. _____

E. _____

F. _____

EXERCISE 12.15:

Endocrine System—Testis and Spermatic Cord (Isolated), Lateral View

SELECT TOPIC	SELECT VIEW	
Testis and Spermatic ▶ Cord (Isolated)	Lateral	GO

- *Click* **LAYER 1** *in the* **LAYER CONTROLS** *window, and you will see the following image:*

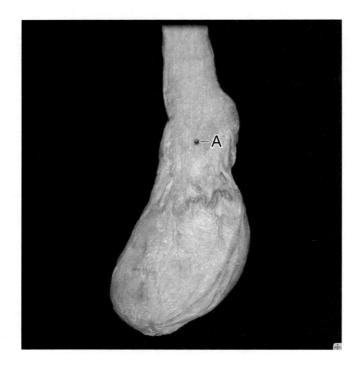

- *Mouse-over the blue pin on the screen to find the information necessary to identify the following non-endocrine system structure:*

A. _____

- *Click* **LAYER 2** *in the* **LAYER CONTROLS** *window, and you will see the following image:*

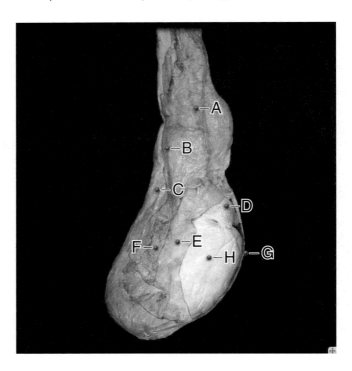

- *Mouse-over the blue pins on the screen to find the information necessary to identify the following non-endocrine system structures:*

A. _____

B. _____

C. _____

D. _____

E. _____

F. _____

G. _____

H. _____

- *Click **LAYER 3** in the **LAYER CONTROLS** window, and you will see the following image:*

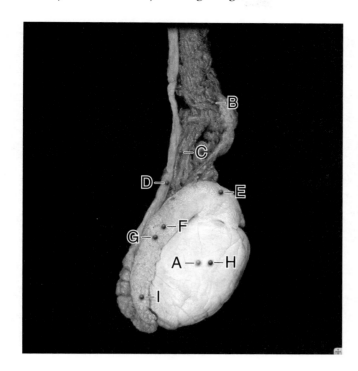

- *Mouse-over the green pin on the screen to find the information necessary to identify the following structure:*

A. _____

Non-endocrine System Structures (blue pins)

B. _____

C. _____

D. _____

E. _____

F. _____

G. _____

H. _____

I. _____

- *Click **LAYER 4** in the **LAYER CONTROLS** window, and you will see the following image:*

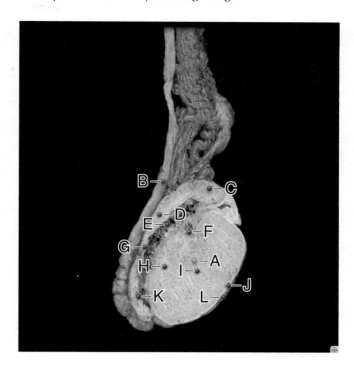

- *Mouse-over the green pin on the screen to find the information necessary to identify the following structure:*

A. _____

Non-endocrine System Structures (blue pins)

B. _____

C. _____

D. _____

E. _____

F. _____

G. _____

H. _____

I. _____

J. _____

K. _____

L. _____

The Seminiferous Tubule

EXERCISE 12.16:

Endocrine System—Seminiferous Tubule, Histology

SELECT TOPIC	SELECT VIEW	
Seminiferous Tubule	➤ **High Magnification**	**GO**

- *Click the* **TURN TAGS ON** *button, and you will see the following image:*

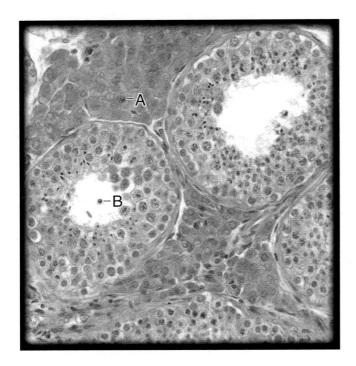

- *Mouse-over the pins on the screen to find the information necessary to identify the following structures:*

A. _____

B. _____

CHECK POINT:

Seminiferous Tubule, Histology

1. Name the structures responsible for testosterone production.
2. What is another name for these cells?
3. Name the structures that these cells surround. What correlation do they have with these structures?

Self Test
Take this opportunity to quiz yourself by taking the **SELF TEST.** See page 15 for a reminder on how to access the self-test for this section.

IN REVIEW

What Have I Learned?

The following questions cover the material that you have just learned, the endocrine system. Apply what you have learned to answer these questions on a separate piece of paper.

1. The pituitary gland rests in what structure of which bone?

2. Name the fossa that surrounds the pituitary gland.

3. Describe the structure and function of the mammillary body.

4. Name the cell that represents a chromophil depleted of hormone.

5. List the hormones released by the anterior pituitary gland.

6. List the hormones released by the posterior pituitary gland.

7. Name the function of the following hormones: T_3 and T_4 and calcitonin. (These may require a little research on your part.)

Continued

Continued

8. Which division of the autonomic nervous system is responsible for the fight-or-flight response? What two hormones regulate this response? What is the origin of these hormones?

9. Name two disorders of suprarenal cortex hormone secretion.

10. Name the layer of the suprarenal gland responsible for mineralocorticoid production.

11. Name the layer of the suprarenal gland responsible for glucocorticoid production.

12. In the following chart, list the hormones produced by the suprarenal gland, where they are produced, and their function.

HORMONE	WHERE PRODUCED	FUNCTION